Microneedles for Drug and Vaccine Delivery and Patient Monitoring

Microneedles for Drug and Vaccine Delivery and Patient Monitoring

Edited by
Ryan F. Donnelly, Thakur Raghu Raj Singh, Eneko Larrañeta, and Maelíosa T.C. McCrudden
Queen's University Belfast, UK

WILEY

This edition first published 2018

Registered Offices
John Wiley & Sons, Inc., 111 River Street, Hoboken, NJ 07030, USA
John Wiley & Sons Ltd, The Atrium, Southern Gate, Chichester, West Sussex, PO19 8SQ, UK

Editorial Office
The Atrium, Southern Gate, Chichester, West Sussex, PO19 8SQ, UK

For details of our global editorial offices, customer services, and more information about Wiley products visit us at www.wiley.com.

Wiley also publishes its books in a variety of electronic formats and by print-on-demand. Some content that appears in standard print versions of this book may not be available in other formats.

Library of Congress Cataloging-in-Publication Data applied for

Hardback: 9781119305149

Cover design: Wiley
Cover image: Courtesy of Eneko Larrañeta

Set in 10/12pt WarnockPro by SPi Global, Chennai, India

10 9 8 7 6 5 4 3 2 1

Contents

List of Contributors

Marco Abbate
Community Pharmacist, Belfast, UK

Sarah Brown
Community Pharmacist, Belfast, UK

Grace Cole
School of Pharmacy, Queen's University Belfast, UK

Aaron J. Courtenay
School of Pharmacy, Queen's University Belfast, UK

Ryan F. Donnelly
School of Pharmacy, Queen's University Belfast, UK

Nicholas J. Dunne
School of Mechanical and Manufacturing Engineering, Dublin City University, Dublin, Ireland

Iman Hamdan
School of Pharmacy, Queen's University Belfast, UK and School of Pharmacy, Middle East University, Amman, Jordan

Mary-Carmel Kearney
School of Pharmacy, Queen's University Belfast, UK

Eneko Larrañeta
School of Pharmacy, Queen's University Belfast, UK

Emma McAlister
School of Pharmacy, Queen's University Belfast, UK

Helen McCarthy
School of Pharmacy, Queen's University Belfast, UK

Maelíosa T.C. McCrudden
School of Pharmacy, Queen's University Belfast, UK

Helen Quinn
School of Pharmacy, Queen's University Belfast, UK

Thakur Raghu Raj Singh
School of Pharmacy, Queen's University Belfast, UK

Ismaiel Tekko
School of Pharmacy, Queen's University Belfast, UK and Vice-Dean for Scientific and Research Affairs, Aleppo University, Syria

Lalit Vora
School of Pharmacy, Queen's University Belfast, UK

Ahlam Zaid-Alkilani
School of Pharmacy, Zarqa University, Jordan

About the Editors

Professor Ryan F. Donnelly holds the Chair in Pharmaceutical Technology in the School of Pharmacy at Queen's University Belfast. A registered Pharmacist, his research is centred on design and physicochemical characterisation of advanced polymeric drug delivery systems for transdermal and topical drug delivery, with a strong emphasis on improving patient outcomes. He is currently developing a range of novel microneedle technologies through independent research, but also in collaboration with several major pharma partners. He has obtained substantial Research Councils UK, charity and industrial funding and authored over 500 peer-reviewed publications, including four patent applications, four textbooks, 23 book chapters and more than 150 full papers. He has been an invited speaker at numerous national and international conferences. Professor Donnelly is Editor-in-Chief of *Recent Patents on Drug Delivery & Formulation* and a member of the Editorial Advisory Boards of several leading pharmaceutical science journals. He won the Controlled Release Society's Young Investigator Award in 2016, BBSRC Innovator of the Year and the American Association of Pharmaceutical Scientists *Pharmaceutical Research* Meritorious Manuscript Award in 2013, the GSK Emerging Scientist Award in 2012 and the Royal Pharmaceutical Society's Science Award (2011).

Dr Thakur Raghu Raj Singh is Senior Lecturer in Pharmaceutics in the School of Pharmacy at Queen's University Belfast. He obtained his PhD in Drug Delivery from the School of Pharmacy at Queen's in 2009, his MSc in Pharmaceutical Sciences from University Science Malaysia in 2006 and his BPharm from Jawaharlal Nehru Technological University, India in 2002. Dr Thakur's research interests are in the design and physicochemical characterisation of advanced polymeric drug delivery systems for ocular, transdermal and topical applications. In particular, his current research involves fabrication and design of novel long-acting injectable and implantable drug delivery systems for treating back-of-the-eye disorders. Dr Thakur's ocular drug delivery research has led to formation of a university spinout company, Re-vana Therapeutics Ltd, for which he is the Co-founder and Chief Scientific Officer. He has authored over 140 scientific publications, including one patent, 46 peer-reviewed research papers, eight book chapters and three textbooks and has been an invited speaker at a number of national and international meetings. He is currently an Editorial Board member of the *Journal of Pharmacy & Pharmacology*, *Chronicles of Pharmacy* and *Science Domain International*, and acts as a scientific advisor to *The Journal of Pharmaceutical Sciences*, in addition to regularly acting as a reviewer for many other international scientific journals.

Dr Eneko Larrañeta works as Lecturer in Pharmaceutical Sciences in the School of Pharmacy at Queen's University Belfast. His main fields of expertise are drug delivery and pharmaceutical materials. During the last five years, he has been working on several projects to develop more efficient drug delivery systems for the oral and transdermal routes. For this purpose, he worked on the development of advanced systems, such as nanoparticles and microneedles. He worked on the development of nanoparticles for oral delivery of difficult-to-deliver drugs and nutraceuticals at the University of Navarra (Spain) under the supervision of Professor Juan Manuel Irache, an internationally renowned expert on mucosal drug delivery. In addition to oral drug delivery, he has done extensive work on microneedle-mediated transdermal drug delivery with Professor Donnelly's Group. He has worked on a range of projects funded by the European 7th Framework Programme and the UK's Biotechnology & Biological Sciences Research Council. Moreover, he has also worked on projects sponsored by global not-for-profit organisations, as well as the pharmaceutical and cosmeceutical industry. Using his previous research experience, the main objective of his current research is to develop materials that can be easily transferred to industry for ultimate patient benefit. He has published 20 scientific articles in indexed journals and has been co-author of two book chapters.

Dr Maelíosa T.C. McCrudden is a Senior Research Fellow in the School of Pharmacy at Queen's University Belfast. Having received her PhD from Queen's in 2008, she first carried out postdoctoral research in the laboratory of Dr Fionnuala Lundy, determining the roles of specific innate host peptides in wound healing and defence mechanisms in the oral cavity. She then moved her research focus to the field of pharmaceutical sciences and has, since 2012, worked in the Microneedles Research Group of Professor Donnelly. Her research has centred on transdermal delivery of drugs and intradermal vaccine administration using a range of novel microneedle systems. She has worked on Research Council and pharmaceutical industry funded projects and has most recently worked with the international not-for-profit organisation, PATH, to deliver an anti-HIV drug in a sustained fashion. Over and above these research interests, she is passionately committed to science communication and was recognised nationally for her work by receipt of both the Mendel Gold Medal and the Eric Wharton Medal at the Westminster SET for Britain awards in 2016. Dr McCrudden has published 25 articles in peer-reviewed journals, in addition to seven review articles and one book chapter. She is a regular invited speaker at both national and international conferences.

Preface

The worldwide transdermal patch market approached $32 billion in 2015, despite limited innovation in this area over the previous five years. Indeed, the entire market is still based on only 20 drugs. This rather limited number of drug substances is attributed to the excellent barrier function of the skin, which is accomplished almost entirely by the outermost 10–15 μm (in the dry state) of tissue, the *stratum corneum* (*SC*). Before being taken up by blood vessels in the upper dermis and prior to entering the systemic circulation, substances permeating the skin must cross the *SC* and the viable epidermis. There are three possible pathways leading to the capillary network: through hair follicles with associated sebaceous glands, via sweat ducts, or across continuous *SC* between these appendages. As the fractional appendageal area available for transport is only about 0.1%, this route usually contributes negligibly to apparent steady state drug flux. The intact *SC* thus provides the main barrier to exogenous substances, including drugs. The corneocytes of hydrated keratin are analogous to "bricks," embedded in a "mortar" composed of highly organised, multiple lipid bilayers of ceramides, fatty acids, cholesterol and its esters. These bilayers form regions of semicrystalline gel and liquid crystal domains. Most molecules penetrate through skin via this intercellular micro-route. Facilitation of drug penetration through the *SC* may involve by-pass or reversible disruption of its elegant molecular architecture. The ideal properties of a molecule that can penetrate the intact *SC* well are:

- Molecular mass less than 600 Da.
- Adequate solubility in both oil and water so that the membrane concentration gradient, which is the driving force for passive drug diffusion along a concentration gradient, may be high.
- Partition coefficient, such that the drug can diffuse out of the vehicle, partition into, and move across, the *SC*, without becoming sequestered within it.
- Low melting point, correlating with good solubility, as predicted by ideal solubility theory.

Clearly, many drug molecules do not meet these criteria. This is especially true for biopharmaceutical drugs, which are becoming increasingly important in therapeutics and diagnostics of a wide range of illnesses. Drugs that suffer poor oral bioavailability or susceptibility to first-pass metabolism, and are thus often ideal candidates for transdermal delivery, may fail to realise their clinical application because they do not meet one or more of these conditions. Examples include peptides, proteins and vaccines, which, due to their large molecular size and susceptibility to acid destruction in the stomach, cannot be given orally and, hence, must be dosed parenterally. Such agents are currently precluded from successful transdermal administration, not only by their large sizes, but also by their extreme hydrophilicities. Several approaches have been used to enhance the transport of drugs through the *SC*. However, in many cases, only moderate success has been achieved and each approach is associated with significant problems. Chemical penetration enhancers allow only a modest improvement in penetration. Chemical modification to increase lipophilicity is not always possible and, in any case, necessitates additional studies for regulatory approval, due to generation of new chemical entities. Significant enhancement in delivery of a large number of drugs has been reported using iontophoresis. However, specialised devices are required and the agents delivered tend to accumulate in the skin appendages. The method is presently best-suited to acute applications, with several commercialised products intended for regular at-home use by patients being withdrawn relatively quickly after market approval for a range of reasons. Electroporation and sonophoresis are known to increase transdermal delivery. However, they both cause pain and local skin reactions and sonophoresis can cause breakdown of the therapeutic entity. Techniques aimed at removing the *SC* barrier, such as tape-stripping and suction/laser/thermal ablation are impractical, while needle-free injections have so far failed to replace conventional needle-based insulin delivery. Clearly, a robust alternative strategy is required to enhance drug transport across the *SC* and thus widen the range of drug substances amenable to transdermal delivery.

Microneedle arrays are minimally invasive devices that can be used to by-pass the *SC* barrier and thus achieve transdermal drug delivery. Microneedles (MN) (50–900 μm in height, up to 2000 MN/cm^2) in various geometries and materials (silicon, metal, polymer) have been produced using recently developed microfabrication techniques. Silicon MN arrays are prepared by modification of the dry- or wet-etching processes employed in microchip manufacture. Metal MN are produced by electrodeposition in defined polymeric moulds or photochemical etching of needle shapes into a flat metal sheet and then bending these down at right angles to the sheet. Polymeric MN have been manufactured by micromoulding of molten/dissolved polymers. MN are applied to the skin surface and pierce the epidermis (devoid of nociceptors), creating microscopic holes through which drugs diffuse to the dermal microcirculation. MN are long enough to penetrate to the dermis but are short and narrow enough to avoid stimulation of dermal nerves. Solid MN puncture the skin prior to application of a drug-loaded patch or are pre-coated with drug prior to insertion. Hollow bore microneedles allow diffusion or pressure-driven flow of drugs through a central lumen, while polymeric drug-containing microneedles release their payload as they biodegrade in the viable skin layers. *In vivo* studies using solid MN have demonstrated delivery of oligonucleotides, desmopressin and human growth hormone, reduction of blood glucose levels from insulin delivery, increase of skin transfection with DNA and enhanced elicitation of immune response from delivery of DNA and protein antigens. Hollow MN have also been shown to deliver insulin and reduce blood glucose levels. MN arrays do not cause pain on application and no reports of development of skin infection currently exist.

MN have been considered for a range of other applications, in addition to transdermal and intradermal drug/vaccine delivery. These include minimally invasive therapeutic drug monitoring, as a stimulus for collagen remodelling in anti-ageing strategies and for delivery of active cosmaceutical ingredients. MN technology is likely to find ever-increasing utility in the healthcare field as further advancements are made. However, some significant barriers will need to be overcome before we see the first MN-based drug delivery or monitoring device on the market. Regulators, for example, will need to be convinced that MN puncture of skin does not lead to skin infections or any long-term skin problems. MN will also need to be capable of economic mass production.

In this book, we review the work that has been carried out on MN to date in both the academic and industrial sectors. We have looked in detail at both *in vitro* and *in vivo* studies and covered the important area of MN-based vaccines. We also consider safety and public perception aspects of MN and discuss newer applications of this exciting technology, such as delivery of gene therapies, photodynamic therapy, ocular delivery and enhanced administration of nanomedicines.

The MN field continues to expand, with ever-increasing numbers of publications that are increasingly being cited (Figure 1). However, the number of commercialised products on the market remains disappointing. Indeed, no true MN array drug delivery system is currently available to patients. Research work is largely confined to universities and specialised drug delivery companies, many of which have been spun-out from universities. It is hoped that the exciting data generated and published will encourage large pharmaceutical and medical device forms to make the considerable financial investment necessary for scaled-up manufacture and comprehensive clinical trials over the coming years so that MN technology will finally deliver the impact envisioned by research scientists.

We took a very different approach to the production of this book, as compared with how we wrote the first text on microneedles published by Wiley in 2012. The first book was written by four authors, with Desmond Morrow and David Woolfson coming on board with Raj and myself. Given the expansion of the field, Desy's move to New Zealand and David's well-deserved retirement, I asked Eneko Larrañeta and Maelíosa McCrudden to join Raj and I as Editors of this new book. Instead of writing the entire book ourselves, we asked members of my research Group and close collaborators to each co-author a chapter in an area of their specialisation. This approach worked extremely well, with our hectic schedules of research over the past year only slightly delaying delivery of the final text.

Editing this text took quite a considerable amount of time and I would like to thank my wife Johanne for her patience and support throughout the project. I am also grateful to Raj, Eneko and Maelíosa for agreeing to assist me in the editing of the book and for their work in co-authoring several chapters. This is now the fourth book I have worked on with Raj and it has been every bit as enjoyable collaborating with him again this time. This has been a new experience for Eneko and Maelíosa and I hope that this is the first of many books for them.

I am highly appreciative of the past and present members of the Microneedles Group at Queen's for their hard work and imagination in the lab. Special mention must go to the current Group members: Maelíosa, Aaron Courtenay, Ismaiel Tekko, Lalit Vora, Patricia Gonzalez-Vazquez, Helen Quinn, Mary-Carmel Kearney, Aoife Rodgers, Bridie Dutton, Sara Cordeiro, Emma McAlister, Eman Migdadi, Rehan Al-Kasasbeh, Iman Hamdan, Kurtis Moffatt, Dian Permana, Fabiana Volpe-Zanutto, Boonnada Pamornpathomkul, Michelle Barreto-Requena, Yadira Pastor-Garcia, Ke Peng, Delly Ramadon, Sarah Stewart, Inken Ramöller, Heba Abdelazim and Alvaro

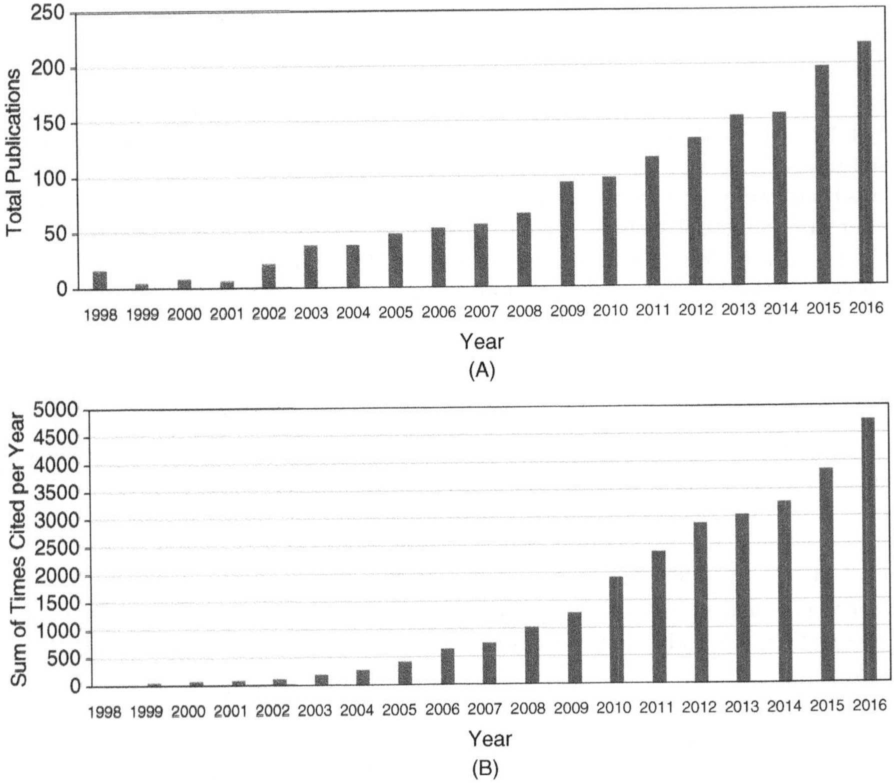

Figure 1 (A) Total number of journal articles published on microneedles, by year, since the first publication in 1998 and (B) total number of citations of microneedles articles, by year, since 1998.

Carcamo-Martinez. I would also like to acknowledge BBSRC, EPSRC, MRC, The Wellcome Trust, PATH, USAID, Invest Northern Ireland, Action Medical Research, Prostate Cancer UK, Arthritis Research UK and The Royal Society for funding my work in this area. I have been fortunate to have many excellent industrial collaborators who continue to support the translation of my research towards commercialisation and patient benefit, but confidentiality prevents direct acknowledgement. You know who you are – thank you! Emma Strickland and Elsie Merlin from Wiley provided considerable help and encouragement as we completed this project and their support and guidance are greatly appreciated. Judith Egan-Shuttler is also thanked for her work as the copyeditor. It is my hope that this book will serve as a comprehensive overview of the field and, hence, that it will be of use to those new to microneedles, as well as people already engaged in work in this area in both industry and academia.

Belfast, October 2017 *Ryan F. Donnelly*

1

Genesis of Transdermal Drug Delivery

Ahlam Zaid-Alkilani

School of Pharmacy, Zarqa University, Jordan

On the basis of the drug delivery systems, the market can be segmented into eight categories: oral, pulmonary, transdermal, injectable, ocular, nasal, implantable and transmucosal drug delivery. The most common routes of drug delivery are oral, which represent the largest market share (more than 50%), followed by transmucosal (26.2%) and transdermal delivery (12%) [1]. The conventional routes of oral drug delivery have many inherent limitations that could potentially be overcome by advanced drug delivery methodologies, such as transdermal drug delivery (TDD).

The administration of chemical agents to the skin surface has long been practiced, whether for healing, protective or cosmetic reasons. Historically, the skin was thought to be totally impervious to exogenous chemicals [2]. Thus, topical drug therapy typically involved the localized administration of medicinal formulations to the skin, generally when the skin surface was breached by disease or infection and a route of drug absorption into the deeper cutaneous layers was consequently open. However, once it was understood that the skin was a semi-permeable membrane rather than a totally impermeable barrier, new possibilities arose for the use of this route as a portal for systemic drug absorption.

In the early twentieth century, it was recognised that lipophilic agents had increased skin permeability, and in 1919 the barrier properties of the skin were attributed specifically to the outermost layers [3]. Scheuplein and Blank thoroughly investigated skin permeability to a wide range of substances *in vitro* [2]. They modelled skin as a three-layer laminate of *stratum corneum*, epidermis and dermis, with drug permeation driven by Fickian diffusion. By digesting the epidermal layer, the *stratum corneum* was separated from the lower layers of the skin and was determined to be the principal barrier to drug absorption.

Transdermal drug delivery refers to the delivery of the drug across intact, healthy skin into the systemic circulation [4]. The diffusive process by which this is achieved is known as percutaneous absorption. The drug initially penetrates through the *stratum corneum* and then passes through the deeper epidermis and dermis, without drug accumulation in the dermal layer [5]. When the drug reaches the dermal layer, it becomes available for systemic absorption via the dermal microcirculation. Thus, classical topical formulations can be distinguished from those intended for transdermal drug delivery in that,

Microneedles for Drug and Vaccine Delivery and Patient Monitoring, First Edition.
Edited by Ryan F. Donnelly, Thakur Raghu Raj Singh, Eneko Larrañeta, and Maelíosa T.C. McCrudden.

whilst the former are generally applied to a broken, diseased or damaged integument, the latter are used exclusively on healthy skin where the barrier function is intact.

It is, indeed, fortuitous for all of us that the skin is a self-repairing organ. This ability, together with the barrier protective properties associated with the integument, is a direct function of skin anatomy. Therefore, in order to develop an effective approach to transdermal drug delivery, it is necessary to be aware of how skin anatomy restricts the percutaneous absorption of exogenously applied chemicals.

1.1 Skin Anatomy

As the largest, and one of the most complex, organs in the human body, the skin is designed to carry out a wide range of functions [6]. Thus, the skin forms a complex membrane with a non-homogenous structure and a surface area of 1.7 m^2, compromising 16% of the total body mass of an average person (Figure 1.1). It contains and protects the internal body organs and fluids, and exercises environmental control over the body with respect to temperature and, to some extent, humidity. In addition, the skin is a communicating organ, relaying the sensations of heat, cold, touch, pressure and pain to the central nervous system.

1.1.1 The Epidermis

The multilayered nature of human skin can be resolved into three distinct layers. These consist of the outermost layer, the epidermis, beneath which lies the much larger dermis and, finally, the deepest layer, the subcutis. The epidermis, which is essentially a stratified epithelium, lies directly above the dermo–epidermal junction. The viable epidermis is often referred to as the epidermal layers below the *stratum corneum*. This provides mechanical support for the epidermis and anchors it to the underlying dermis. The junction itself is a complex glycoprotein structure about 50 nm thick [7].

Directly above the undulating ridges of the dermo–epidermal junction lies the basal layer of the epidermis, the *stratum germinativum*. This layer is single cell in thickness with columnar-to-oval shaped cells, which are actively undergoing mitosis. As the name implies, the *stratum germinativum* generates replacement cells to counterbalance the constant shedding of dead cells from the skin surface. In certain disease states, such as psoriasis, the rate of mitosis in this layer is substantially raised in order to compensate for a diminished epidermal barrier, the epidermal turnover time being as fast as four days. As the cells of the basal layer gradually move upwards through the epidermis, they undergo rapid differentiation, becoming flattened and granular and the ability to divide by mitosis is lost. Directly above the *stratum germinativum* is a layer, several cells in thickness, in which the cells are irregular and polyhedral in shape. This layer is the *stratum spinosum*, and each cell has distinct spines or prickles protruding from the surface in all directions. Although they do not undergo mitosis, the cells of this layer are metabolically active. The prickles of adjacent cells interconnect via desmosomes or intercellular bridges. The increased structural rigidity produced by this arrangement increases the resistance of the skin to abrasion [7].

As the epidermal cells migrate upwards towards the skin surface they become flatter and more granular in appearance, forming the next epidermal layer, the *stratum*

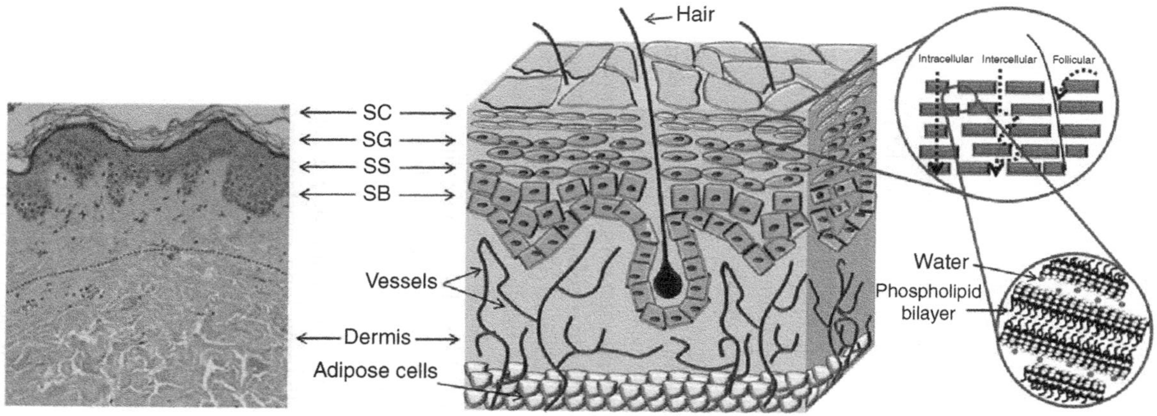

Figure 1.1 Schematic representation of the skin. Comparison with a histological section of mammary skin from a 19 year old patient (left). From inside to outside, the adipose cells, the dermis (the only vascular layer; thickness 0.3–4 mm; fibroblasts, sweat glands and hair follicles are present), the epidermis (thickness 100–150 μm) composed of *stratum basale* (*SB*), *stratum spinosum* (*SS*), *stratum granulosum* (*SG*), and *stratum corneum* (*SC*). Cells differentiate from *SB* to *SC* up to lose their nuclei. In the *SC* cells (corneocytes) are embedded in a matrix of lipid bilayers (brick and mortar model) (right). Penetration routes of molecules are reported with dotted arrows. The number of *SC* layers depends on the body site, age, skin condition and skin hydration (generally 6–20 but also 86 in the heel). Total thickness of *SC* is 10–30 μm. Reproduced with permission from [30] R.J. Scheuplein (1972) Properties of the skin as a membrane. *Adv. Biol. Skin*. 12: 125–152 .

granulosum, consisting of a few layers of granular cells. Their appearance is due to the actively metabolising cells producing granular protein aggregates of keratohyalin, a precursor of keratin [8]. As cells migrate through the *stratum granulosum*, cell organelles undergo intracellular digestion and disappear. The cells of the *stratum granulosum* die due to degeneration of the cell nuclei and metabolic activity ceases towards the top of this layer. A further differentiation of cells above the *stratum granulosum* can be seen in sections taken from thick skin, such as on the palm of the hand or the sole of the foot. This distinct layer of cells is termed the *stratum lucidum*. The cells of this layer are elongated, translucent and anuclear.

1.1.2 The *Stratum Corneum*

The *stratum corneum* is the most superficial layer of the epidermis, and thus the skin. It is in direct contact with the external environment and its barrier properties may be partly related to its very high density (1.4 g/cm^3 in the dry state) [9]. It is now well accepted that this layer constitutes the principal barrier for penetration of most drugs. The horny layer represents the final stage of epidermal cell differentiation. The thickness of this layer is typically 10 μm, but a number of factors, including the degree of hydration and skin location, influence this. The *stratum corneum* consists of 10–25 rows of dead keratinocytes, now called corneocytes, embedded in the secreted lipids from lamellar bodies [10]. The corneocytes are flattened, elongated, dead cells, lacking nuclei and other organelles [11]. The cells are joined together by desmosomes, maintaining the cohesiveness of this layer [12]. The heterogeneous structure of the *stratum corneum* is composed of approximately 75–80% protein, 5–15% lipid and 5–10% other substances on a dry weight basis [13, 14].

The majority of protein present in the *stratum corneum* is keratin and is located within the corneocytes [14]. The keratins are a family of α-helical polypeptides. Individual molecules aggregate to form filaments (7–10 nm diameter and many microns in length) that are stabilised by insoluble disulfide bridges. These filaments are thought to be responsible for the hexagonal shape of the corneocytes and provide mechanical strength for the *stratum corneum* [15]. Corneocytes possess a protein-rich envelope around the periphery of the cell, formed from precursors, such as involucrin, loricrin and cornifin. Transglutaminases catalyse the formation of γ-glutamyl crosslinks between the envelope proteins that render the envelope resistant and highly insoluble. The protein envelope links the corneocytes to the surrounding lipid enriched matrix [12].

The main lipids located in the *stratum corneum* are ceramides, fatty acids, cholesterol, cholesterol sulfate and sterol/wax esters [14]. These lipids are arranged in multiple bilayers called lamellae. Phospholipids are largely absent, a unique feature for a mammalian membrane. The ceramides are the largest group of lipids in the *stratum corneum*, accounting for approximately half of the total lipid mass [16], and are crucial to the lipid organisation of the *stratum corneum* [12].

The bricks and mortar model of the *stratum corneum* (Figure 1.1) is a common representation of this layer [17]. The bricks correspond to parallel plates of dead keratinised corneocytes, and the mortar represents the continuous interstitial lipid matrix. It is important to note that the corneocytes are not actually brick shaped, but rather are polygonal, elongated and flat (0.2–1.5 μm thick and 34.0–46.0 μm in diameter) [11]. The "mortar" is not a homogenous matrix. Rather, lipids are arranged in the lamellar phase (alternating layers of water and lipid bilayers), with some of the lipid bilayers in the gel or crystalline state [18]. The extracellular matrix is further complicated by the presence of intrinsic and extrinsic proteins, such as enzymes. The barrier properties of the *stratum corneum* have been assigned to the multiple lipid bilayers residing in the intercellular space. These bilayers prevent desiccation of the underlying tissues by inhibiting water loss and limit the penetration of substances from the external environment [18].

1.1.3 The Dermis

This region, also known as the corium, underlies the dermo–epidermal junction and varies in thickness from 2 to 4 mm. Collagen, a fibrous protein, is the main component of the dermis and is responsible for the tensile strength of this layer. Elastin, also a fibrous protein, forms a network between the collagen bundles and is responsible for the elasticity of the skin and its resistance to external deforming forces. These protein components are embedded in a gel composed largely of mucopolysaccharides. The skin appendages, such as the sebaceous and sweat glands, together with hair follicles, penetrate this region. Since these open to the external environment they present a possible entry point into the skin [19, 20].

The dermis has a rich blood supply extending to within 0.2 mm of the skin surface and derived from the arterial and venous systems in the subcutaneous tissue. This blood supply consists of microscopic vessels and does not extend into the epidermis. Thus, a drug reaching the dermis through the epidermal barrier will be rapidly absorbed into the systemic circulation, a key advantage of the use of microneedles to by-pass the barrier to drug penetration offered by the *stratum corneum* [6, 11].

1.1.4 Skin Appendages

The skin appendages comprise the hair follicles and associated sebaceous glands, together with the eccrine and apocrine glands. Hairs are formed from compacted plates of keratinocytes, with the hair shaft housed in a hair follicle formed as an epidermal invagination. Associated flask-like sebaceous glands are formed as epidermal outgrowths. The sebaceous gland secretes an oily material (sebum), which plays a role in lubricating the skin surface and maintaining skin pH at around 5 [21]. Hairs can be pigmented or non-pigmented and can extend more than 3 mm into the hypodermis [22]. In humans, the skin density of these units varies with body region. For example, on the face, follicular openings can account for up to 10% of the surface area, whilst on other parts of the body, these orifices make up only 0.1% of the surface area [22]. Thus, a transfollicular route may be important for certain veterinary transdermal drug delivery applications, where the hair follicle density is much higher, but not in humans.

The eccrine glands respond to increased temperature and stress by exuding a dilute salt solution (sweat), where its evaporation plays an important thermoregulatory role. The coiled and tubular eccrine glands are located in the dermal tissue, and are connected to a duct that ascends towards the surface. They are distributed throughout the body surface, being particularly concentrated in the hands and feet [21]. Humans have approximately 3–4 million eccrine glands on their skin, which produce as much as 3 litres of sweat per hour [23]. The apocrine glands are found closer to the epidermal–dermal boundary and are associated with the axillae and ano-genital regions [21]. Apocrine ducts exit to the skin surface via the hair follicle [23].

1.2 Routes to Percutaneous Drug Absorption

It is now well established that the *stratum corneum* is the principal barrier to the percutaneous absorption of exogenous substances, including drugs seeking to use the skin as a portal via transdermal drug delivery. There are three routes by which a drug can, in theory, breach the *stratum corneum* barrier, thus reaching viable tissue and, ultimately, the skin microcirculation (Figure 1.1). From here, entry is made into the systemic circulation to complete the drug absorption process. The available routes are: trans-appendageal, via the hair follicles and sweat glands (sometimes referred to as the shunt route); transcellular, by diffusion through and across the corneocytes; and intercellular, by diffusion through the ordered domains of intercellular skin lipids. The relative contributions of the pathways to the overall drug flux are governed by the physicochemical properties of the permeating molecule, the fractional area of the route and whether drug permeation is facilitated in any way by disruption of the skin barrier.

An elegant model for the percutaneous absorption of a topically applied drug has been proposed [24] based on an analogy between the flow of electrons in an electrical circuit through series and parallel resistors, and the passive diffusional flow of a drug through the resistances offered by the various skin components. The current flow is driven by an electrical potential gradient whereas the diffusional drug flow, in contrast, is driven by a concentration gradient across the skin (Figure 1.2).

Skin diffusional resistances can be thought of as the transepidermal and transappendageal routes, in parallel. The transepidermal resistance is essentially that offered by the *stratum corneum*. As with the ohmic magnitude of an electrical resistance, the chemical magnitude (R) of a membrane resistor with respect to drug diffusion through

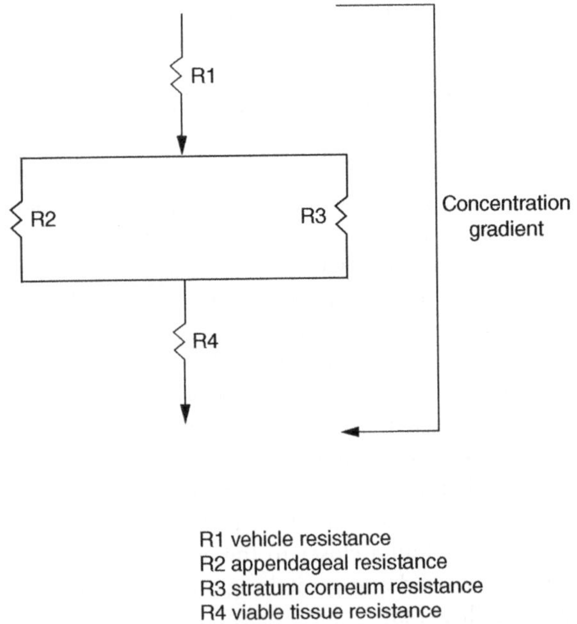

Figure 1.2 Series and parallel resistances to percutaneous drug penetration.

that membrane can be expressed as:

$$R = \frac{h}{FDK} \tag{1.1}$$

where h is the thickness of the resistor membrane, F is the fractional area of the route (where there is more than one pathway involved), D is the diffusion coefficient of the drug through that resistor (the ease of movement of the drug through the tissue) and K represents the capacity of a particular tissue for the drug (in effect, the partition coefficient of the drug between one tissue phase and that immediately preceding it). It follows that the rate of skin penetration of a given drug is inversely proportional to the total diffusional resistance due to the various skin layers and components.

The transepidermal route has a fractional area approaching unity. In the percutaneous absorption process the total diffusional resistance offered by this route would consist of the sum of resistances due to the *stratum corneum*, viable epidermis and dermis. However, any diffusional resistance due to the dermis is minimal compared with that of the *stratum corneum* and can be neglected.

The *stratum corneum* is a narrow layer; hence the value of h in Equation (1.1) is small, thus tending to reduce the diffusional resistance of this layer. However, the main factor to consider is the densely packed, organised anatomical characteristics of this layer, ensuring that its overall resistance to chemical penetration is substantial, notwithstanding the reduced thickness of the horny layer compared with that of the viable epidermis.

The transappendageal route has a very low fractional area [25]. Shunt diffusion of penetrants through the skin appendages appears to be of significance only during the initial phase following application of the drug. The higher diffusion coefficients through the appendages compared with the *stratum corneum* leads to an excess initial penetration via this route, with an exponential relationship with time compared with the linear time dependency of drug penetration that characterises the establishment of steady-state diffusion [2, 25, 26]. Thus, although the transappendageal route may be important initially, its small fractional area suggests that it is of no great significance in the overall percutaneous penetration of most topically applied drugs [27]. Given the tortuous nature of the skin ducts and glands, and the upwards flow of material towards the skin surface opposing the downwards diffusion of an applied drug, it is not surprising that the shunt route is unimportant in steady state drug diffusion through the skin [28]. However, the initial build-up of drug achieved by rapid diffusion along the appendageal route, probably the hair follicles, prior to the establishment of steady state transepidermal diffusion, may explain the appearance of vasoactive phenomena associated with nicotinates (erythema) and steroids (skin blanching), both effects rapidly seen following topical administration of these agents [29].

Since the transappendageal route can be neglected as a major contributor to the overall penetration of non-electrolytes, the overall resistance to the drug reaching its target site of action can be seen as analogous to the flow of current through electrical resistors in series. Thus, the total resistance (R) of the skin to the percutaneous absorption of a diffusing molecule can be described by:

$$R = \frac{h}{F_{sc}D_{sc}K_{sc}} + \frac{h}{F_{c}D_{c}K_{c}} \tag{1.2}$$

where the denominator subscripts *sc* and c refer to the *stratum corneum* and viable epidermis, respectively.

The *stratum corneum* has been shown to have approximately 10^3 times greater resistance to water penetration than the dermis, and is thus even more resistant to the passage of polar solutes [30]. For non-polar lipophilic solutes the *stratum corneum* has a lower resistance than to the passage of water. Although the viable epidermis and the dermis are more resistant to the passage of non-polar compared with polar materials, as might reasonably be expected, this effect is relative and minimal, with only 4% of the total skin resistance being ascribed to these viable layers [30]. It is clear, therefore, that the passage of the drug through the *stratum corneum* is the rate-limiting step for the percutaneous absorption of both polar and non-polar molecules. The decreased resistance of the horny layer to lipophilic drugs dictates the use of lipophilic molecules for conventional transdermal delivery, that is, where diffusion is driven by the drug concentration gradient across the barrier.

Although numerous mathematical models are available to describe the process of percutaneous absorption, that proposed by Flynn and coworkers [31] provides a good description of the overall process involved in the percutaneous absorption of a drug. Where that drug is a relatively low molecular weight, lipophilic molecule, the model can be considerably simplified. Thus, the resistance to drug penetration of the dermis can be neglected since it is minimal compared with that of the *stratum corneum*. The transappendageal route is largely insignificant, and the resistance due to the viable epidermis is so small compared with that due to the *stratum corneum* that it approaches zero. Thus, the *stratum corneum* fractional area can, in this case, be taken as unity. When steady-state diffusion of the drug across the *stratum corneum* barrier has been established, the amount of material passing through the barrier per unit area of vehicle coverage per unit time, that is, the drug flux, J, is given by

$$J = \left(\frac{D_{sc}.K_{sc/w}}{h_{sc}}\right)\Delta C \tag{1.3}$$

where $K_{sc/w}$ represents the partition coefficient between the *stratum corneum* and the formulation vehicle and ΔC is the drug concentration gradient across the *stratum corneum*, which, assuming sink conditions is the effective drug concentration in the vehicle. This equation, which is essentially Fick's first law for a steady state [11, 32] can be simplified to:

$$J = P(\Delta C) \tag{1.4}$$

where P is the permeability coefficient of the drug through the skin; P is described by the term in parentheses in Equation (1.3).

Equation (1.3) provides a guide to those factors that can be acted upon to maximise the efficiency of the percutaneous absorption of a drug through the *stratum corneum* barrier. Clearly, little can be done to reduce the value of h, the barrier thickness, unless an adhesive tape stripping technique is employed [28]. The barrier thickness may be reduced in the event of an existing clinical disease state but otherwise it can be regarded as a constant.

The drug diffusivity in the *stratum corneum*, as measured by D_{sc}, is a physicochemical parameter of the chosen drug or drug combination. Although the barrier characteristics may be altered by the use of a chemical penetration enhancer [11, 14], the relative values of D_{sc} for different drug molecules will retain their same comparative ranking. An increase in the value of $K_{sc/w}$, the vehicle/*stratum corneum* partition coefficient, therefore represents the best available means to ensure that an adequate concentration of drug can penetrate through the *stratum corneum* barrier. In practice, therefore, a conventional approach to transdermal drug delivery via drug diffusion through the *stratum corneum* along a concentration gradient is highly dependent on the physico-chemical properties of the drug, with some limited influence exerted by formulation factors. Hence, for water-soluble or large, particularly macromolecular, actives, other approaches are needed if the transdermal route is to be used to its full potential.

1.3 Facilitated Transdermal Drug Delivery

Transdermal delivery has potential advantages over other conventional routes of drug delivery. It can provide a non-invasive and painless alternative to parenteral routes [19, 32]. Furthermore, the pharmacokinetic profiles of drugs are more uniform with lower variability, resulting in compliance and adherence with drug use [19, 33, 34]. Transdermal administration can be stopped by removal of the patch, which puts the patient in control in the event of an adverse reaction [35]. The skin is the largest organ in the body, which allows many placement options for transdermal absorption and ease of access. TDD is used when there is a significant first-pass effect of the liver, since it avoids pre-systemic metabolism, thus improving bioavailability [20, 34]. In addition, it is also the best route for paediatric patients, and a suitable route for unconscious or vomiting patients or those who rely on self-administration. The skin is known to be a highly immunogenic site for vaccination, because this organ is known to be crowded with dendritic cells in both the epidermal and dermal layers, which play a central role in immune responses [36]. Therefore, TDD is an attractive and novel vaccination route for therapeutic proteins and peptides.

A three-day patch that delivers scopolamine to treat motion sickness was the first transdermal patch to be approved, in 1979 [34]. The biggest challenge for transdermal delivery is that only a limited number of drugs are suitable to be administered transdermally. The number of commercially available transdermal patches that are approved by the US Food and Drug Administration (FDA) is less than 20.

The use of this route is severely limited by the restrictions imposed by the lipophilic *stratum corneum* barrier, which allows a limited number of drug molecules with certain physicochemical properties to be delivered transdermally [20, 34]. These approved molecules are only up to a few hundred Daltons, with octanol–water partition coefficients that heavily favour lipids ($\log P = 1\text{–}3$). Moreover, reasonable potency with doses of milligrams per day or less is required for candidates to become suitable for conventional TDD [15, 34]. Therefore, the transdermal delivery of hydrophilic drugs, peptides and macromolecules, for example, DNA or small-interfering RNA, has posed challenges [34, 36]. Penetration enhancement technology would broaden the range of drugs available for transdermal administration [20]. Technologies used to modify the barrier properties of the *stratum corneum* are classified into passive methods (chemical penetration enhancement) and physical methods.

Passive methods include modulation of formulation excipients and addition of chemical enhancers, in order to temporarily alter the barrier properties of the *stratum corneum* [13, 16]. Passive methods are inexpensive, available and easily incorporated into transdermal patches, such as chemical penetration enhancers

[14, 37]. Ideal penetration enhancers should be non-toxic, non-allergic, inert and work unidirectionally [14]. Chemical penetration enhancers facilitate drug permeation across the skin by various mechanisms without long-term damage to the skin [14]. The mechanisms of action of penetration enhancers are complex, so it is difficult to classify them accordingly. They have several mechanisms of action such as: enhancing solubility, improving partitioning between the formulation and the *stratum corneum*, fluidising the *stratum corneum* lipid bilayers, interaction with intercellular proteins, causing dissolution and disruption of *stratum corneum* lipids [11, 14, 26, 32]. In addition, they may enhance diffusion across the skin by increasing the diffusion coefficient of the drug in the *stratum corneum* through disruption of the barrier properties of the *stratum corneum* and increasing the drug's thermodynamic activity [14, 25, 32]. The most widely studied penetration enhancers are water, polyols, sulfoxides, azone, pyrrolidones, essential oils, surfactants, terpenes, fatty acids and urea [14]. Nonetheless, the major drawbacks of passive methods are that only modest degrees of increased flux can be achieved in practice and there is a time lag in drug release. Moreover, unacceptable skin irritation could occur, especially when using more than one enhancer in the formulation.

The recent trend in TDD is to use a novel vehicle. Recently, researchers have designed suitable drug delivery vehicles for transdermal patches, such as emugel, proniosomes, microemulsions and nanoemulsions, into the field of penetration enhancers [38]. Ammar and coworkers [39] designed a new transdermal formulation for tenoxicam and showed that proniosomal gels act as penetration enhancers that enhance the drug permeation from the skin barrier. Proniosomal gel formulations showed a significantly higher therapeutic compared with the oral tenoxicam tablets of the same dose on the market, thus revealing a more promising tenoxicam dosage form [39]. Barakat and coworkers [40] showed that nanoemulsions can be used as potential vehicles for improved transdermal delivery of indomethacin as an approach to eliminate the side effects of the oral dose.

Modulation of formulation excipients and addition of chemical enhancers can increase drug flux, but not sufficiently to ensure delivery of a pharmacologically effective concentration of the drug. Therefore, several new active rate-controlled transdermal drug delivery technologies (electrically-based, structure-based, velocity-based, etc.) have been developed for the transdermal delivery of wide ranges of drugs [41]. This is particularly of interest given the high economic value of the transdermal delivery market, despite the relatively small number of actives that can be delivered by this route [42]. Broadly, facilitated delivery falls into two categories: technological [42], of which microneedles, the subject of this text, is a good example; and formulation approaches, most notably the focus on nanoscale delivery systems [43]. The following are some of the technologies presently being considered as aids to transdermal drug delivery.

1.3.1 Electrical-based Devices

1.3.1.1 Iontophoresis

Perhaps the oldest method in use for facilitated transdermal delivery, this technique employs the application of physiologically acceptable electrical currents (0.1–1.0 mA/cm^2) to drive charged drugs into the skin through electrostatic effects and make ionic drugs pass through the skin into the body by its potential gradient [20, 33,

44–46]. Unlike other transdermal enhancement methodologies, drug transport across the skin is facilitated by two primary mechanisms: electrorepulsion and electroosmosis. Using electrorepulsion, whereby like charges repel each other, delivery of a positively charged drug can be achieved by dissolving the drug in a suitable vehicle in contact with an electrode of similar polarity (anode). The most frequently used electrodes are aluminium foil, platinum and silver/silver chloride electrodes [44, 45, 47]. The one that is most preferred is the Ag/AgCl, because the electrode materials used for iontophoretic delivery need to be harmless to the body and flexible in order to be applied close to the body surface [44]. It also resists the changes in pH.

Application of a small direct current (approximately 0.5 mA/cm^2), causes the drug to be repelled from the anode, and it is attracted towards to the oppositely charged electrode (cathode) [20, 44, 48]. This process is termed anodal iontophoresis. Conversely, cathodal iontophoresis occurs when anions are repelled from the cathode towards the anode. Importantly, iontophoresis is not reserved for just charged drugs, delivery of small neutral molecules may also be enhanced through electroosmosis.

Many factors influence iontophoresis, including pH of the donor solution, electrode type, drug concentration, buffer concentration, current intensity and duration, patients' anatomical factors and the type of current employed [20, 47–49]. At pH values above 4, the skin is negatively charged, due to ionisation of carboxylic acid groups within the membrane. Positively charged ions, such as Na^+, are more easily transported, as they attempt to neutralise the charge in the skin, hence there is a flow of Na^+ to the cathode [13]. Owing to a net build-up of NaCl at the cathodal compartment, osmotic flow of water is induced from the anode to the cathode. This net flow of water facilitates transfer of neutral molecules across the skin [45, 48].

Regarding current intensity and duration factor, there is a proportional relationship between the current and drug flux across the skin, but the current is limited to 1 mA due to sensitivity, safety concerns and tolerance of the patients. As with increasing current, the risk of nonspecific vascular reactions (vasodilatation) also increases [47]. In addition, the duration of current that can be applied is up to 3 min, in order to prevent local skin irritation or burns. The maximum physiologically acceptable iontophoretic current is 0.5 mA/cm^2 [50]. Therefore, the current should be sufficiently high to deliver more drugs but it should not produce harmful effects to the skin [51]. The use of continuous direct current (DC) can reduce the drugs flux due to its polarization effect on the skin. Owing to its polarization effect on the skin, pulsed or modulated current has been used [49]. However, a limited number of studies have been carried out comparing pulsed direct current iontophoresis versus conventional direct current iontophoresis. Recently, Kotzki *et al.*, in 2015, concluded that pulsed iontophoresis of treprostinil significantly enhanced cutaneous blood flow compared with continuous iontophoresis [49]. The transappendageal route is thought to offer the path of least electrical resistance across the skin and is suggested to be the principal pathway taken by a permeant during iontophoresis [52].

The use of iontophoresis as a technique to enhance the delivery of low as well as high molecular weight compounds has not been studied extensively, although there was a study on whether molecules with a molecular weight of less than 12 000 Da could be successfully delivered across skin via iontophoresis [50]. However, small proteins, such as cytochrome c (12.4 kDa) [53, 54] and ribonuclease A, (13.6 kDa) were successfully delivered across porcine and human skin [47, 55]. Recently, biologically active human basic fibroblast growth factor (hbFGF; 17.4 kDa) was also able to be delivered in therapeutically relevant amounts, corresponding to those used in clinical trials and animal studies via iontophoresis [56–58].

Iontophoresis can be used in the drug delivery of therapeutic and diagnostic applications, for example, diagnosing cystic fibrosis [59] and recently for monitoring blood glucose levels [45]. The major benefit of using iontophoresis in diagnostic applications is that there is no mechanical penetration or disruption of the skin involved in this approach [48, 51, 52]. There are a number of commercially available FDA approved iontophoretic delivery systems, such as Phoresor®, Lidosite® and Zecuity®. The first commercially marketed iontophoretic patch system was LidoSite, which was developed to deliver lidocaine for fast dermal anaesthesia. The LidoSite system is comprised of a patch loaded drug, re-usable battery-powered controller and a flexible interconnect module.

1.3.1.2 Electroporation

In contrast to iontophoresis, which uses small voltages (<10 V), electroporation employs relatively high voltage pulses (10–1000 V) for brief periods of time (less than a few hundred milliseconds) [60, 61]. Electroporation has been shown to increase transport across the skin for different molecular weight drugs ranging from small, such as fentanyl, timolol [61, 62] or orcalcein [63], to high molecular weight drugs such as luteinizing hormone releasing hormone (LHRH), calcitonin, heparin or fluorescein isothiocyanate (FITC)–dextran, with molecular weights up to 40 kDa [64–67]. The technique was first described by Neumann *et al.* in 1982 [60]. When applied to the *stratum corneum*, cells are temporarily exposed to high intensities of electrical pulses, which are thought to induce formation of aqueous pores in the lipid bilayers, thus allowing the diffusion of drugs across the skin [64]. The aqueous pores may facilitate drug transport by passive diffusion, electroosmosis or iontophoresis during the pulse. Transdermal delivery of charged molecules can be further enhanced by iontophoretic transport through the transfollicular pathway during pulsation [68]. Most recently, a laser microporation technology has been described and successfully demonstrated *ex vivo* [69]. However, the major drawbacks are the lack of quantitative delivery and potential damage to labile drugs, such as those of a protein origin [70]. Another problem is the lack of safety profile evaluations, since some studies showed that cell death may be occurring with high-voltage pulses [61, 63, 70].

1.3.1.3 Ultrasound

Ultrasound, sonophoresis or phonophoresis can be defined as physical enhancers for drugs across the skin by application of longitudinal pressure waves at frequencies of 20 kHz–16 MHz, which reduce the resistance of the *stratum corneum* [25, 71]. However, the mechanisms of action are less clearly understood or characterised [72–75]. Many mechanisms have been proposed, including thermal effects and cavitation effects [33]. Researchers initially focused on high-frequency ultrasound (≥1 MHz) to enhance the transdermal drug delivery [25]. Recently, research has focused on the use of low-frequency ultrasound (≤100 kHz), since cavitation is believed to be the predominant mechanism in the enhancement of TDD at these frequencies [1, 72, 76]. The use of low-frequency ultrasound for the transdermal delivery of drugs, referred to as low-frequency sonophoresis, has enabled the delivery of a wide range of drugs, regardless of their electrical characteristics, including both hydrophilic molecules and macromolecules [33, 72, 76, 77].

Fellinger and Schmidt introduced the concept of the use ultrasound in TDD in 1950 for the successful treatment of polyarthritis using hydrocortisone ointment combined with sonophoresis [see 72, 74]. Nevertheless, the first FDA approved ultrasound device for transdermal application was in 2004 for the delivery of local dermal anaesthesia, developed by Sontra Medical, the SonoPrep®. Since that time, it has become widely used in the field of TDD in the treatment of many other diseases, including bone joint diseases and bursitis [71]. Recent research has demonstrated the feasibility of delivering proteins, hormones, vaccines, liposomes and other nanoparticles through treated skin [75, 78]. *In vivo* studies have also established that low-frequency sonophoresis can act as a physical immunisation adjuvant. Low-frequency ultrasound (frequencies below 100 kHz) has been used to enhance delivery of a range of low and high molecular weight drugs across the skin [75, 79]. *In vitro* studies using human *stratum corneum* demonstrate enhanced transport (by several orders of magnitude) of the macromolecules insulin, interferon-γ and erythropoietin using low-frequency ultrasound [78, 80]. Park and collaborators [81] reported the use of a compact, lightweight, low-frequency transducer to enhance transdermal insulin delivery. The ultrasound-treated group showed a significant reduction in blood glucose, compared with the control. The authors proposed that the device was capable of safely reducing blood glucose to within a normal range.

In vitro and *in vivo* studies have demonstrated the efficacy of sonophoresis, with some studies reporting up to 1000-fold better penetration compared with simple topical application. However, challenges remain in terms of gaining a full understanding of how the technology operates, broadening the range of drugs that can be delivered and fully evaluating its safety profile [1, 80]. Singer *et al.* [76] demonstrated that low-intensity ultrasound induced only minor skin reactions in dogs, but high-intensity ultrasound was capable of causing second-degree burns. Wang and coworkers [56] showed that with the application of ultrasound, the delivery of yellow–green fluorescent nanoparticles and high molecular weight hyaluronic acid (HA) in the skin samples could be raised above its passive diffusion permeability.

1.3.1.4 Cryopneumatic and Photopneumatic Technologies

Novel approaches to facilitated transdermal delivery have recently been reported [82] using cryopneumatic technology and photopneumatic technology to enhance the permeation of the *stratum corneum*. Cryopneumatic technology produces micro-cracks at the skin surface by successively freezing and stretching the skin with vacuum suction. Photopneumatic technology combines stretching of the skin by vacuum suction with intense pulsed light. The enhancing effects of both methods were studied on *ex vivo* porcine skin and *in vivo* human skin models using fluorescent hydrophilic macromolecules as drug surrogates. It was shown that the enhancing effect of cryopneumatic technology is due to drug permeation through the micro-cracks produced by freezing–stretching cycles, while photopneumatic technology could promote drug permeation through sweat glands.

1.3.1.5 Velocity-based Devices

Velocity-based devices use a high-velocity jet ranging from 100 to 200 m/s to puncture the skin and deliver drugs using a power source, which can be powder or liquid jet injections [83]. Transdermal powder delivery is where the therapeutic compound is formulated as a fine powder (20–100 μm diameter) and is accelerated in a supersonic flow of helium gas to penetrate the skin [84]. The PMED® (Pfizer) device, formerly known as PowderJect®, has been reported to successfully deliver, for example, vaccines [85, 86] and lidocaine [87]. Dry-powder formulations are generally more stable than solutions and can eliminate the need for the "cold chain" to be maintained when using vaccines. This would be particularly advantageous for large-scale immunisation in developing countries with hot climates.

Liquid jet injectors consist of a power source (compressed gas or spring), piston, drug-loaded compartment and a nozzle with an orifice size typically ranging between 150 and 300 μm [84]. Upon triggering the actuation mechanism, the power source pushes the piston that impacts the drug-loaded compartment, thereby leading to a quick increase in pressure [88]. This forces the drug solution through the nozzle orifice as a liquid jet with a velocity ranging between 100 and 200 m/s.

It is claimed that needle-free injection has several potential benefits. The fear of needles can be avoided, specific skin strata can be targeted and accidental needle stick injuries can also be avoided [89]. However, there are some drawbacks, including the issue that dosing accuracy may vary due to skin variability between patients and also the risk of cross contamination cannot be excluded, since splashback of interstitial liquid from the skin may contaminate the nozzle. So the use of multi-use nozzle jet injectors has been discontinued and multi-dose drug deliveries are now only used to the same individual, for example the Tjet® device, which delivers somatropin (human growth hormone (hGH)). The long-term side effects of high-speed particles or liquids on the skin are not known and some jet injection technologies have resulted in reports of adverse reactions [90].

1.3.1.6 Thermal Approaches (Lasers and Radiofrequency Heating)

Thermal- or energy-based ablation is a minimally invasive method used to deliver drugs systemically through the skin by application of heat to the surface of the skin, which depletes the *stratum corneum* selectively at that site of heating only, without damaging deeper tissues [91]. Several techniques can be used to cause thermal ablation, such as lasers [92] or radiofrequencies (RF) [93]. In order to generate the high temperatures needed to ablate the *stratum corneum* without damaging the underlined epidermis, the exposure time should be short. With radiofrequency, thermal ablation involves the placement of a needle electrode directly into the skin and exposing skin cells to a high frequency (100–500 kHz), causing ionic vibrations within the tissue, which attempts to localize the heating to a specific area of the skin, resulting in increased drug transport across the skin [93].

1.3.1.7 Microneedles

Of all the available facilitated transdermal drug delivery technologies, microneedles are presently attracting the most interest. A search of the scientific literature over the past five years using the term "transdermal delivery technologies" reveals that around 30% of published studies involve microneedles. The first report of microneedle assisted topical drug delivery was in the late 1990s, whereby puncturing the skin using micron-sized needles was shown to increase permeability of human skin to a model drug, calcein [94]. Subsequently, there has been intense interest in this technology, with significant developments being made both in the fields of microneedle fabrication and drug delivery.

Microneedle arrays are manufactured based on etching methods used by the microelectronics industry, to create arrays of micron sized needles [95, 96]. The majority of studies to date have used silicon or metal microneedles, although devices have also been made from dextrin [97, 98], glass [99], maltose [100] and various polymers [101, 102].

Microneedles can be made of varying length, as short as 25 μm and as long as 2000 μm. In addition, the base diameter of the needle and needle density can also be altered. These devices have been shown to penetrate across the *stratum corneum* and into the viable epidermis, avoiding nerve fibres and blood vessels that reside primarily in the dermal layer. The overriding benefit of using microneedles is the promise of pain-free injection of active pharmaceutical ingredients of both small and large molecular weight [103]. Therefore, in the present text, the focus is on emerging microneedle technologies [104] and the possibilities that they offer for the future in widening the scope and applications of transdermal drug delivery.

References

1 R. Liuzzi, A. Carciati, S. Guido and S. Caserta (2016). Transport efficiency in transdermal drug delivery: What is the role of fluid microstructure? *Colloids Surf., B: Biointerfaces* 139: 294–305.

2 R.J. Scheuplein and I.H. Blank (1971). Permeability of the skin. *Physiol. Rev.* 51: 702–747.

3 H.W. Smith, G. Clowes and J. Marshal (1919). On dichloroethyl sulfide (mustard gas). IV. The mechanism of absorption by the skin. *J. Pharmacol. Exp. Ther.* 13: 1–30.

4 A.Z. Alkilani, M.T. McCrudden and R.F. Donnelly (2015). Transdermal drug delivery: innovative pharmaceutical developments based on disruption of the barrier properties of the stratum corneum. *Pharmaceutics* 7: 438–470.

5 R.F. Donnelly, T.R.R. Singh, D.I. Morrow and A.D. Woolfson (2012). *Microneedle-Mediated Transdermal and Intradermal Drug Delivery*. Wiley.

6 C.M. Chuong, B.J. Nickoloff, P.M. Elias, *et al.* (2002). What is the 'true' function of skin? *Exp. Dermatol.* 11: 159–187.

7 M. Aumailley and P. Rousselle (1999). Laminins of the dermo–epidermal junction. *Matrix Biol.* 18: 19–28.

8 E.P. Reaven and A.J. Cox (1965). Histidine and keratinization. *J. Invest. Dermatol.* 45: 422–431.

9 G. El Maghraby, B. Barry and A. Williams (2008). Liposomes and skin: from drug delivery to model membranes. *Eur. J. Pharm. Sci.* 34: 203–222.

10 J.W. Wiechers (1989). The barrier function of the skin in relation to percutaneous absorption of drugs. *Pharm. World Sci.* 11: 185–198.

11 H.A. Benson (2005). Transdermal drug delivery: penetration enhancement techniques. *Curr. Drug Delivery* 2: 23–33.

12 G.K. Menon (2002). New insights into skin structure: scratching the surface. *Adv. Drug Deliv. Rev.* 54: S3–S17.

13 B.W. Barry (2001). Novel mechanisms and devices to enable successful transdermal drug delivery. *Eur. J. Pharm. Sci.* 14: 101–114.
14 A.C. Williams and B.W. Barry (2012). Penetration enhancers. *Adv. Drug Deliv. Rev.* 64: 128–137.
15 J.M. Jensen and E. Proksch (2009). The skin's barrier. *G. Ital. Dermatol. Venereol.* 144: 689–700.
16 C.S. Asbill, A.F. El-Kattan and B. Michniak (2000). Enhancement of transdermal drug delivery: chemical and physical approaches. *Crit. Rev. Ther. Drug Carrier Syst.* 17.
17 A. Michaels, S. Chandrasekaran and J. Shaw (1975). Drug permeation through human skin: Theory and in vitro experimental measurement. *AIChE J.* 21: 985–996.
18 J.A. Bouwstra, G.S. Gooris, J.A. van der Spek and W. Bras (1991). Structural investigations of human stratum corneum by small-angle X-ray scattering. *J. Invest. Dermatol.* 97: 1005–1012.
19 A. Naik, Y.N. Kalia and R.H. Guy (2000) Transdermal drug delivery: overcoming the skin's barrier function. *Pharm. Sci. Technol. Today* 3: 318–326.
20 T. Gratieri, I. Alberti, M. Lapteva and Y.N. Kalia (2013). Next generation intra-and transdermal therapeutic systems: using non-and minimally-invasive technologies to increase drug delivery into and across the skin. *Eur. J. Pharm. Sci.* 50 (5): 609–622.
21 S. Singh and J. Singh (1993). Transdermal drug delivery by passive diffusion and iontophoresis: a review. *Med. Res. Rev.* 13: 569–621.
22 V.M. Meidan (2010). Methods for quantifying intrafollicular drug delivery: a critical appraisal. *Expert Opin. Drug Delivery* 7: 1095–1108.
23 D.J. Tobin (2006). Biochemistry of human skin–our brain on the outside. *Chem. Soc. Rev.* 35: 52–67.
24 G.L. Flynn (1990). Topical drug absorption and topical pharmaceutical systems. *Drugs Pharm. Sci.* 40: 263–325.
25 D. Brambilla, P. Luciani and J. Leroux (2014) Breakthrough discoveries in drug delivery technologies: The next 30 years. *J. Controlled Release* 190: 9–14.
26 M.B. Brown, M.J. Traynor, G.P. Martin and F.K. Akomeah (2008) Transdermal drug delivery systems: skin perturbation devices. *Anonymous Drug Delivery Systems*, 119–139. Springer.
27 R.J. Scheuplein (1976) Percutaneous absorption after twenty-five years: or "old wine in new wineskins", *J. Invest. Dermatol.* 67: 31–38.
28 I.H. Blank (1965) Cutaneous barriers** from the research laboratories of the Department of Dermatology of the Harvard Medical School at the Massachusetts General Hospital, Boston, Massachusetts 02114. *J. Invest. Dermatol.* 45: 249–256.
29 R.B. Stoughton (1972) Some bioassay methods for measuring percutaneous absorption. *Adv. Biol. Skin.* 12: 535–546.
30 R.J. Scheuplein (1972) Properties of the skin as a membrane. *Adv. Biol. Skin.* 12: 125–152.
31 G. Flynn, S.H. Yalkowsky and T. Roseman (1974) Mass transport phenomena and models: theoretical concepts. *J. Pharm. Sci.* 63: 479–510.
32 L. Brown and R. Langer (1988) Transdermal delivery of drugs. *Annu. Rev. Med.* 39: 221–229.

33 C.M. Schoellhammer, D. Blankschtein and R. Langer (2014) Skin permeabilization for transdermal drug delivery: recent advances and future prospects. *Expert Opin. Drug Delivery* 11: 393–407.

34 R.H. Guy (2010). Transdermal drug delivery. *Anonymous Drug Delivery*, 399 410. Springer.

35 M.N. Pastore, Y.N. Kalia, M. Horstmann and M.S. Roberts (2015). Transdermal patches: history, development and pharmacology. *Br. J. Pharmacol.* 172: 2179–2209.

36 Y. Kim and M.R. Prausnitz (2011). Enabling skin vaccination using new delivery technologies. *Drug Delivery Transl. Res.* 1: 7–12.

37 K. Ita (2016). Transdermal iontophoretic drug delivery: advances and challenges. *J. Drug Target.* 24: 386–391.

38 K. Rehman and M.H. Zulfakar (2014). Recent advances in gel technologies for topical and transdermal drug delivery. *Drug Dev. Ind. Pharm.* 40: 433–440.

39 H. Ammar, M. Ghorab, S. El-Nahhas and I. Higazy (2011). Proniosomes as a carrier system for transdermal delivery of tenoxicam. *Int. J. Pharm.* 405: 142–152.

40 N. Barakat, E. Fouad and A. Elmedany (2011). Formulation design of indomethacin-loaded nanoemulsion for transdermal delivery. *Pharm. Anal. Acta* 10: 1–8.

41 R. Kumar and A. Philip (2007). Modified transdermal technologies: Breaking the barriers of drug permeation via the skin. *Trop. J. Pharm. Res.* 6: 633–644.

42 R.K. Subedi, S.Y. Oh, M. Chun and H. Choi (2010). Recent advances in transdermal drug delivery. *Arch. Pharm. Res.* 33: 339–351.

43 B. Baroli, M.G. Ennas, F. Loffredo, *et al.* (2007). Penetration of metallic nanoparticles in human full-thickness skin. *J. Invest. Dermatol.* 127: 1701–1712.

44 A. Khan, M. Yasir, M. Asif, *et al.* (2011). Iontophoretic drug delivery: history and applications. *J. Appl. Pharm. Sci.* 1: 11–24.

45 E. Krueger, J.L. Claudino Junior, E.M. Scheeren, *et al.* (2014). Iontophoresis: principles and applications. *Fisioterapia em Movimento* 27: 469–481.

46 S. Lakshmanan, G.K. Gupta, P. Avci, *et al.* (2014). Physical energy for drug delivery; poration, concentration and activation. *Adv. Drug Deliv. Rev.* 71: 98–114.

47 N. Dixit, V. Bali, S. Baboota, *et al.* (2007). Iontophoresis-an approach for controlled drug delivery: a review. *Curr. Drug Deliv.* 4: 1–10.

48 Y.N. Kalia, A. Naik, J. Garrison and R.H. Guy (2004). Iontophoretic drug delivery. *Adv. Drug Deliv. Rev.* 56: 619–658.

49 S. Kotzki, M. Roustit, C. Arnaud, *et al.* (2015). Effect of continuous vs pulsed iontophoresis of treprostinil on skin blood flow. *Eur. J. Pharm. Sci.* 72: 21–25.

50 M. Roustit, F. Gaillard-Bigot, S. Blaise, *et al.* (2014) Cutaneous iontophoresis of treprostinil in systemic sclerosis: a proof-of-concept study. *Clin. Pharmacol. Ther.* 95: 439–445.

51 O. Pillai, V. Nair and R. Panchagnula (2004). Transdermal iontophoresis of insulin: IV. Influence of chemical enhancers. *Int. J. Pharm.* 269: 109–120.

52 B. Priya, T. Rashmi and M. Bozena (2006). Transdermal iontophoresis. *Expert Opin. Drug Deliv.* 3: 127–138.

53 T. Gratieri and Y.N. Kalia (2014). Targeted local simultaneous iontophoresis of chemotherapeutics for topical therapy of head and neck cancers. *Int. J. Pharm.* 460: 24–27.

54 S. Dubey and Y. Kalia (2010). Non-invasive iontophoretic delivery of enzymatically active ribonuclease A (13.6 kDa) across intact porcine and human skins. *J. Controlled Release* 145: 203–209.

55 S. Dubey, R. Perozzo, L. Scapozza and Y. Kalia (2011). Non-invasive electrically-assisted transdermal delivery of human basic fibroblast growth factor. *Mol. Pharm.* 8: 1322–1331.

56 H. Wang, P. Fan, X. Guo, *et al.* (2016). Ultrasound-mediated transdermal drug delivery of fluorescent nanoparticles and hyaluronic acid into porcine skin in vitro. *Chin. Phys. B* 25: 124314.

57 Y. Wang, R. Thakur, Q. Fan and B. Michniak (2005). Transdermal iontophoresis: combination strategies to improve transdermal iontophoretic drug delivery. *Eur. J. Pharm. Biopharm.* 60: 179–191.

58 S. Dubey and Y. Kalia (2014). Understanding the poor iontophoretic transport of lysozyme across the skin: When high charge and high electrophoretic mobility are not enough. *J. Controlled Release.* 183: 35–42.

59 V.A. LeGrys, J.R. Yankaskas, L.M. Quittell, *et al.* (2007). Diagnostic sweat testing: the Cystic Fibrosis Foundation guidelines. *J. Pediatr.* 151: 85–89.

60 E. Neumann, M. Schaefer-Ridder, Y. Wang and P.H. Hofschneider (1982). Gene transfer into mouse lyoma cells by electroporation in high electric fields. *EMBO J.* 1: 841–845.

61 A.R. Denet, R. Vanbever and V. Préat (2004). Skin electroporation for transdermal and topical drug delivery. *Adv. Drug Delivery Rev.* 65 (5): 659–674.

62 A. Denet and V. Preat (2003). Transdermal delivery of timolol by electroporation through human skin. *J. Controlled Release* 88: 253–262.

63 M.R. Prausnitz, V.G. Bose, R. Langer and J.C. Weaver (1993). Electroporation of mammalian skin: a mechanism to enhance transdermal drug delivery. *Proc. Natl. Acad. Sci. U. S. A.* 90 10504–10508.

64 M.R. Prausnitz, E. Edelman, J. Gimm, *et al.* (1995). Transdermal delivery of heparin by skin electroporation, *Biotechnology* 13: 1205–1209.

65 C. Lombry, N. Dujardin and V. Préat (2000). Transdermal delivery of macromolecules using skin electroporation. *Pharm. Res.* 17: 32–37.

66 D.B. Bommannan, J. Tamada, L. Leung and R.O. Potts (1994). Effect of electroporation on transdermal lontophoretic delivery of luteinizing hormone releasing hormone (LHRH) in vitro. *Pharm. Res.* 11: 1809–1814.

67 S. Chang, G.A. Hofmann, L. Zhang, *et al.* (2000). The effect of electroporation on iontophoretic transdermal delivery of calcium regulating hormones, *J. Controlled Release.* 66: 127–133.

68 K. Sung, J. Fang, J. Wang and O.Y. Hu (2003). Transdermal delivery of nalbuphine and its prodrugs by electroporation. *Eur. J. Pharm. Sci.* 18: 63–70.

69 Y. Bachhav, S. Summer, A. Heinrich, *et al.* (2010). Effect of controlled laser microporation on drug transport kinetics into and across the skin. *J. Controlled Release.* 146: 31–36.

70 J. Yi, A.J. Barrow, N. Yu and B.E. O'Neill (2013). Efficient electroporation of liposomes doped with pore stabilizing nisin. *J. Liposome Res.* 1–6.

71 T. Han and D.B. Das (2015). Potential of combined ultrasound and microneedles for enhanced transdermal drug permeation: A review. *Eur. J. Pharm. Biopharm.* 89: 312–328.

72 D.M. Skauen and G.M. (1984). Zentner, Phonophoresis. *Int. J. Pharm.* 20: 235–245.
73 S.T. Sonis (2016) Ultrasound-mediated drug delivery. *Oral Dis.* 23: 135–138.
74 J. Simonin (1995). On the mechanisms of in vitro and in vivo phonophoresis. *J. Controlled Release.* 33: 125–141.
75 I. Lavon and J. Kost (2004). Ultrasound and transdermal drug delivery. *Drug Discov. Today* 9: 670–676.
76 A.J. Singer, C.S. Homan, A.L. Church and S.A. McClain (1998). Low-frequency sonophoresis: Pathologic and thermal effects in dogs. *Acad. Emerg. Med.* 5: 35–40.
77 J. Lepselter, A. Britva, Z. Karni and M.C. Issa (2016) Ultrasound-assisted drug delivery in fractional cutaneous applications. *Anonymous Lasers, Lights and Other Technologies*, 1–16. Springer.
78 S. Mitragotri, D. Blankschtein and R. Langer (1995). Ultrasound-mediated transdermal protein delivery. *Science* 269: 850.
79 G. Merino, Y.N. Kalia and R.H. Guy (2003). Ultrasound-enhanced transdermal transport. *J. Pharm. Sci.* 92: 1125–1137.
80 S. Mitragotri (2005). Healing sound: the use of ultrasound in drug delivery and other therapeutic applications. *Nat. Rev. Drug Discovery* 4: 255–260.
81 E. Park, J. Werner and N.B. Smith (2007). Ultrasound mediated transdermal insulin delivery in pigs using a lightweight transducer. *Pharm. Res.* 24: 1396–1401.
82 F. Sun, R. Anderson and G. Aguilar (2010). Stratum corneum permeation and percutaneous drug delivery of hydrophilic molecules enhanced by cryopneumatic and photopneumatic technologies. *J. Drugs Dermatol.* 9: 1528–1530.
83 J.C. Stachowiak, T.H. Li, A. Arora, *et al.* (2009). Dynamic control of needle-free jet injection. *J. Controlled Release* 135: 104–112.
84 A. Arora, I. Hakim, J. Baxter, *et al.* (2007). Needle-free delivery of macromolecules across the skin by nanoliter-volume pulsed microjets. *Proc. Natl. Acad. Sci. U. S. A.* 104: 4255–4260.
85 H.J. Dean and D. Chen (2004). Epidermal powder immunization against influenza. *Vaccine* 23: 681–686.
86 L.K. Roberts, L.J. Barr, D.H. Fuller, *et al.* (2005). Clinical safety and efficacy of a powdered Hepatitis B nucleic acid vaccine delivered to the epidermis by a commercial prototype device. *Vaccine* 23: 4867–4878.
87 A.R. Wolf, P.A. Stoddart, P.J. Murphy and M. Sasada (2002). Rapid skin anaesthesia using high velocity lignocaine particles: a prospective placebo controlled trial. *Arch. Dis. Child.* 86: 309–312.
88 S. Mitragotri (2006). Current status and future prospects of needle-free liquid jet injectors. *Nat. Rev. Drug Discovery* 5: 543–548.
89 K. Benedek, E. Walker, L.A. Doshier and R. Stout (2005). Studies on the use of needle-free injection device on proteins. *J. Chromatogr. A* 1079: 397–407.
90 C. Houtzagers, A.P. Visser, P. Berntzen, *et al.* (1988). The Medi-Jector II: Efficacy and acceptability in insulin-dependent diabetic patients with and without needle phobia. *Diabetic Med.* 5: 135–138.
91 S. Mitragotri (2013). Devices for overcoming biological barriers: The use of physical forces to disrupt the barriers. *Adv. Drug Deliv. Rev.* 65: 100–103.
92 S. Giannos (2014). Skin microporation: strategies to enhance and expand transdermal drug delivery. *J. Drug Delivery Sci. Technol.* 24: 293–299.

93 J. Park, J. Kim, H.J. Park and W. Kim (2016). Evaluation of safety and efficacy of noninvasive radiofrequency technology for submental rejuvenation. *Lasers Med. Sci.* 31: 1599–1605.

94 S. Henry, D.V. McAllister, M.G. Allen and M.R. Prausnitz (1998). Microfabricated microneedles: a novel approach to transdermal drug delivery. *J. Pharm. Sci.* 87: 922–925.

95 J.Z. Hilt and N.A. Peppas (2005). Microfabricated drug delivery devices. *Int. J. Pharm.* 306: 15–23.

96 S.L. Tao and T.A. Desai (2003). Microfabricated drug delivery systems: from particles to pores. *Adv. Drug Deliv. Rev.* 55: 315–328.

97 Y. Ito, E. Hagiwara, A. Saeki, *et al.* (2006). Feasibility of microneedles for percutaneous absorption of insulin. *Eur. J. Pharm. Sci.* 29: 82–88.

98 Y. Ito, J. Yoshimitsu, K. Shiroyama, et al. (2006). Self-dissolving microneedles for the percutaneous absorption of EPO in mice. *J. Drug Target.* 14: 255–261.

99 P.M. Wang, M. Cornwell, J. Hill and M.R. Prausnitz (2006). Precise microinjection into skin using hollow microneedles. *J. Invest. Dermatol.* 126: 1080–1087.

100 C.S. Kolli and A.K. Banga (2008). Characterization of solid maltose microneedles and their use for transdermal delivery. *Pharm. Res.* 25: 104–113.

101 J. Park, M.G. Allen and M.R. Prausnitz (2006). Polymer microneedles for controlled-release drug delivery. *Pharm. Res.* 23: 1008–1019.

102 J. Park, M.G. Allen and M.R. Prausnitz (2005). Biodegradable polymer microneedles: fabrication, mechanics and transdermal drug delivery. *J. Controlled Release* 104: 51–66.

103 S. Kaushik, A.H. Hord, D.D. Denson, *et al.* (2001). Lack of pain associated with microfabricated microneedles. *Anesth. Analg.* 92: 502–504.

104 R.F. Donnelly, T.R.R. Singh and A.D. Woolfson (2010). Microneedle-based drug delivery systems: microfabrication, drug delivery, and safety. *Drug Deliv.* 17: 187–207.

2

Microneedle Manufacturing and Testing

Eneko Larrañeta and Thakur Raghu Raj Singh

School of Pharmacy, Queen's University Belfast, Belfast BT9 7BL, UK

2.1 Introduction

Techniques that are principally to assist in skin microporation, to desired depths, are gaining importance to enhance transdermal drug delivery. The use of microneedles (MNs) is included in this group of techniques. MNs consist of a plurality of microprojections and the pain-free delivery is due to the micron dimensions of the needles, which are attached to a base support and generally range from 25 to 2000 μm in height and are available in different shapes and materials [1]. MNs have been shown to penetrate the skin across the *stratum corneum* and into the viable epidermis, avoiding contact with nerve fibres and blood vessels, which reside primarily in the dermal layer. Application of MN arrays to the skin surface can create transport pathways with dimensions of the order of microns. Once created, these micropores should readily allow the transport of drugs, vaccines and macromolecules, as well as possibly supramolecular complexes, microparticles and nanoparticles [2–4]. Therefore, the principal benefit of using MNs is the promise of pain-free penetration and delivery of both low and high molecular weight active pharmaceutical ingredients (APIs). In addition, MNs could also be used for alternative purposes, such as sampling interstitial body fluids for monitoring purposes, for example, measuring of the blood glucose levels in diabetic therapy.

The ALZA Corporation, as described in a 1976 patent [1], appears to be the first to have conceived the use of plurality of MNs to penetrate the skin in order to overcome the barrier function of the *stratum corneum*. However, it was not possible to make such microstructured devices until the 1990s; this delay was due to the fact that the technology needed to design feasible micron or submicron structures only became available with the advent of high-precision microelectronic industrial tools during the 1990s. The first paper, as far as the authors are aware, to demonstrate MNs for transdermal delivery was not published until 1998 [5]. Since then, a plethora of work has been published, by both research academia and industry, which has demonstrated the applicability of MNs, from the basic concepts of fabrication to application in human clinical trials. An extensive number of research papers have been published and a relatively higher number

Microneedles for Drug and Vaccine Delivery and Patient Monitoring, First Edition.
Edited by Ryan F. Donnelly, Thakur Raghu Raj Singh, Eneko Larrañeta, and Maelíosa T.C. McCrudden.

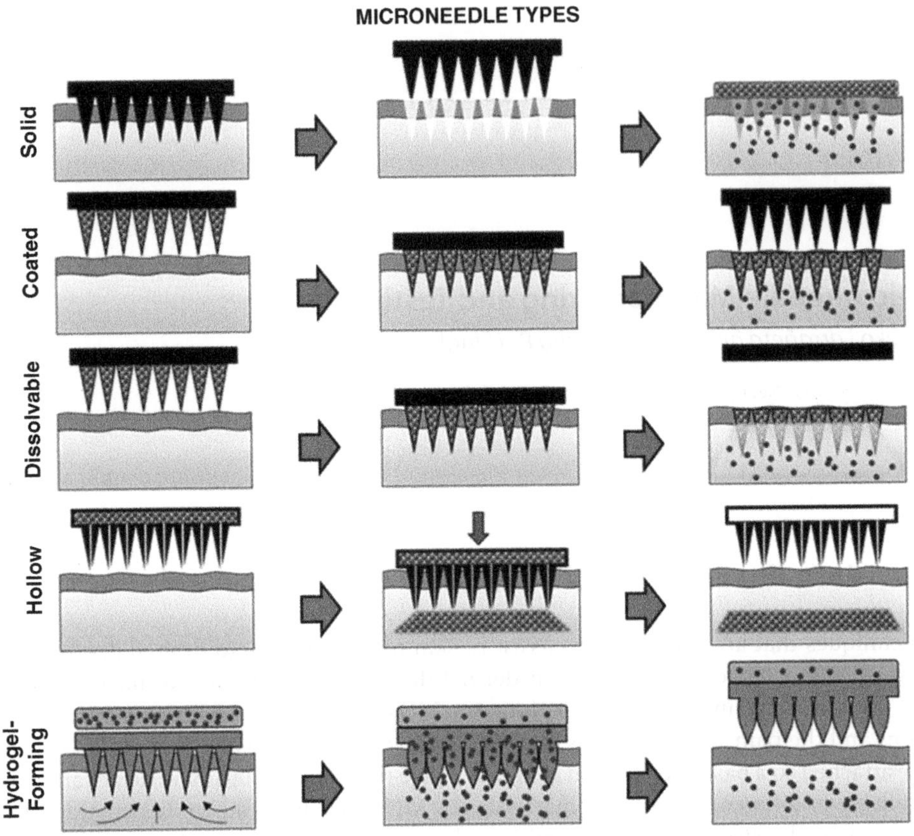

Figure 2.1 A schematic representation of the different types of MN used for transdermal drug delivery. Solid MNs: for increasing the permeability of a drug formulation by creating micro-holes across the skin. Coated MNs: for rapid dissolution of the coated drug into the skin. Dissolvable MNs: for rapid or controlled release of the drug incorporated within the microneedles. Hollow MNs: used to puncture the skin and enable release of a liquid drug following active infusion or diffusion of the formulation through the needle bores. Hydrogel-forming MNs: take up interstitial fluids from the tissue inducing diffusion of the drug located in a patch through the swollen microprojections. Reproduced with permission from: [9] E. Larrañeta, R.E.M. Lutton, A.D. Woolfson and R.F. Donnelly (2016). Microneedle arrays as transdermal and intradermal drug delivery systems: Materials science, manufacture and commercial development. *Mater. Sci. Eng. R-Rep.* 104: 1–32.

of patents have been issued concerning the concept, design and manufacturability of various types of MNs (Figure 2.1) from different techniques.

Importantly, with the introduction of the latest microfabrication technologies, it is now possible to microfabricate MNs in three-dimensional designs and from a variety of materials. In *in vitro, ex vivo* and *in vivo* experiments these MNs have shown enhanced transdermal drug delivery of compounds with a wide variety of physicochemical properties. For effective penetration into the epidermal skin layers, the MNs should be sharp and robust and should withstand lateral forces (i.e. without much bending) experienced by the skin tissue during penetration. However, a certain degree of flexibility of the MN shaft is desired, to counteract the elasticity of the skin. While the needle length within the MN device should be controlled to avoid nerve contact, the needle density should be optimised in order to reduce pain, skin damage or insertion forces [6]. All these factors are dependent on the material selected for the manufacture, the design and the manufacturing technology of the MN.

In addition to the manufacturing of MN arrays, another key issue during the production of these devices is the testing. The need for universally accepted tests that can be used as quality control measurements to characterise microneedles is one of the considerations that has not been properly addressed in the scientific literature. Only a few research papers have been published describing these aspects [7, 8].

This chapter aims to give a detailed account of a variety of microfabrication technologies presently being used in fabricating different types of MNs from various materials, such as silicon, metals and polymers. We have also attempted to demonstrate the need for optimised MN design, in anticipation of fabricating MNs with effective skin penetration to be able to achieve enhanced transdermal drug delivery. Finally, the chapter will describe the testing of MN arrays.

2.2 Material Types, Properties and Biocompatibility

2.2.1 Silicon

The development of high precision microelectronic industrial tools during the 1990s [1, 10–12] facilitated the development of silicon MNs [5, 13, 14]. Silicon presents an important number of interesting properties that make it extremely attractive in the field of microstructures and microelectromechanical systems (MEMS) [12]. The main one is the flexibility of the process to provide a diverse series of shapes using monocrystalline or polycrystalline silicon. Furthermore, silicon possesses many attractive physical properties making it an ideal candidate for a versatile sensor material. Lastly, the manufacturing processes that exist for this material are precise and allow batch production at low cost [10].

As an anisotropic material, the properties of silicon depend on orientation relative to the crystal lattice [14]. Consequently, the elastic modulus of silicon ranges between 50 and 180 GPa [15]. Regardless of this variability, silicon MN can successfully pierce the skin [5, 16, 17] allowing transdermal delivery of different biologically active compounds [13, 18]. Therefore, silicon has been used widely to produce an extensive number of MN arrays with varying heights, densities and shapes by means of microfabrication methods [19–21]. Silicon has been used to produce solid [5, 12, 18, 22, 23], hollow [13, 24] and coated MNs [25] (Figure 2.2).

Despite all of these beneficial properties, the main limitations to the use of silicon are: its high cost and the complex multi-step manufacturing processes [10, 26]. In addition to all these factors, biocompatibility should be taken into account. There have been some concerns over the biocompatibility of silicon [27, 28]. It is a brittle material [28] and some silicon MNs could fracture after insertion in the skin [29]. Owing to the normal turnover of the epidermis, any broken MN pieces should be discarded naturally within four weeks [30]. However, some cases of silicon-related granulomas have been reported [31]. In order to ascertain the reliability of this type of MN, several studies were performed evaluating the failure forces and failure mechanisms of the needle tips for silicon MNs [16, 17, 29] (Figure 2D–E).

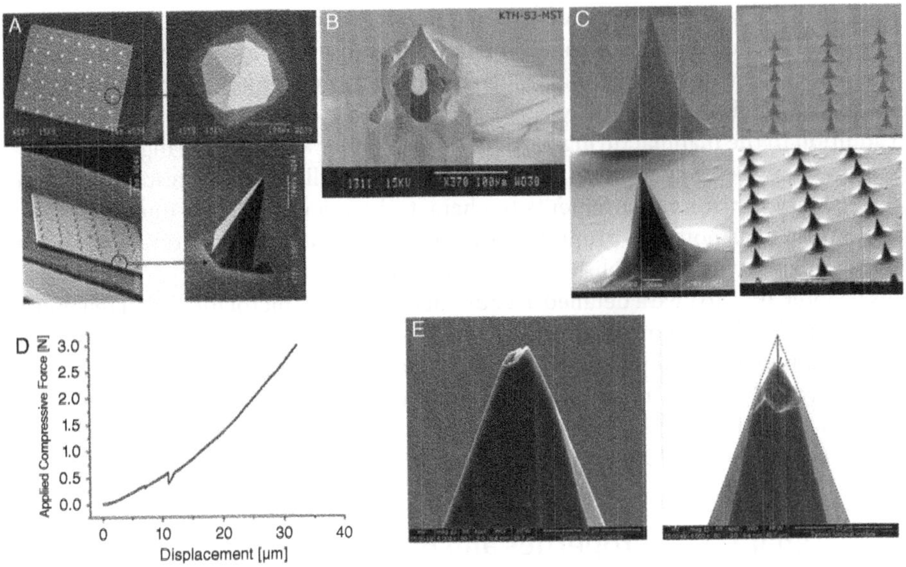

Figure 2.2 Scanning electron microscopy (SEM) images of (A) solid, (B) hollow and (C) coated silicon MN arrays. (D) Force-displacement graph for a silicon MN compression test showing discontinuities characteristic of structural failure. (E) SEM images of damage to needle after application of 0.3 and 0.5 N compressive loads to the needle tip. l_c is the distance from the needle apex to the intersection of the longitudinal axis of the MN and the broken plane. In this case, l_c is 14 μm. Reproduced with permission from: [18] R.F. Donnelly, D.I. Morrow, P.A. McCarron, *et al.* (2009). Microneedle arrays permit enhanced intradermal delivery of a preformed photosensitizer. *Photochem. Photobiol.* 85: 195–204; [24] N. Roxhed, P. Griss and G. Stemme (2008). Membrane-sealed hollow microneedles and related administration schemes for transdermal drug delivery. *Biomed. Microdevices* 10: 271–279; [25] M.G. McGrath, A. Vrdoljak, C. O'Mahony, *et al.* (2011). Determination of parameters for successful spray coating of silicon microneedle arrays. *Int. J. Pharm.* 415: 140–149; [29] C. O'Mahony (2014). Structural characterization and in-vivo reliability evaluation of silicon microneedles. *Biomed. Microdevices* 16: 333–343.

2.2.1.1 Biocompatibility of Silicon

Despite the number of studies carried out during the last 20 years [32], the biocompatibility of silicon is not totally clear. There have been a wide range of studies assessing the biocompatibility of silicon MEMS devices for brain and subcutaneous implants [33–35]. In a paper published by Bayliss *et al.*, nanocrystalline silicon did not exhibit significant cytotoxicity [36]. However, several research workers have described some biocompatibility problems, such as granulomas in subcutaneous tissue [31] or the formation of nodules on periodontal ligament fibroblasts [37]. Despite all of these concerns about the biocompatibility of silicon, there is a silicon MN-based product on the market, Micronjet®, which was approved by the FDA in 2010 [38]. However, in this case the system consists of silicon MNs attached to a syringe and consequently the product is not designed to be inserted inside the skin for prolonged periods of time.

2.2.2 Metals

Metals are used extensively in medicine. Hypodermic needles made of stainless steel or prostheses made from titanium are a couple of examples. Consequently, the use of metals for MN manufacturing could be a good option for smoothing the way towards commercial and regulatory approval, as these materials should not present any safety issues.

Stainless steel, titanium, palladium, palladium–cobalt alloys and nickel are the main types of metals used for MN production [39]. These metals provide good biocompatibility and mechanical properties. The most widely used stainless steel for medical implants (SUS316L) presents a Young's modulus of around 180 GPa [40]. Titanium and its alloys generally present smaller Young's moduli than those of stainless steels. As an example: two common materials used for implant devices, titanium and its alloy, Ti-6Al-4V ELI, have a Young's modulus of around 110 GPa [40]. A comparison of the mechanical properties of these types of metals can be found in Table 2.1. When compared with silicon, these types of materials present comparable elastic moduli to the highest ones of silicon (up to 180 GPa), higher fracture toughness and similar values of yield strength [9] (Figure 2.3). Accordingly, metals present better properties for MN manufacturing when compared with silicon.

The first metal used in the production of MN arrays was stainless steel [9]. The simplest way to obtain stainless-steel MNs is by using conventional hypodermic needles and exposing defined lengths of the needles held in a supporting material [1]. This strategy has been followed to form MN arrays [42]. In addition to this simple strategy, microfabrication technology can be used to produce stainless-steel MNs [43, 44]. Consequently, this type of metal has then been used to produce different types of MN arrays: solid [42–46], hollow [42, 45] and coated [47, 48] (Figure 4A–B).

Titanium is a good alternative to stainless steel; the mechanical properties are worse than for stainless steel, but it is strong enough for biomedical applications [40]. MNs made from titanium have been designed to be used as transdermal delivery systems [49] and as bio-sensors [50, 51]. In addition to stainless steel and titanium, palladium,

Table 2.1 Strengths of materials used to make microneedles. *Source*: data from Monteiro-Riviere (2010). *Source*: adapted from [41] N.A. Monteiro-Riviere (2010). *Toxicology of the Skin*. Boca Raton: CRC Press.

Material	Young's modulus (GPa)	Ultimate tensile strength (MPa)
Silicon	110	7000
Glass	85	50
Nickel	214	586
Palladium	117	186
Platinum	147	117
Titanium	110	241
Stainless steel	200	1000
Ormocer®	17	30
PMMA	3	170
Maltose	31.1	—
SU-8	3	—

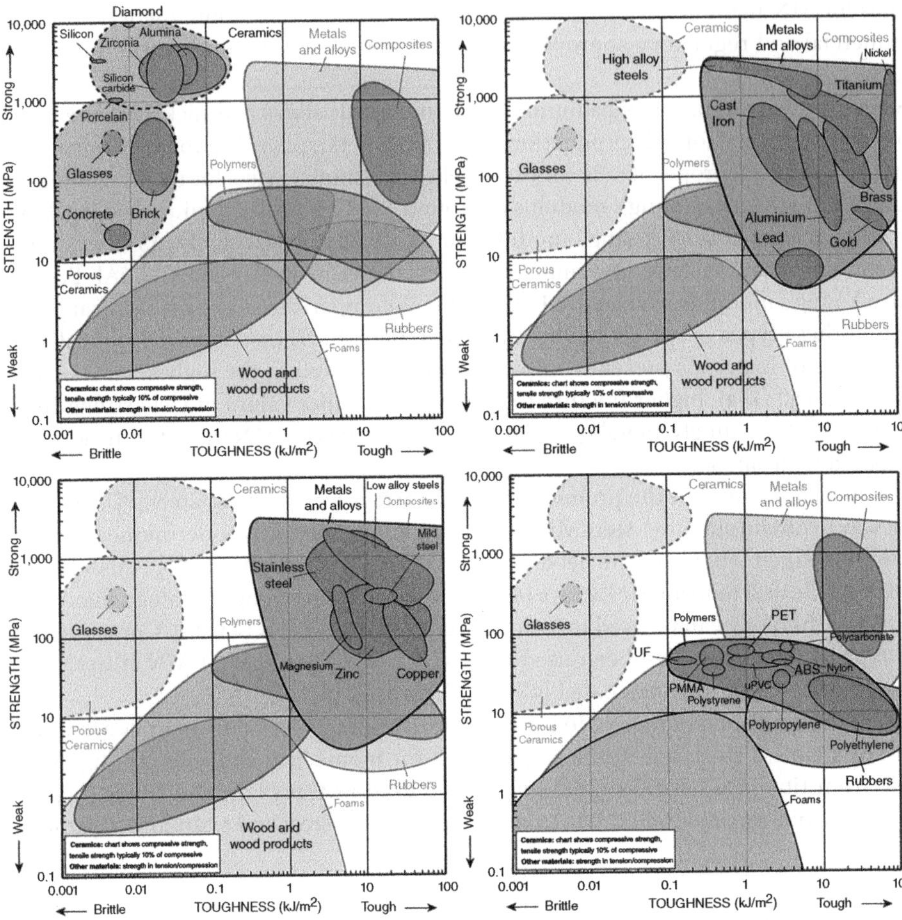

Figure 2.3 Strength versus toughness graphs for different types of materials. Adapted from: www-materials.eng.cam.ac.uk/mpsite (Lovatt A.M., Shercliff H.R. and Withers P.J. (2000), "Material selection and processing"). *Source*: data courtesy of Granta Design Ltd, Cambridge, UK.

palladium–cobalt alloys [52, 53], nickel [52–54] and platinum [55, 56] have been used in the manufacture of MNs.

2.2.2.1 Biocompatibility of Metals

All the types of metals described thus far are typically biocompatible and have been used predominantly as structural biomaterials for reconstructive surgery (mainly orthopaedics) and recently in non-osseous tissues, for example, in blood vessels [59]. Stainless steel is an iron alloy that contains a high percentage of chromium (11–30 wt%) and varying amounts of nickel [59]. Consequently, the biocompatibility of stainless steel is strongly affected by its composition [59]. The most commonly used surgical stainless steel is 316L, which provides relatively good biocompatibility. Although titanium possesses better resistance to corrosion [59], this is not critical for MN arrays, as they do not require long application times. The application time for MN arrays will be similar to that for stainless-steel hypodermic needles, which are widely used and considered to be biocompatible [60].

Titanium alloys present excellent corrosion resistance [59] and several studies have shown where they do not presented significant mutagenicity. Consequently, these alloys are considered safe for humans and animals [59]. However, the first generation of titanium alloys were reported to cause an allergic reaction in the human body [61]. Although the second generation of titanium alloys appear to be relatively safe, more studies need to be undertaken as there is a lack of long-term clinical application data [59].

Palladium and platinum have been used to a lesser extent to produce MNs and they present good biocompatibility [62, 63]. In contrast, nickel is carcinogenic and can cause adverse allergic reactions when used in biomaterials [64].

2.2.3 Ceramics

Alumina (Al_2O_3) is the main type of ceramic used to produce MNs [57, 65] (Figure 2.4C). This material presents some interesting properties: high resistance to corrosion or adverse conditions [66] and good mechanical compression resistance [66] (Table 2.1). Despite all of these properties, alumina shows lower strength resistance to tension than other materials such as metals (Table 2.1) (Figure 2.3). Moreover, it is brittle under tensile strengths [66]. Thus, Al_2O_3 MNs may fracture after manual insertion, as reported by Bystrova *et al.* [65]. Alumina is a porous material and consequently such MNs provide nanoporosity [57]. Because of these interesting surface properties, Al_2O_3 MNs have been coated with drugs for transdermal drug delivery. The porous nature of the material holds the active drug in the surface for controlled release into the skin [57].

Alternative types of ceramic that have been used to prepare MNs are calcium sulfate dihydrate (gypsum ($CaSO_4{\cdot}2H_2O$)) and calcium phosphate dihydrate (brushite ($CaHPO_4{\cdot}2H_2O$)) [58] (Figure 2.4D). These ceramics provide good mechanical and drug loading properties, as shown by their use as drug delivery bone cements [67, 68]. In addition to alumina, gypsum and brushite, the organic–ceramic hybrid material Ormocer® has been used to prepare MNs [69, 70]. Ormocer contains a three-dimensional network of organically modified silicon alkoxides and organic monomers [69]. By modifying the composition and the synthesis parameters the final properties of the material can be adjusted [71].

2.2.3.1 Biocompatibility of Ceramics

Owing to the biocompatibility and high strength, ceramics have been used as replacement parts for the musculoskeletal system [72]. However, the biocompatibility of the material depends on the type of ceramic. Alumina has been used extensively as an orthopaedic material for bone or dental implants for nearly 25 years [72] and consequently its biocompatibility has been studied extensively [73–75]. Despite its good biocompatibility, some studies have questioned the stability of alumina under physiological conditions and whether there is a risk of Al release from long-term bone implants [76]. However, this is not a problem for MN arrays as this phenomenon has not been observed for such short-term implants [76].

Calcium phosphate ceramics are used as bone substitutes [77, 78] and consequently they are considered biocompatible, bioactive in the sense of osteoconduction and

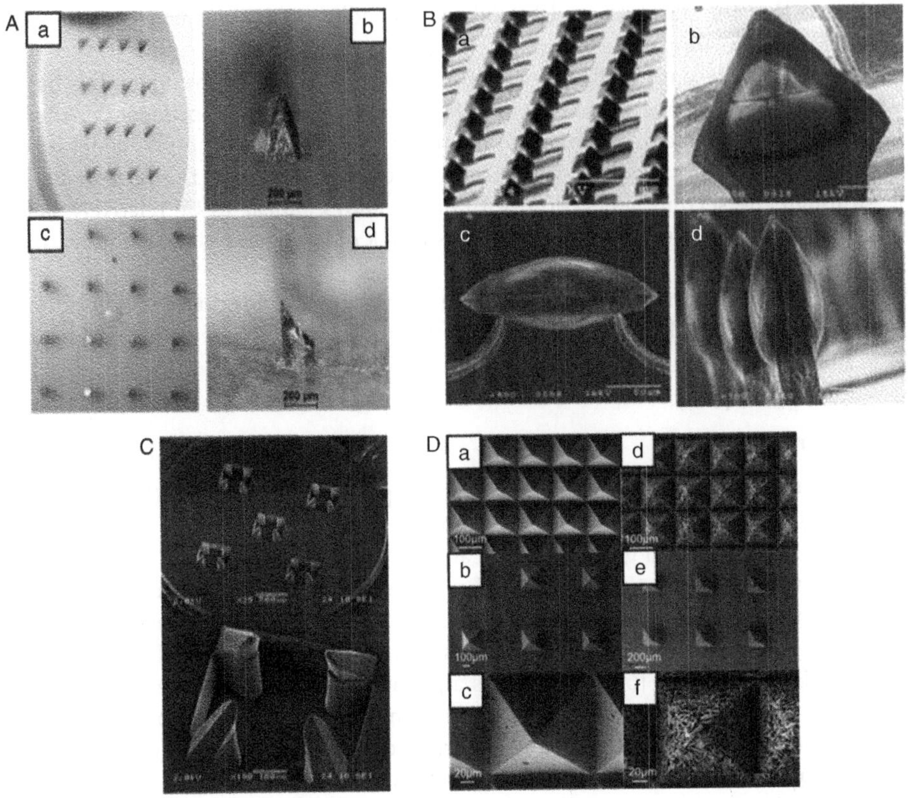

Figure 2.4 (A) Different types of stainless-steel MNs: (a) hollow stainless-steel 4×4 MN array; (b) higher magnification of a single hollow stainless-steel needle; (c) solid stainless-steel 4×4 MN array; and (d) a higher magnification of a single solid needle. (B) Scanning electron micrographs of a titanium microneedle array coated with 80 μg of desmopressin per array: (a) bar = 1 mm; (b–d) bar = 50 μm. (C) Scanning electron micrographs of an alumina ceramic. (D) a–c, gypsum; d–f, brushite. Reproduced with permission from: [45] F.J. Verbaan, S.M. Bal, D.J. van den Berg, *et al.* (2008). Improved piercing of microneedle arrays in dermatomed human skin by an impact insertion method. *J. Control. Release* 128: 80–88; [49] M. Cormier, B. Johnson, M. Ameri, *et al.* (2004). Transdermal delivery of desmopressin using a coated microneedle array patch system. *J. Control. Release* 97: 503–511; [57] M. Verhoeven, S. Bystrova, L. Winnubst, *et al.* (2012). Applying ceramic nanoporous microneedle arrays as a transport interface in egg plants and an ex-vivo human skin model. *Microelectron. Eng.* 98: 659–662; [58] B. Cai, W. Xia, S. Bredenberg and H. Engqvist (2014). Self-setting bioceramic microscopic protrusions for transdermal drug delivery. *J. Mater. Chem. B.* 2: 5992–5998 .

bioresorbable [79]. Calcium sulfates present similar characteristics [80], so they are good candidates for the manufacture of MN. Finally, the biocompatibility of Ormocer has been studied and shown to be safe for use as a medical material [81–83]. In a study published in 2007, Ovsianikov *et al.* demonstrated that this material does not adversely affect the growth of one of the major cellular components of the skin, human epidermal keratinocytes [81].

2.2.4 Silica Glass

Silica glass has been used to manufacture MN arrays as it can be produced quickly for small-scale laboratory use in various geometries [9]. This material can be fabricated with

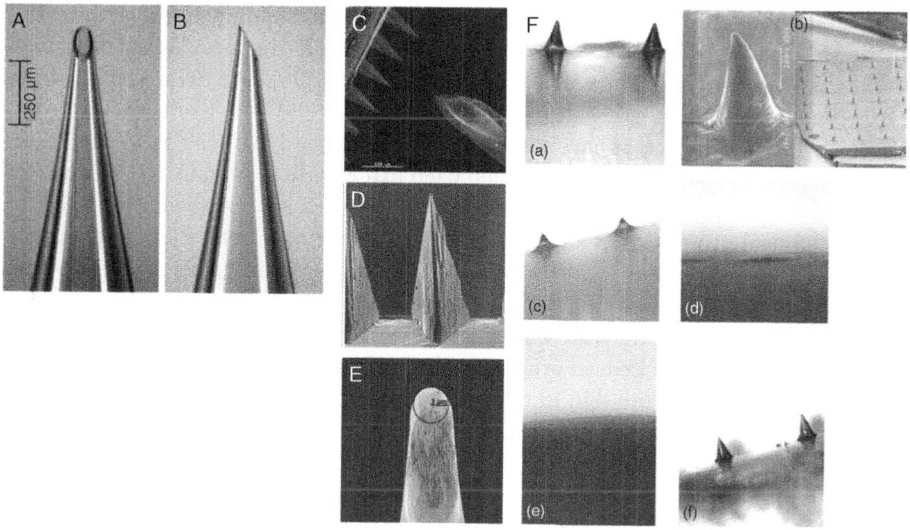

Figure 2.5 (A) Front and (B) side views of a representative hollow, glass microneedle. (C) Scanning electron micrograph image of 500 μm long solid maltose microneedles shown opposite to a tip of a 26G hypodermic needle, (D) in an individual array and (E) a magnified view that shows the radius of the tip. (F) Influence of the storage conditions for galactose MN arrays: (a) light micrograph of galactose microneedles upon preparation; (b) scanning electron micrographs of the same array; (c) light micrograph of galactose microneedles after storage at a relative humidity of 43% for 1 hour and (d) 6 hours; (e) light micrograph of galactose microneedles after storage at a relative humidity of 75% for 1 hour; (f) light micrograph of galactose microneedles after storage at a relative humidity of 0% for 3 weeks. Reproduced with permission from: [85] W. Martanto, J.S. Moore, O. Kashlan, *et al.* (2006). Microinfusion using hollow microneedles. *Pharm. Res.* 23: 104–113; [88] C.S. Kolli and A.K. Banga (2008). Characterization of solid maltose microneedles and their use for transdermal delivery. *Pharm. Res.* 25: 104–113; [89] R.F. Donnelly, D.I. Morrow, T.R. Singh, *et al.* (2009). Processing difficulties and instability of carbohydrate microneedle arrays. *Drug Dev. Ind. Pharm.* 35: 1242–1254.

dimensions similar to those of microfabricated microneedles, it allows easy visualisation of fluid flow and is physiologically inert [84–86] (Figure 2.5). The main application of glass MNs is to by-pass the *stratum corneum* and the injection of medicines [84, 85].

Borosilicate glass is a brittle material, with similar fracture toughness to silicon and silica glass [9] (Figure 2.3). Consequently, this material presents similar limitations to silicon when used for MN manufacture. The manufacturing process of glass MNs is carried out by hand so it is not time efficient [87]. Accordingly, glass microneedles are still used today, but only for laboratory experimental purposes and are not viable for commercial use as a drug delivery system [84].

2.2.4.1 Biocompatibility of Silica Glass

The biocompatibility of glass is not fully understood, as silica glass can cause granulomas in the skin [90] but borosilicate glass apparently seems to be biocompatible when used for cortical implants [91]. Therefore, there needs to be a realistic study on the use of glass for MN manufacture.

2.2.5 Carbohydrates

Micromoulding techniques can be used to manufacture MNs from carbohydrates [89, 92–95]. In order to prepare carbohydrate MN arrays, a slurry or a solution of the carbohydrate containing a drug substance is placed into a mould to form the MNs [89, 92–95]. Carbohydrates are good alternatives to all the previously described materials as they are cheap and safe for use with humans [95]. The main carbohydrate used to prepare MN arrays is maltose [88, 92, 93] (Figure 5C–E). Trehalose, sucrose, mannitol, xylitol and galactose have also been used to prepare MNs [89, 96]. In addition to simple sugars, polysaccharides have been used extensively to prepare MN arrays. As these are macromolecules they will be discussed in the next section (Section 2.2.6).

The mechanical properties of all these sugars have not been well studied. Donnelly *et al.* showed that there was significant reductions in height in galactose MNs when compressed against an aluminium block using relatively strong forces [89]. However, the same study described that they can be successfully inserted into the heat-stripped epidermis. Therefore, the mechanical properties do not limit the efficacy of galactose MNs as they can be inserted in the skin to release their cargoes [88, 92]. On the other hand, this material is not ideal for the preparation of MN arrays as there are some limitations during their manufacture, storage and use [89]. The main limitation is that during the manufacture of this type of MNs there is a thermal treatment, which limits the amount of compound that can be loaded inside the arrays because of stability issues [89]. In addition to this, after insertion of the needles, they should dissolve to release their cargo but the dissolved sugar seals the holes, thus limiting the drug release [89]. Lastly, the storage conditions strongly affect the integrity of this type of MN arrays (Figure 2.5F). Consequently, all these limitations are expected to preclude clinical applications of sugar MN arrays.

2.2.5.1 Biocompatibility of Carbohydrates

Natural sugars have been used extensively as ingredients in drug delivery systems as they are safe for use in humans [95]. Maltose and galactose have been used in different parenteral formulations approved by the FDA [97]. Consequently, maltose and galactose can be considered safe for MN production. However, it is important to note that those products which contain sugar can interfere with blood glucose monitoring [97]. In addition to maltose and galactose, sucrose, mannitol, trehalose and xylitol are present in several parenteral formulations as cryoprotectants, stabilisation agents or as parenteral nutrition products [98, 99].

2.2.6 Polymers

Polymers are good candidates for the manufacture of MNs. Within this group of materials there are a wide variety of macromolecules, such as synthetic polymers, proteins or polysaccharides, which may show excellent biocompatibility, degradability, mechanical properties at low cost [1, 100]. These materials do not possess the strength of silicon, metals, ceramics and glass. However, they show higher toughness than ceramics and glass [9].

The main types of polymers that have been used to produce MNs includes: poly-L-lactic acid (PLA) [101], poly(methyl methacrylate) (PMMA) [6, 102, 103], poly(lactic-*co*-glycolic acid) (PLGA) [104], polyglycolic acid (PGA) [87], polycarbonate [105], cyclic olefin copolymer [106, 107], poly(vinylpyrrolidone) (PVP) [108, 109], poly(vinyl alcohol) (PVA) [109], polystyrene (PS) [110], poly(methyl vinyl ether-*alt*-maleic anhydride) [4, 111], poly(methyl vinyl ether-*alt*-maleic acid) [112],

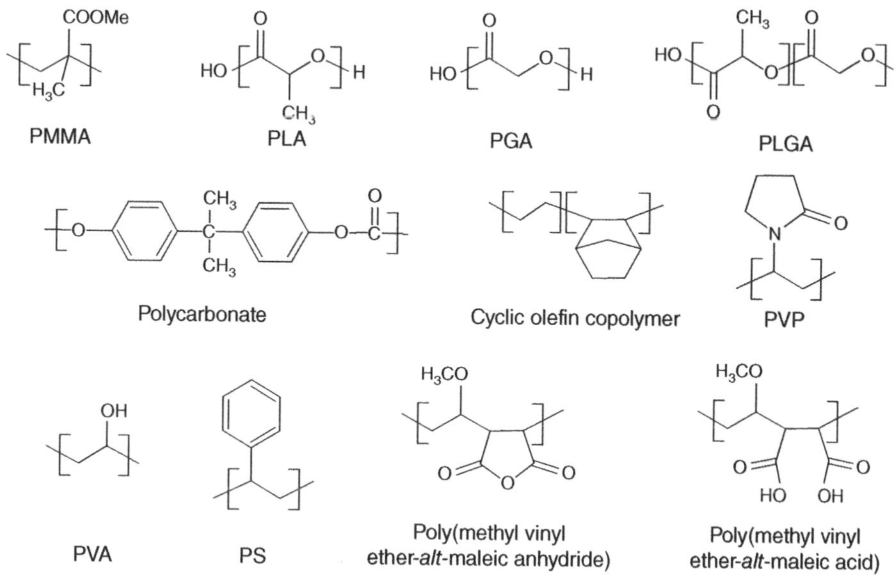

Figure 2.6 Chemical structure of some of the polymers used to produce MNs.

SU-8 photoresist [113] and polysaccharides [94, 114–116]. Figure 2.6 shows the structures of some of these polymers.

Primarily, polymers have been used to produce dissolving/biodegradable and hydrogel-forming MN arrays. In addition, but to a lesser extent, solid [102, 117], hollow [19, 103] and coated MN arrays [118] have also been made from polymers.

Dissolving/biodegradable polymeric MNs have been manufactured using various types of polymers [100]. The drug cargo is inside the MN, and after insertion into the skin, the drug can be released as the polymer slowly dissolves/biodegrades for local or systemic delivery (Figure 2.1 and Figure 2.7A) [1]. Polysaccharides have been used extensively to prepare dissolving MN arrays including: carboxymethylcellulose (CMC), amylopectin, dextrin, hydroxypropyl cellulose, alginate and hyaluronic acid (Figure 2.7B) [100, 116]. MNs prepared with polysaccharides showed adequate mechanical properties for appropriate insertion into the skin and delivery of their payloads (model molecules, lysozyme and insulin) [94, 114, 115].

Synthetic polymers have been used widely to prepare dissolving MN arrays. The main polymers used for this purpose are: PVP [119], PVA [109] and poly(methyl vinyl ether-*alt*-maleic anhydride) (Gantrez AN-139®) (Figure 2.7A) [111, 120, 121]. MNs prepared with all these polymers provide sufficient strength to pierce the skin and deliver their cargoes [109, 111, 119–121]. Alternatively, biodegradable MN arrays have been manufactured using PLGA, PLA, PGA and chitosan [100].

Dissolving/biodegradable MN arrays are not the only types of polymeric MN arrays. The second main type of polymeric MN arrays is the hydrogel-forming or swelling MN array [122]. This type of MN allows drug release as a result of the swelling of the polymeric matrix on absorption of body fluids (Figure 2.1 and Figure 2.7C–D) [122]. The swollen needle tips produce aqueous conduits. In this case the drug is located in a separated patch reservoir and after the swelling of the needle tips the drugs can diffuse into

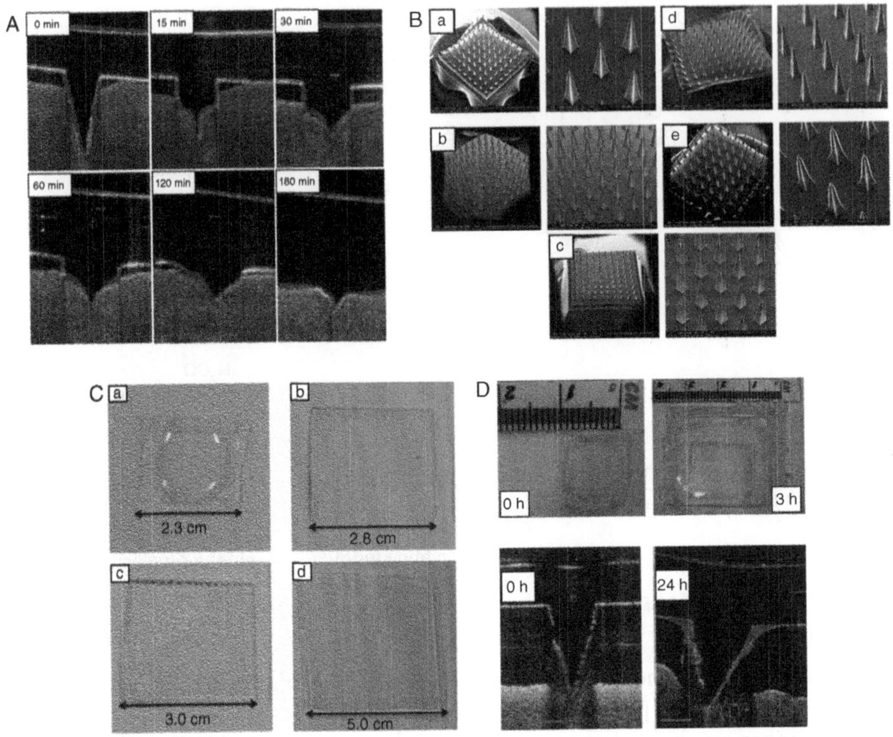

Figure 2.7 (A) False colour optical coherence tomography images of the *in vitro* dissolution profile of Gantrez-AN 139® microneedles in porcine skin over a 3-hour period. (B) SEM photographs of parts from 10×10 MN arrays made from: (a) alginate, (b and c) hydroxypropyl cellulose, (d) crosslinked PVA-gelatine and (e) chitosan. (C) Digital photographs of equilibrium swollen hydrogels (with initial dimensions of 1×1 cm^2) prepared from aqueous blends of 15% w/w PMVE/MA and 7.5% w/w PEG 10 000 containing (a) 0, (b) 1, (c) 2 and (d) 5% w/w of $NaHCO_3$ showing changes in respective dimensions. (D) Gantrez S-97® MN before and after swelling for 3 hours in PBS at pH 7.4. (E) Representative optical coherence tomography images of hydrogel forming MN array before and after insertion into neonatal porcine skin for a period of 24 hours. Scale bar represents a length of 300 μm. Reproduced with permission from: [112] R.F. Donnelly, M.T.C. McCrudden, A. Zaid Alkilani, *et al.* (2014). Hydrogel-forming microneedles prepared from "super swelling" polymers combined with lyophilised wafers for transdermal drug delivery, *PLoS ONE* 9: e111547; [116] Y.K. Demir, Z. Akan and O. Kerimoglu (2013). Characterization of polymeric microneedle arrays for transdermal drug delivery. *PLoS ONE*. 8: e77289; [121] R.F. Donnelly, M.J. Garland, D.I. Morrow, *et al.* (2010). Optical coherence tomography is a valuable tool in the study of the effects of microneedle geometry on skin penetration characteristics and in-skin dissolution. *J. Control. Release* 147: 333–341; [125] .R.R. Singh, M.J. Garland, K. Migalska, *et al.* (2012). Influence of a pore-forming agent on swelling, network parameters, and permeability of poly(ethylene glycol)-crosslinked poly(methyl vinyl ether-co-maleic acid) hydrogels: Application in transdermal delivery systems. *J. Appl. Polym. Sci.* 125: 2680–2694 .

the skin through the aqueous conduits [122]. This type of MN arrays were developed by Donnelly *et al.* using poly(methyl vinyl ether-*alt*-maleic anhydride) (Gantrez AN-139) or poly(methyl vinyl ether-*alt*-maleic acid) (Gantrez S-97®) crosslinked with polyethylene glycol (PEG) [4, 112, 123]. The mechanical properties of these MN arrays provide mechanical resistance to fracture and successful insertion into the skin [4, 112, 124]. The swelling of this type of device can be modified by adding $NaHCO_3$, a pore-forming agent [125]. Hydrogel-forming MN arrays for drug delivery have been prepared using PVA and mixtures of polysaccharides (dextran, gelatin, etc.) [116, 122, 126]. Yang *et al.* described the use of PVA, dextran and CMC to manufacture hydrogel-forming MN arrays [126]. The crosslinking process was carried out with a freeze–thaw method. In a similar crosslinking method, Demir *et al.* prepared swelling MN arrays using PVA crosslinked with gelatine [116].

2.2.6.1 Biocompatibility of Polymers

The biocompatibility of polymers used to prepare MNs has been studied extensively. The majority of the polymers selected to prepare MNs have been used previously for other medical applications. Polysaccharides have been widely used as biomaterials for a series of biomedical applications such as drug delivery or regenerative medicine [127]. They are chemically similar to human extracellular matrix components and consequently are recognised and accepted by the body [127]. If they are absorbed, the elimination process will be dependent on the size and the shape of the polysaccharide. When the molecular weight is below the glomerular threshold, they can be easily eliminated in the kidney by glomerular filtration [128]. This applies to non-degradable synthetic polymers.

Cellulose derivatives such as carboxymethyl cellulose and hydroxypropyl cellulose are good candidates for MN manufacture as they are biocompatible and biodegradable [129, 130]. They have been used in drug delivery formulations and in various therapies to prevent postsurgical adhesion [131]. Another type of polysaccharide that is widely used is chitosan, which has been reported to be biocompatible and biodegradable, yielding non-toxic residues [132]. This molecule is degraded by lysozyme and its degradation process is highly dependent on its molecular weight and the degree of deacetylation [128]. Alginates present interesting properties as they are biocompatible [133]. This type of macromolecules has been used extensively for microencapsulation, wound dressing or tissue repair. Consequently, their biocompatibility has been extensively proven [133]. One of the main disadvantages of alginates is the inability to undergo enzymatic degradation by mammals [9]. However, some procedures to prepare alginate-based materials with enhanced biodegradability can be found in the literature [134]. In a similar way, starch-based polymers (i.e. amylopectin) are not ideal for biomedical applications, as in spite of being biocompatible they are not easily biodegraded within the human body [135].

There has been considerable investigation into, and use of, hyaluronic acid for biomedical and drug delivery applications [136]. Consequently, this material is considered biodegradable and biocompatible [136]. Inside the body, hyaluronic acid is degraded by free radicals present in the extracellular matrix before endocytosis.

Another example of an ideal type of polysaccharides for MN manufacturing in terms of biocompatibility are dextrins. They have not been widely used for biomedical applications, but it has been demonstrated that they are biocompatible, degradable by α-amylases, non-immunogenic and can be eliminated from the body by renal clearance, preventing accumulation [137, 138].

In addition to polysaccharides, and as discussed, the majority of the synthetic polymers used to manufacture MN arrays are biocompatible and some of them are also biodegradable.

There have been many studies on, and broad use of, PC and PMMA for biomedical purposes, showing their biocompatibility [139–141]. PMMA can be found in bone cements or intraocular lenses [140]. On the other hand, PC can be found in a wide variety of medical devices, such as cannulas or syringes and for dental brackets [141]. The main drawback of these materials is that PMMA [142] is not biodegradable but PC can be biodegraded releasing bisphenol-A [143–145], which induces hormone-related effects and therefore its use should be controlled [143].

The use of PS should be restricted as its biocompatibility is limited [146, 147]. However, Vesel *et al.* showed that the biocompatibility of this polymer can be improved by non-equilibrium oxygen plasma treatment [146].

PVA and PVP are widely used for biomedical applications [148, 149]. PVA is another example of an ideal candidate for MN manufacturing as it is biocompatible, non-toxic and non-carcinogenic [148]. In a similar way, PVP presents an extremely low cytotoxicity [150] and consequently it is considered biocompatible [149]. In addition, it is biodegradable but to a lesser extent than PVA [150].

SU-8, an epoxy-based polymer, has been used to produce MN arrays. It has been demonstrated that this type of polymer is biocompatible [151]. This was established after *in vitro* and *in vivo* testing of SU-8 as an implant material [151]. In a similar way, cyclic olefin copolymers have been proven to be suitable materials for tissue engineering and consequently are biocompatible [152, 153]. However, the biocompatibility can be affected by surface modification [153].

PLA, PGA and PLGA are widely used as drug delivery systems and consequently show good biocompatibility [154, 155]. They are also biodegradable, making them appealing materials to be used in MN manufacturing as they will degrade after being inserted into the skin (Figure 2.8) [104]. In the case of PLGA, the degradation rate can be controlled by modifying the PGA/PLA ratio in the copolymer [41, 154].

Finally, poly(methyl vinyl ether-*alt*-maleic anhydride) and its acid form have been used for MN manufacturing, mainly through the ability to form hydrogels. These

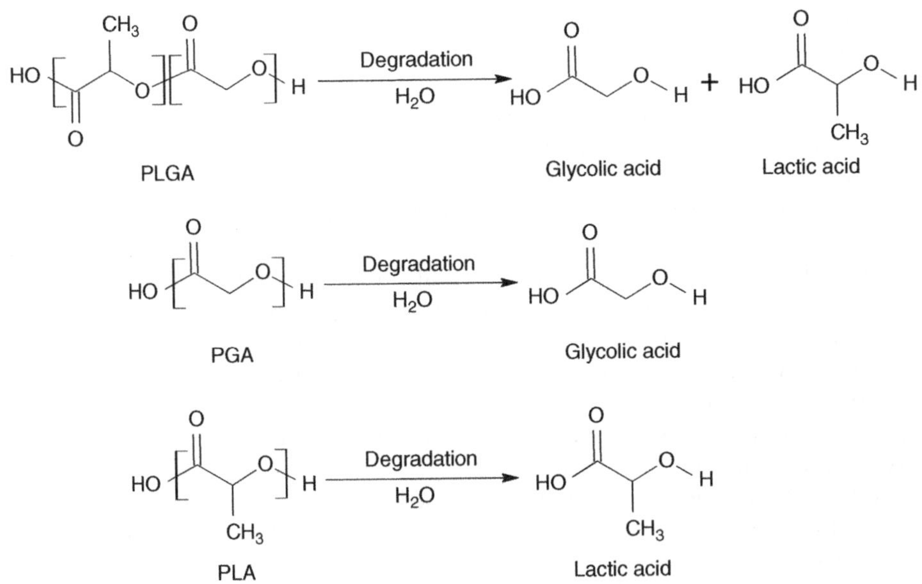

Figure 2.8 Hydrolysis of PLGA, PGA and PLA.

polymers can be found in bioadhesives for dentures and thickening/suspending agents for topical ointments [156]. They have been used to prepare nanoparticulate drug delivery systems [157–160] and in several studies have been proven to be biocompatible [120, 161, 162].

2.3 Microneedles Manufacturing and Design

The first MN devices were fabricated from silicon [5], but in the last decade the number of practicable materials for the fabrication of MNs has significantly increased. As discussed earlier, other materials employed in the fabrication of MN are metals, polymers and ceramics, such as stainless steel, dextrin, glass, maltose, galactose, and so on. A multiplicity of methods has been employed in the manufacturing of a wide variety of MNs, predominantly conventional microelectronics fabrication technologies, including chemical isotropic etching, injection moulding, reactive ion-etching, surface/bulk micromachining, polysilicon micromoulding, lithography-electroforming-replication, laser-drilling and two-photon polymerisation [6, 39, 163, 164]. Furthermore, MNs have been fabricated with a wide range of designs (various sizes and shapes) and of different types (solid, hollow, sharp or flat). However, the two basic designs are: in-plane and out-of-plane designs. Combinations of both in-plane and out-of-plane MN designs have also been reported. In the in-plane designs, the MNs are parallel to the fabrication surface, whereas in out-of-plane designs the MNs are perpendicular to the fabrication surface [165]. The following sections proved detailed descriptions of the basics of microfabrication technology and the manufacturing of silicon, metals and polymer MNs.

2.3.1 Basics of Microfabrication

Microfabrication technology (i.e. micro-machining or micro-electromechanical systems, MEMS) was traditionally utilised in the production of microprocessors, but subsequently it has been applied in the manufacture of a variety of micron-scale devices, such as micropumps, microreactors, accelerometers, micromirrors and so on [166]. Consequently, exploitation of these MEMS techniques has led to potential applications in biomedical fields (termed "BioMEMS"), such as in drug delivery, DNA sequencing devices, biosensors and chemical analysis systems [167, 168]. MEMS technology has been applied effectively in the fabrication of MN arrays [169]. Although these tools offer the potential for mass production of MNs, production is often highly specialised and includes complex multi-step processes [170].

MEMS technology utilises a number of tools and methodologies to create small three-dimensional (3D) structures, with dimensions ranging from sub-centimetre to sub-micrometre. MEMS technology is adapted from that of integrated circuit (IC) technology. In a MEMS process, before the actual device is generated, a series of sequential operations are necessary. The three basic techniques in MEMS technology are: (1) application of a patterned mask on top of a film by photolithograpic imaging; (2) the deposition of thin films of material on a substrate; and (3) selective etching of the films to the mask [171, 172]. A discussion on the microfabrication of silicon MNs is not complete without understanding the basics of microfabrication. Therefore, the following sections provide details on the basics of silicon microfabrication.

2.3.1.1 Lithography and Etching

Both in microelectronics and micromachining, fabrication starts with lithography (lithos "stone" and gráphein "to write"); this is the technique used to transfer the master pattern onto the surface of a substrate (e.g. a silicon wafer), previously coated with a photosensitive material, by selective exposure to a radiation source (e.g. UV light). The most widely used type of lithography is photolithography. Figure 2.9a–g shows a schematic representation of basic photolithography. In general, the following sequential steps are involved in the mask transfer onto the photosensitive-coated substrate [171]:

- The first step, with a silicon (Si) wafer as a substrate, is to grow a thin layer of oxide by heating between 900 and 1150 °C in the presence of steam or humidified oxygen.
- A thin layer of an organic polymer, known as a photosensitive/photoresist or resist material (sensitive to UV radiation), is spin coated onto the oxide surface of the silicon wafer, which is then spun at a high speed, between 1500 and 8000 rpm, to yield a resist of defined thickness [171].
- After spin coating, the solvent in the resist layer is removed by soft baking at 75 to 100 °C for 10 min. This also promotes adherence to the wafer. After soft baking, the resist-coated wafers are exposed to illumination. The simplest form is the use of a UV lamp, through a mask, allowing a nearly perfect transfer (in other words "printing") of the mask image onto the resist-coated wafers. The UV wavelengths employed range from far-UV (i.e. 150–300 nm) to near-UV (i.e. 350–500 nm) [171].

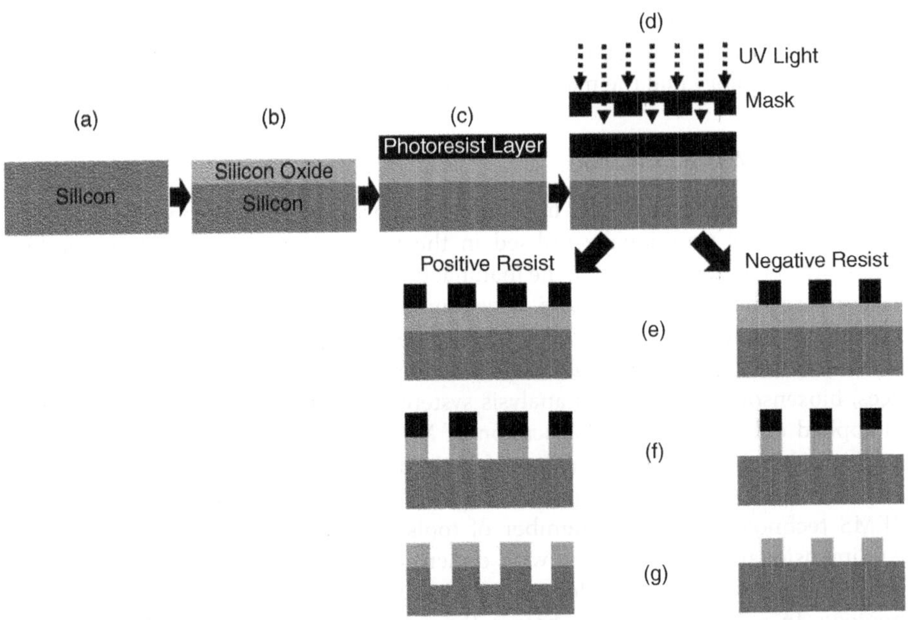

Figure 2.9 Sequential processes in the transfer of a pattern to the substrate surface: (a) Si wafer; (b) Si wafer with oxide coating; (c) spin-coated photoresistive material; (d) mask guided UV light exposure on the photoresistive material; (e) development process to remove the soluble resist material; (f) etching of SiO_2 film; and (g) photoresist removal. Reproduced with permission from: [9] E. Larrañeta, R.E.M. Lutton, A.D. Woolfson and R.F. Donnelly (2016). Microneedle arrays as transdermal and intradermal drug delivery systems: Materials science, manufacture and commercial development. *Mater. Sci. Eng. R-Rep.* 104: 1–32 .

- The radiation treatment induces a chemical reaction in the exposed regions of the resist, altering the solubility of the exposed resist in a particular solvent. During the development process, selective dissolving of the resist takes place, either by wet (using solvents) or dry (using vapour phase or plasma) developments [173].

The unwanted resist left behind after development is removed by oxygen-plasma treatment, called descumming. Finally, before moving to the next process (i.e. before etching the substrate or adding a material) the wafers are post-baked to remove the residual developing solvents and improve adhesion between the resist–substrate interfaces. There are two basic types of resists: positive resists and negative resists. In positive resists, the chemical bonds within the resists are weakened when exposed to UV light and, subsequently, the exposed resists become more soluble in the development solutions. Conversely, the chemical bonds in negative resists are strengthened when exposed to UV light. In photolithography, the mask used is typically an optically flat glass or quartz plate (transparent to near-UV) with a metal (e.g. 800 Å thick chromium player) absorber layer. Alternatively, other techniques, such as, X-ray and charged particle beam lithography can be used to create the desired patterns on the substrates [171, 172].

2.3.1.2 Thin-film Deposition on Substrates

One of the basic steps in the MEMS process is the deposition of a thin film on the substrate surface (e.g. Si wafer). These films can then be patterned using photolithographic techniques or suitable etching techniques. Common materials include silicon dioxide (oxide) and silicon nitride (nitride) [172]. Thin-film deposition on the substrate can also be achieved by using a wide range of materials, including noble metals such as gold [171].

Table 2.2 summarises a variety of different methods used to deposit thin-films on substrates. In general, in physical vapour deposition (PVD) based techniques, the raw materials (solid, liquid or vapour) are released from the source (material to be coated) and deposited on the substrate surface. For instance, in thermal evaporation, the Si wafer is placed inside a vacuum chamber and the source (e.g. aluminium) is heated by electron-beam or radiofrequency. The heating causes the source to boil and the vapours are then condensed onto the substrate surface to form a film. In the sputtering technique,

Table 2.2 Different MEMS deposition techniques. *Source*: adapted from [171] M.J. Madou (2002). *Fundamentals of Microfabrication: The Science of Miniaturization*, 2nd edn. Boca Raton: CRC Press.

Physical vapour deposition (PVD) techniques	Chemical vapour deposition (CVD) techniques	Other deposition techniques
Thermal evaporation Sputtering	Plasma-enhanced CVD (PECVD)	Eptiaxy
Molecule-beam epitaxy Ion plating	Atmospheric pressure CVD	Casting
Laser ablation deposition	Low pressure CVD (LPCVD)	Electrochemical deposition
Cluster-beam deposition	Very low pressure CVD (VLPCVD)	Silk-screen printing
	Metallorganic CVD	Plasma spraying
	Spray pyrolysis	Casting

the substrate and source are placed in a chamber containing an inert gas (e.g. argon or Xe) at low pressure. Using a power source, the gas plasma is ionised, which then accelerates the ions promoting them to impinge on the substrate surface [171].

In chemical vapour deposition (CVD), thin-films are produced by a chemical reaction between the hot substrate and inert carrier gases in the chamber. The CVD method is versatile and works at low or atmospheric pressure at relatively low temperatures. The two most common CVD technologies in MEMS are the LPCVD and PECVD. The LPCVD allows coating of large numbers of wafers without detrimental effects on film uniformity at higher temperatures (>600 °C). On the other hand, PECVD operates at low temperatures and films grow at a faster rate [174].

Finally, the choice of a specific deposition process is dependent upon a variety of considerations, such as substrate structure, source (e.g. chemistry, purity or thickness), apparatus, operating temperature, rate of deposition and total production time. After deposition, the thin-film can be locally etched using lithographic or etching processes to create the final device [172].

2.3.1.3 Etching

Following the lithography, and to create the final functional form of the MEMS structure on a substrate, it is necessary to etch the thin films previously deposited and/or the substrate itself (Figure 2.9f). In general, there are two classes of etching processes: wet etching and dry etching.

Wet Etching In this process, the material is removed by immersing the wafer in a liquid bath containing a chemical etchant. The two main wet etching techniques are isotropic and anisotropic etching. Isotropic etchants attack the material, such as oxide, nitride, aluminium, polysilicon, gold or silicon, at the same rate and in all directions. They remove material horizontally under the etch mask (undercutting). In contrast, anisotropic etchants attack the material (silicon wafer) at different rates in different directions, to produce more controlled shapes. The crystal planes in silicon limit anisotropic wet etching.

Hydrofluoric and nitric acids, in combination with either methanol or water, are used as isotropic silicon etchants [175]. For anisotropic wet etching, potassium hydroxide (KOH) and tetramethyl ammonium hydroxide (TMAH) are used as etchants [176]. However, the mask should not dissolve or at least it should etch at a much slower rate than the material to be etched [172].

Dry Etching This form of etching is carried out at low pressure in the presence of inert or reactive gases. Dry etching is categorised into two main types: reactive ion etching (RIE), which involves chemical processes; and ion-beam milling, which involves purely physical processes. In RIE, a plasma of reactive ions is created in a chamber and these ions are accelerated towards the material to be etched. A deep RIE (DRIE) process, in combination with CVD, called the BOSCH process, can be used to create high aspect ratio (height-to-width ratio) structures. In the case of ion-beam milling (IBM), inert ions are accelerated from a source to physically remove the material to be etched. There are two forms of IBM: showered-IBM (SIBM) and focused-IBM (FIBM). In SIBM, the entire substrate is showered with energetic ions, whereas in FIBM, the ions are focused to a spot on the material [172].

2.3.2 Microfabrication MNs

2.3.2.1 Fabrication of Silicon MNs

Hypodermic needle mediated injections are still painful and also remain a common cause of medical device mediated infections [177]. In addition, one must dispose of hypodermic needles after use in special bins, which requires further expensive treatment for their disposal. However, the design of hypodermic needles allows both drug delivery and biological fluid sampling. This has guided researchers to convert the dimensions into those for use in micron-sized needles, called hollow MNs. Hollow MNs resemble the sharp hypodermic needles, but are intended to be used as minimally invasive means of overcoming the outermost barrier of the skin, the *stratum corneum*, for both drug delivery and/or biological fluid sampling [178]. In contrast, solid MNs, without bores, have also been developed for minimally invasive means of penetrating the skin. A variety of microfabrication techniques have been used effectively to manufacture both solid and hollow MNs, based on the basic principles of silicon fabrication, as discussed previously.

Prausnitz's research group, from the Georgia Institute of Technology, Atlanta, USA, were the first to publish the fabrication of solid silicon MNs by using an RIE process with a chromium mask, which is a dry-etch process [5]. Similarly, different research groups have used either wet- or dry-etching techniques in the fabrication of solid silicon MNs, of varying shapes, heights and densities. Morrissey's research group from the Tyndall National Institute, Cork, Ireland, fabricated solid silicon MNs using a wet-etch process, where a standard silicon wafer (p-type) was deposited with an oxide layer (300 Å) and a layer of nitride (1000 Å) using a standard LPCVD process. The double layer was then patterned with a plasma-etch and 280 μm high MN arrays (with aspect ratio 3:2) were fabricated using 29% v/v KOH at 79 °C at an etch rate of 1.12 (±0.02) μm/min [28]. Apart from wet etching, they have also demonstrated both solid and hollow MNs, fabricated by a dry-etching technique. Using a standard wafer of 525 μm thickness, conical-shaped solid MNs with an aspect ratio of 4.5:1 (height to base diameter) were fabricated by a modified RIE process. A typically undercut etch rate to vertical etch (using SF_6/O_2) was used alternately with the BOSCH-DIRE process. For hollow MNs, an additional etching from the backside of the silicon wafer was performed using the DIRE process [20]. The major disadvantage of this method of fabrication was the limitation of MN height and density. Owing to the thickness of the original silicon wafer (i.e. approximately 500 μm), fabrication of MNs with heights of 500 μm or greater was impossible. In addition, as a result of the wet-etching technique employed, the minimum inter-spacing possible between the needles was approximately three times the needle base diameter. This further limited the density of MNs per array.

Using a dry-etching technique Paik *et al.* (2004) fabricated in-plane single-crystal silicon MN arrays consisting of a microchannel (Figure 2.10a) [179]. The MN shafts, 2 mm in length, 100 μm wide and of 100 μm thickness, were strong enough to endure a 0.248 mN out-of-plane bending moment and 6.28 N of in-plane buckling load. These MN arrays, when integrated with a polydimethylsiloxane (PDMS) microfluid chip, demonstrated efficient delivery of model solutes, namely Rhodamine B dye and black ink, in *in vitro* (agarose gel) and *ex vivo* (chicken breast) models through the microchannel. Similarly, Roxhed's research group from the Royal Institute of Technology, Sweden, described an interesting concept of sharp hollow silicon MN tips

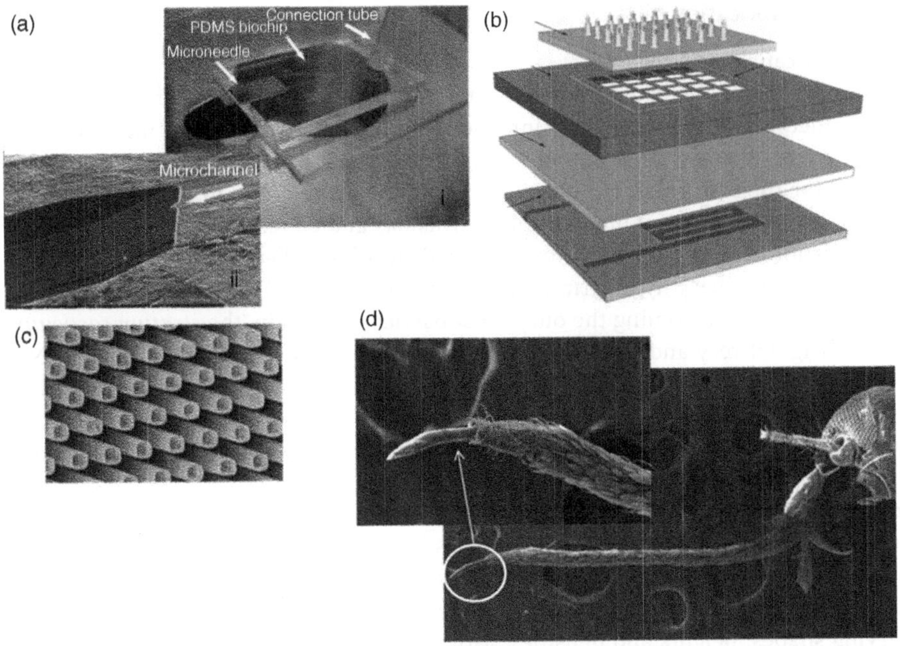

Figure 2.10 SEM micrographs of a variety of hollow silicon MNs. (a) i-PDMS microfluid chip integrated MN array with a ii-microchannel entrance at the tip. Reproduced with permission from: [179] S.J. Paik, S. Byun, J.M. Lim, *et al.* (2004). In-plane single-crystal-silicon microneedles for minimally invasive microfluid systems. *Sens. Actuator A-Phys.* 114: 276–284. (b) Patch-like MN integrated dispensing unit. Reproduced with permission from: [180] N. Roxhed, B. Samel, L. Nordquist, *et al.* (2008). Painless drug delivery through microneedle-based transdermal patches featuring active infusion. *IEEE Trans. Biomed. Eng.* 55: 1063–1071. (c) Array of MNs with 70 nm wall thickness and square arrangement. Reproduced with permission from: [181] A. Rodriguez, D. Molinero, E. Valera, *et al.* (2005). Fabrication of silicon oxide microneedles from macroporous silicon. *Sens. Actuator B-Chem.* 109: 135–140. (d) Mosquito head and proboscis, the inset shows the magnified view of fascicle tip with labella retracted. Reproduced with permission from: [182] M.K. Ramasubramanian, O.M. Barham and V. Swaminathan (2008). Mechanics of a mosquito bite with applications to microneedle design. *Bioinspiration Biomimetics* 3: 046001 .

with side-openings. The hollow MNs tips were sealed with thin leak-tight membranes to provide a closed-package system, thereby preventing drug degradation, evaporation on leaking-out and damage due to handling, thus improving the shelf life of MN integrated patch devices. MNs within the integrated device were etched on 600 μm thick monocrystalline silicon wafers using a two-mask process, an anisotropic DRIE etch through the BOSCH process and an isotropic (unbiased) SF_6 plasma etching. The needle bore was first etched from the backside of the silicon wafer, after which needle geometry was shaped from the front side by mixed isotropic and anisotropic etching. An intersection between the two-directional etchings resulted in a side opening of the needles. The two designs of MN created were a 310 μm long cross-shaped MN and 400 μm long circular-shaped MN. The side opening, previously coated with a 20 nm thick adherent chromium layer, was sealed with a layer of gold coating of three different thicknesses (150, 300 and 450 nm). Three different approaches to opening the seals were studied: burst opening, electrochemical opening and opening upon insertion into the skin [24]. All three methods were found to be efficient for opening the side-bores. However, the biocompatibility of the residual coating material left after its application was not demonstrated. On the other hand, the advantage of the side-opening MNs, unlike the central opening, is the resistance to the blockage due to dermal tissue, which is one of the major concerns for hollow MNs. In another study by Roxhed's research group, the previously described 400 μm long circular-shaped MNs were integrated with a low-cost electrically controlled liquid dispensing unit to form a complete patch-like drug delivery system [180]. The dispenser consisted of three distinct layers: a 500 μm thick printed circuit board (PCB) heater layer, a 500 μm thick expandable layer (a mixture of silicone elastomers) and a liquid reservoir (total volume of 12 μl) (Figure 2.10b). The integrated hollow MN device successfully demonstrated the delivery of insulin in diabetic rats, where active infusion at 2 μl/h showed consistent control over blood glucose levels when compared with passive (unactuated) delivery.

By using an electrochemical-etching process on an n-type Si wafer, novel hollow silicon dioxide MNs were fabricated [181]. MNs with different geometries and lengths (30–140 μm), wall thickness (70–110 nm) and pore diameters (2–5 μm) were etched on the macroporous silicon (Figure 2.10c). To control the fluidic properties of the microstructures, the MNs were glued to a tube, which, in turn, was attached to a syringe. This, in theory, should allow a controlled delivery of the loaded dose with high precision. In another study, an array of hollow silicon MNs were integrated with a piezoelectric (PZT) micro-pump [183]. In this case, the hollow MNs were fabricated on a flexible silicon substrate by inductively coupled plasma (ICP) and anisotropic wet-etching techniques with dimensions: 200 μm height, 30 μm internal diameter and 90 μm outer diameters. The PZT pump is connected to a flexible MN array to create a prototype device, in which the total size of the packaged micropump is claimed to be smaller than any other packaged insulin pumps, and which can perform precise fluid sampling, programmable drug delivery, remote operation and autonomy. Even though the devices with such automation may revolutionise future drug delivery applications, the practicality of such devices is of major concern, therefore thorough investigations in clinical trials are required before widespread application.

Another fully integrated microfluidic drug delivery device was recently demonstrated for treatment of a cardiovascular disorder (hypertension) [184]. The device consisted of blood pressure sensors integrated with hollow silicon MNs, a PZT actuator unit, drug reservoir and a flow sensor to measure the fluid dynamics in real-time. The hollow silicon MNs used within this device were fabricated by the DIRE technique followed by a series of combined isotropic and anisotropic etching processes, as described earlier, using inductively coupled plasma etching technology. The device is capable of sensing the rise in the blood pressure of the patient, it then actuates through the integrated PZT unit to release the drug, which is present in the reservoir, and the flow sensors enable real-time monitoring of the volume delivered.

Arrays of hollow out-of-plane silicon MNs have been fabricated and then integrated with a reservoir made of PDMS. In this method, the silicon MNs were microfabricated through a two-mask MEMS process, consisting of a 40 μm wide lumen that was generated through a DIRE step through the back of the silicon substrate [185]. The needle density was 600 needles/cm^2 with a shaft length of 200 μm. A micromoulding technique was used to fabricate the PDMS reservoir, with a volume capacity of 12.5 μl [185]. The PDMS surface was treated with an oxygen plasma and bonded with the silicon MN base. It was shown that the drug suspension could be loaded into the PDMS reservoir by an assembled MEMS syringe that uses a 28G (gauge) hypodermic needle. The device can be placed on the skin and, by applying gentle finger pressure on the PDMS, the contents of the reservoir can be delivered to the epidermal layer.

Unlike the hypodermic type of design of hollow integrated MN arrays, the approach used by Ramasubramanian *et al.* (2008) [182] was found to be very interesting. The authors conceptualised the naturally existing mosquito fascicle behaviour, which painlessly withdraws blood from the human skin surface (Figure 2.10d). Firstly, anatomical details of mosquito fascicle penetration and its dynamics were investigated with the help of scanning electron microscopy (SEM) and high-speed video imaging, respectively. Secondly, mathematical models were used to determine the role of the lateral support of the fascicle (i.e. labium) and non-conservative force application. Results showed that the lateral support of the labium is critical and helps the mosquito to penetrate the skin. In general, the fascicle is typically 1.8 mm long and has an internal radius of 11 μm, with a blood withdrawal volume of 4.2 ml taking just approximately 141 s. Thus, a careful understating of the anatomical dimensions and dynamics of mosquito bites may help researchers to design novel hollow MNs that resemble natural phenomena. Interestingly, the volume of blood withdrawn is significantly high with only one needle-like fascicle structure; we believe that no other MN system has demonstrated such a high withdrawal volume in such a short period of time. Finally, the microfabrication of silicon, as the substrate material, in the design and application of MNs is a dominant technology due to certain excellent features of silicon, such as the mechanical properties, electrical properties and the possibility to directly integrate circuits on the transducer's substrate, as reported in various publications.

2.3.2.2 Fabrication of Metal and Glass MNs

Fabrication using microelectronics is relatively expensive and requires cleanroom processing [25]. In contrast, metal and glass MNs have been found to be equally effective in skin penetration and can be produced at relatively much lower cost than silicon MNs. Different metals, such as stainless steel, titanium, palladium, palladium–cobalt alloys, nickel and so on, have been used as structural materials for MN fabrication [1, 9, 42]. A number of approaches have been proposed for fabricating metal MNs, such as electroplating (e.g. palladium), photochemical etching (e.g. titanium) and laser cutting (e.g. stainless steel). Metals such as stainless steel have been in medical use for decades, for example as hypodermic needles. Essentially, the use of such materials will effectively reduce the regulatory path of approval, compared with that required for non-approved materials such as silicon. In addition, like silicon MNs, metals can be routinely fabricated into both solid and hollow MNs.

Most metal MNs have been obtained by simply assembling the traditionally available stainless-steel hypodermic needles, to produce hollow MNs. Currently, the smallest hypodermic needles in use are 30 and 31G for conventional syringes and pen injectors (for insulin delivery) with outer diameters of 305 and 254 μm, respectively [169, 186]. For the shortest and thinnest 31G needle the insertion length is 5 mm. These hypodermic needles were translated into MNs by exposing the defined lengths out of a supporting material. Typically, the smallest used hypodermic needles (30/31G) are manually assembled to form an array of MNs. For example, Bouwstra's research group from the Leiden/Amsterdam Center for Drug Research, Leiden, The Netherlands, used commercially available 30G hypodermic needles and assembled them into 4×4 arrays, supported by a polyetheretherketone mould, with needle heights adjusted manually to 300, 550, 700 and 900 μm [42]. These 30G stainless-steel hypodermic needles were chosen due to their low cost and robustness compared with silicon, which is brittle and not biocompatible [27, 28]. Similarly, Mikszta's research group from BD Technologies, Durham, NC, USA, employed 1 mm long hollow MNs using a 34G stainless-steel hypodermic needle fitted to 1 ml syringe for vaccine delivery in rat models [187, 188]. However, in these applications, the role of hollow MNs was similar in application to that of solid MNs, where the drug (in the form of a patch or solution) is applied following the MNs treatment. In such a case, the role of the hollow bores in the MNs becomes irrelevant. Furthermore, these types of hollow MNs are limited in terms of design; the only feasible modifications that can be achieved are the needle dimensions (i.e. height and width) or the needle density.

Bouwstra's research group have also demonstrated the simplest form of solid metal MNs, which were fabricated by simply assembling stainless-steel wires of 200 μm diameter and 300 μm height into 4×4 arrays, where the tips were cut tangentially to obtain sharp tips [42]. Wu *et al.* (2007) [189] proposed use of acupuncture needles that were assembled on a silicon sheet (with dimensions of 1.0×20×20 mm). Each needle tapered over a 400 μm length with a tip angle of 28°, and a base diameter of 200 μm. Badran *et al.* (2009) [190] utilised a novel metal MN device called Dermaroller®, with different stainless-steel needle lengths (150, 500 and 1500 μm) protruding out from a cylindrical assembly containing 24 circular arrays of eight needles each (192 needles in total). Unlike assembling of the stainless-steel wires, stainless-steel needles or acupuncture needles, which are limited in terms of the needle dimensions, Prausnitz research group used an infrared laser technique to fabricate arrays of MN shafts from stainless-steel sheets (75 μm thick). In this technique AutoCAD software was used to draft the shape and orientation of the arrays [191]. The laser was operated at 1000 Hz at an energy density of 20 J/cm^2, and took a total of 4 min to cut an array. Each needle was manually bent at an angle of 90° to create an out-of-plane MN array followed by electropolishing (6:3:1, glycerine–phosphoric acid–water). This approach is promising to achieve MNs of different designs and dimensions, but it requires specialised instruments for fabrication that may add to the cost. Similarly, solid in-plane and raised out-of-plane titanium MNs were fabricated using photolithography [192]. The effect of titanium foil thickness (i.e. 75, 127, 250 μm) on the height and base of the MNs was studied. It was observed that the 127 μm thick titanium foil produced MNs with a length and width of 454±42 μm and 225±11 μm, respectively, after wet etching within a reasonable time [192]. Titanium-based MN arrays (commercially known as Macroflux®) were previously demonstrated by Cormier and Dadonna (2004) [193], from the ALZA Corporation, which consisted of 321 MNs/cm^2, with an area of 2 cm^2, height 200 μm, base-width 170 μm and thickness 35 μm, affixed to an adhesive patch (patch area 5.3 cm^2). The MN arrays were produced using a manufacturing process incorporating an AutoCAD-generated microprojection array design, photochemical etching and forming. As mentioned earlier, after etching the microprojections were bent to an angle of 90° (relative to the sheet plane) using a forming tool [193].

As an alternative to metal MNs, Prausnitz and colleagues fabricated a single hollow glass MN by pulling fire-polished glass pipettes using a micropipette puller. The resulting blunt-tip MN was bevelled (at an angle of 35–38°) and cleaned in a series of solutions, as shown in [85]. These cleaned MNs were attached to a 250 μl or 1 ml glass syringe (containing 1×10^{-3} M sulforhodamine B solution) and then investigated for infusion flow rates into human cadaver skin.

Unlike the methodologies just discussed, the commonly used MEMS fabrication techniques, such as, photolithography, DIRE and deep X-ray lithography of LIGA (Lithographie, Galvanoformung and Abformung) are based on the inherently planar geometries of 2D substrates. These techniques have been reported to create MNs, but with limited heights due to the limitations of substrative projection lithography [194]. Therefore, to overcome this and to create relatively ultra-high aspect ratio (UHAR) metal MNs a "drawing lithography" method was employed (Figure 2.11). The method involves spin coating and curing of a thermosetting polymer (SU-8 2050) following by controlled drawing using pillars of a defined pattern. This process creates 3D solid long polymeric needles, which act as a mould to finally create hollow metallic nickel MNs. The hollow metallic MNs fabricated by this technique produced MNs with heights of 600, 1200 and 1600 μm [194]. Thus, fabrication of metal or glass MNs can be adapted from various different techniques as discussed here; however, the choice of a given technique is in turn dependent upon various factors, such as manufacturing cost, materials biocompatibility, desired depth of penetration, desired drug plasma concentrations and drug dosage if it is aimed to be coated on the MN arrays.

2.3.2.3 Fabrication of Polymeric MNs

They are some concerns surrounding the use of metal and silicon MNs, such as biocompatibility issues, the immuno-inflammatory response of soft tissue and the chance of material deposits within the skin, as discussed earlier. Thus, polymeric MNs have gained significant importance because of their unique biocompatibility and biodegradability properties. Selected polymeric materials can also provide the strength and toughness required for skin penetration. Additionally, some polymeric materials can provide the optical clarity to the MN system [1]. Further advantages of polymeric MNs may extend to the fabrication and reproducibility, which are considerably much more cost-effective compared with typical MEMS processes employed in silicon MNs fabrication. Accordingly, the micro-scale dimensions of polymer MNs have been accurately produced using a variety of mould-based techniques, such as casting [106], injection moulding (Figure 2.11) [107], hot embossing (Figure 2.11) [196], investment moulding [197], laser [198] and X-ray [6] methods.

Micromoulding-based Fabrication Micromoulding processes involve replication of master structures by means of moulds. To date the most commonly used mould-based material is PDMS, due to its flexibility and accurate reproducibility of master structures. The master structures have been produced using different materials (e.g. silicon or metal) and techniques, but they can eventually be micromoulded into polymeric MNs. A wide range of polymers have been used in micromoulding of the respective master structures. Initially, a number of groups investigated carbohydrates as potential MN materials. When carbohydrates are heated to temperatures around their melting points they change into yellow–brown coloured substances, known as caramels. This process, known as caramelisation, involves the removal of water from a carbohydrate followed by complex isomerisaton and polymerisation steps. This leads to a hard, brittle material sometimes referred to as a "candy". Miyano *et al.* (2005) [95] were the first to report the use of natural sugars in the fabrication of MNs. The method involved use of powdered

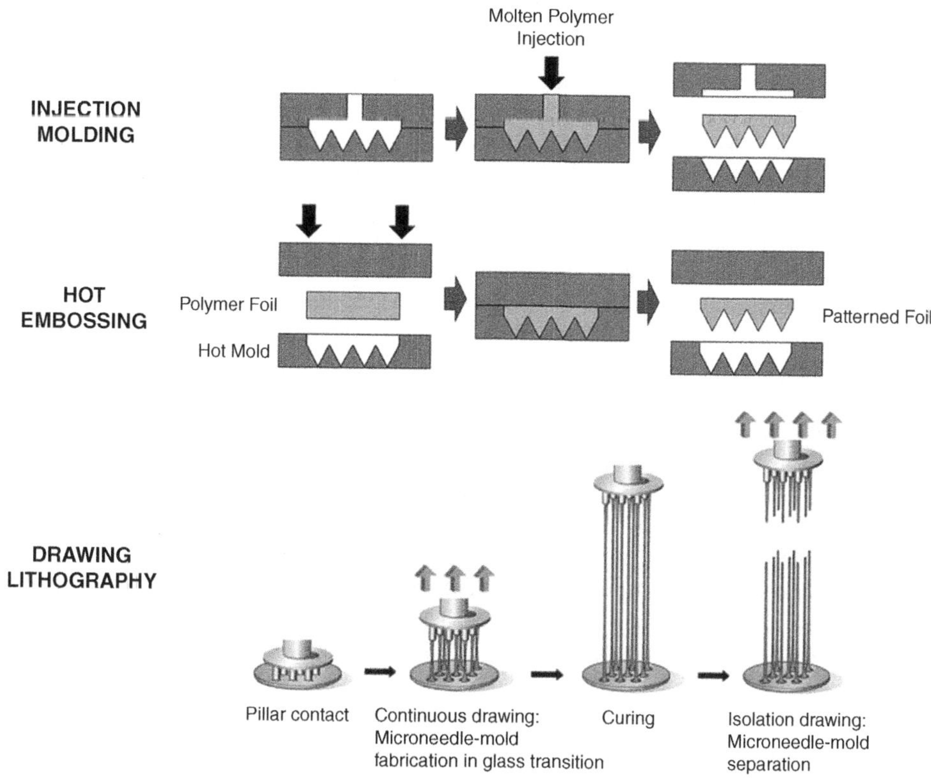

Figure 2.11 Schemes of injection moulding, hot embossing and drawing lithography processes. Reproduced with permission from: [9] E. Larrañeta, R.E.M. Lutton, A.D. Woolfson and R.F. Donnelly (2016). Microneedle arrays as transdermal and intradermal drug delivery systems: Materials science, manufacture and commercial development. *Mater. Sci. Eng. R-Rep.* 104: 1–32; and [195] K. Lee and H. Jung (2012) Drawing lithography for microneedles: a review of fundamentals and biomedical applications. *Biomaterials* 33: 7309–7326 .

maltose, heated to 140 °C for 1 h, to form maltose-candy. The powdered drug was added to this candy and mixed uniformly within a minute and stored in a dry environment at room temperature. A small amount of this maltose-candy containing drug was then placed onto a casting MN mould at 95 °C and cast into MN arrays. Arrays of 500 μm high MNs containing ascorbate-2-glicoside (5% w/w), sodium salicylate (10% w/w) and calcein (10% w/w) have been fabricated by this method. These sugar MNs dissolved within a few hours at humidities of more than 50% and retained their shapes for at least 3 months at 40% humidity. Furthermore, Miyano *et al.* (2007) [199] developed hydrolytic MNs of maltose and poly(ethylene glycol) (PEG, MW = 600 Da). The fast-dissolving maltose MNs were one-dimensional tetrahedral MNs with a length of 500 μm, dissolved within 3 h after insertion. However, the high viscosity and low throughput prohibited the fabrication of maltose MNs deployed in 2D-arrays. In contrast, the low viscosity of PEG not only allowed high throughput production of PEG MNs, but also fabrication of 2D-arrays of PEG MNs with lengths of 1000 or 2000 μm, respectively. Likewise, micromoulded maltose MNs were reported to enhance the flux of nicardipine hydrochloride across full-thickness rat skin fourfold *in vitro*. However, the drug was not incorporated into the MN, but was contained in a liquid reservoir patch that was applied to the skin on top of the inserted MN [88].

A number of reports [88, 95, 199] have suggested the fabrication of hot-melts of carbohydrate materials into MNs using silicon or metal MNs as master templates. Principally, such MNs should dissolve upon skin insertion to release their drug payload. However, Donnelly *et al.* (2009) [89] reported that the carbohydrates require high processing temperatures to produce hot-melts, which are extremely viscous and resistant to flow. In addition, the materials produced upon cooling are extremely hygroscopic, which causes difficulties with storage and handling. In this study, it was shown that due to the high processing temperature (i.e. 160 °C) of galactose MNs, substantial losses of the APIs, namely 5-aminolevulinic acid (ALA) and bovine serum albumin (BSA), were incurred during processing. Furthermore, galactose MNs were found to be unstable at ambient relative humidities (43 and 83% RH) and became adhesive. Finally, it was suggested by the authors that the carbohydrate-based MNs are not the solution to the problems posed by use of silicon and metal MNs. This is especially true as MNs produced from hot polymer melts also suffer from substantial drug loss during processing.

To overcome some of the disadvantages associated with the preparation of carbohydrate MNs by heating, Takada's research group from Kyoto Pharmaceutical University, Kyoto, Japan, proposed that self-dissolving MNs could be fabricated from a thread-forming biopolymer (i.e. dextrin). Unlike micromoulding, poly(propylene) tips were used to dip into the dextrin gel and then withdrawn perpendicularly to form thread-like needles, which were dried at 4 °C in a desiccator [114]. Clearly, this fabrication method has the advantage of not requiring high temperatures. However, at an average length of 3.0 mm and base diameter of 0.55 mm, these devices are considerably larger than most true MN systems studied elsewhere. These MNs loaded with insulin showed a dose-dependent hypoglycaemic effect in mice and insulin was found to be stable in the MNs for at least 1 month at 40 °C. However, this method of MN fabrication may create high inter-individual variability in the MN dimensions. Erythropoietin (EPO) loaded polymeric MNs were also produced using a thread-forming method for percutaneous administration of EPO in mice. The thread-forming materials used in this study were made from dextrin, chondroitin sulfate and albumin [200].

Takada's group also prepared self-dissolving MNs from dextrin, chondroitin sulfate or dextran containing lower molecular weight heparin (LMWH) (MW = 4.5 kDa), where the length and basal diameter of the MN were 1.5 and 0.5 mm, respectively. The LMWH loaded into the dextran-MNs was found to be stable for at least 3 months when stored under three different conditions, namely, 40, 4 and −80 °C [201].

In addition to sugars, various other materials have also been reported. Prausnitz and colleagues demonstrated encapsulated molecules within pyramidal MNs made from carboxymethylcellulose (CMC), amylopectin (AP) and bovine serum albumin (BSA) using casting techniques. These MNs were shown to be sufficiently strong to insert into cadaver skin and to dissolve within minutes [93]. This group also reported an alternative matrix for the preparation of pyramidal polymeric MNs (10×10 arrays of 300×300×600 μm width, length and height, respectively), which was a mixture of PVA (MW 2000)/PVP (BASF, K15), in a 3:1 ratio [109]. PDMS micromoulds were created from the master arrays, and three different MN designs were then produced, namely, solid, bubble and pedestal MNs. Park *et al.* (2006) [163] also fabricated bevelled-tip MNs from PLGA using a micromoulding technique to encapsulate various APIs.

Han *et al.* (2007) [105] demonstrated a casting technique to design a prototype MN array, in which in-plane MNs were transformed into out-of-plane biocompatible polycarbonate MN arrays. The in-plane MNs were then aligned parallel to each other to form an out-of-plane MN array. Using these MN arrays a negative mould of PDMS was fabricated. Finally, the biocompatible out-of-plane MNs were produced using a hot-embossing machine. Furthermore, Han *et al.* (2009) [196] proposed three different types of groove-embedded MN shafts with sharp 3D-tips to reduce insertion force, increase fracture force and increase drug loading within the grooves on the MN. The height, base width and thickness of these polymeric MNs were 880±20, 710±15 and 145±15 μm, respectively. The groove-embedded MNs were used to produce negative PDMS moulds by a hot-embossing process. Biodegradable polymeric PLA MN arrays were then fabricated by replicating the negative moulds. The mould and PLA grains were heated to 190 °C and subjected to a pressure of 20 kg/cm^2 for 10 min, followed by cooling to room temperature. The groove-embedded replica MN arrays were then demoulded. Unlike the soluble MNs that dissolve to release the API, here these authors reported coating of the groove-embedded MNs. One advantage of these types of MNs over the sugar MNs is that the APIs are not exposed to high temperatures.

Lippmann *et al.* (2007) [106] fabricated MNs using a plastic polymer "cyclic olefin copolymer (COC)" for its balance of strength, easy of manufacturing and its biocompatibility. The hollow, in-plane, polymeric MNs were 280 μm in height, with a base width of 130 μm and a bore diameter of 35 μm. Similarly, Sammoura *et al.* (2007) [107] fabricated in-plane, open-channel polymeric MNs via microinjection moulding techniques. The key components were the shank portion (4.7 mm) with an open-channel (cross-sectional area of 0.1×0.1 mm^2) and the base portion consisting of a reservoir. In this study Topas® COC, a copolymer based on cyclic olefins and ethylene, was used as the plastic material for making in-plane MNs by an injection moulding process. Upon application into beef liver, these MNs were observed to draw approximately 0.04 μl of liquid immediately from tissues.

Other Techniques Moon *et al.* (2005) [6] proposed an inclined LIGA process to fabricate polymeric MN arrays for both drug delivery and whole blood sampling. Since the LIGA process is based on MN mould fabrication, it is easier and more convenient than other silicon or glass etching techniques. In this method, a simple change in the X-ray angle can easily control the needle tip and its high aspect ratio. Following exposure to high-energy X-rays, materials will show high surface roughness, which is an advantage for mass production of a mould. A conventional LIGA process uses a planar exposure and layer-by-layer technique to fabricate 3D microstructures [6]. However, a deep X-ray lithography (DXRL) technique employs both vertical and inclined deep X-ray exposure to create high-aspect ratio hollow microstructures with sharp tips. Thus, in order to obtain sharp MN tips with high-aspect ratio out-of-plane hollow MN arrays, a DXRL process was developed [6]. With this design strategy, high-aspect ratio and sharp tip poly(methyl methacrylate) (PMMA) MN arrays were fabricated. The height of these 3D-shaped triangular hollow MNs, ranged from 750 to 1000 μm, with base diameters of from 270 to 400 μm and bore diameters between 70 and 100 μm. Though high aspect ratio polymeric MNs are prone to mechanical failure, especially in the case of hollow polymeric MNs, the authors demonstrated the mechanical performance in human skin through applications at different sites. It was observed that due to the greater thickness and flexibility of fingertips, MNs (750 μm high) could not penetrate the dermis layer to extract blood samples. In contrast, blood samples were easily collected from skin on the back of the hand. However, no method was provided for replicating the MN array technique.

Unlike the conventional silicon, steel or titanium microfabrication processes, Ovsianikov *et al.* (2007) [81] illustrated a two-photon polymerisation (2PP) technique to create in-plane and out-of-plane hollow MN arrays using a covalently bonded organic–inorganic material known as Ormocer® (organically modified ceramic) as a substrate. The 2PP method applies a femtoseconds laser pulse (60 fs, 94 MHz, <450 mW, 780 nm) from a titanium:sapphire laser focused onto a small focal volume within a photosensitive resin; the laser pulse cleaves the chemical bonds and the desired structures are fabricated by moving the laser focus in three dimensions. Using this process three different designs of hollow MNs were produced, with different bore sizes and positions. The fabrication time with hollow MNs for a single MN was 2 min, with needle height of 800 μm and base diameter of 150–300 μm. Interestingly, this is the only study conducted to date to demonstrate the biocompatibility of MN materials by using an MTT (3-(4,5-dimethylthiazol-2-yl)-2,5-diphenyltetrazolium bromide, a tetrazole) assay in human epidermal keratinocyte cell lines. Later, Gittard *et al.*, in 2009 [202] from the same research group, suggested that if Ormocer MNs were present on a substrate made of glass, they could be susceptible to detachment from the substrate. Therefore, a novel material and method for fabrication of polymeric MNs was proposed.

The silicon master MNs, fabricated using a previously reported approach [20], produced MNs of approximately 280 μm in height, with a diameter of 240 μm at the base and an interspacing of 750 μm. The aspect ratio of these MNs was 1.12:1 (height to base diameter). The major disadvantage of this method of fabrication is the limitation of the height and density of the MNs. Owing to the thickness of the original silicon wafer (i.e. 500 μm), fabrication of MNs with heights of 500 μm or greater was impossible. In addition, due to the wet-etching technique employed, the minimum possible interspacing between the needles was approximately three times the needle base diameter. This further limited the density of MNs per array. In order to fabricate high aspect ratio MNs at higher densities, Donnelly's research group have developed a novel laser-engineered technique [121].

Lasers have been used in materials processing for nearly three decades. Laser light is special, in that it is monochromatic in nature (i.e. is only one wavelength), coherent in time and space and has a very low angle of divergence (i.e. it is collimated) when compared with light from a conventional incandescent bulb. Because of its monochromatic and coherent nature, a laser beam can focus its entire power onto a very small diameter spot [172]. In micromachining, lasers are used for a wide variety of applications, such as heat treatment, welding, ablation, deposition, etching, lithography, photopolymerisation, microelectroforming and focused-beam milling of plastics, glasses, ceramics and metals [171]. Therefore, laser micromachining is an ideal technology for the microstructuring of medical devices. The most commonly used types of laser are carbon dioxide (CO_2) and neodymium YAG (Nd:YAG), both of which are infrared (IR) types of lasers. These are commonly used industrial lasers, whereas UV lasers are more typically used for MEMS purposes [172]. The main advantage of a UV excimer laser beam lies in its ability to micromachine organic materials (plastics, polymers, etc.). Unlike IR lasers, excimer lasers do not remove material by burning or vaporising. In fact, they selectively ablate the material by interacting with the chemical bonds within the material. Excimer lasers are operated in the pulsed mode, removing material with each pulse, with the quantity of material removed being dependent upon the type of material, length of pulse and the intensity of the laser light. However, laser fabrication is performed individually for each object, which is time consuming [171]. Finally, the running costs of laser units, in particular diode pumped solid state (DPPS) lasers, are relatively low, with no consumables apart from changing the diode every 20 000 hours.

Traditional methods of excimer based laser micromachining production (e.g. a UV excimer laser at 248 nm, KrF) use chrome on quartz masks, such as the masks produced for photolithography, to allow creation of precise shapes in polymer materials. The time to prototype is limited by the creation of masks, and all but the simplest of shapes, such as holes or squares, cannot be manufactured in a timely manner. As a scalable production process, the use of mask technology is well proven. However, the cost of consumables, such as masks and gas refills, make excimer-based processing a high-cost option. Donnelly's group were the first to use a galvanometer to replace the traditional mask-based micromachining technique [111]. Replacing mask projection beam delivery with a galvanometer-based beam delivery is fast becoming the standard method for high throughput, low cost laser micromachining. Galvanometer based laser micromachining allows the positioning and on/off pulsing of the laser beam to be determined by a standard computer aided design (CAD) file. The galvanometer consists of two mirrors that can either position the beam at a given point within a field of view or, more importantly, scan the laser beam within the field of view. Therefore, the laser beam can trace out a CAD-based design onto the work piece, without the need for masks. This makes it ideal for fast prototyping. It is also scalable to production volume runs. The lack of consumables, such as masks, also makes it a cheaper option. In this study, using silicon sheets, it was shown that the time taken to machine an array of 11×11 holes was ≈5.4 min. It was also shown that the silicone sheets (obtained from a laser-engineering method) could be reused more than 50 times in the micromoulding of poly(methyl vinyl ether-*co*-maleic acid) (PMVE/MA) based soluble polymeric MNs.

Aoyagi *et al.* (2007) [101] was the first to show UV excimer laser based fabrication of high aspect ratio thin holes in different polymeric materials, such as PLA, parylene, PDMS, epoxy and nylon. Subsequently, the UV excimer laser was then used to make hollow PLA MNs. Briefly, the laser was applied to create long thin holes with diameters of 10, 20 or 50 μm and of varying depths in PLA sheets. The long PLA MNs, fabricated by a wet-etching technique, were subjected to laser pulses to create holes (10 μm in diameter) within the central core of the PLA MNs. Blood collection through the hollow MNs was successfully demonstrated. However, this method uses an aperture mask, which is placed in the travel path of the beam. The novel method described by Donnelly's group utilises a galvanometer, which reduces the cost and saves time compared with the UV excimer method. Furthermore, as previously stated, the Donnelly method can drill an array of 11×11 holes in less than 6 min. Unlike this last method in which thin holes were laser-engineered into the silicone sheets, Aoyagi *et al.* (2007) [101] later demonstrated fabrication of PLA MNs by micromoulding from a silicon master MN, which had previous been fabricated by a wet-etching technique. The laser was then used to "drill" holes into the PLA MNs to produce hollow PLA MNs.

Finally, MNs prepared from the polymers discussed here can be easily inserted into the skin and be manufactured at low cost and with high yield; in addition, polymers pose lower potential risks due to their biodegradation and biocompatibility properties. However, insertion of polymeric MNs into any biological membrane requires the necessary strength. Brittle polymeric materials may fracture during application or very hard polymeric materials may be poor at assisting the drug delivery. Also, the stability of the polymers, such as moisture absorbing capacity and many other parameters, requires detailed investigation with respect to the manufacturability of polymeric MNs. Polymeric MNs reported to date mainly rely on micromoulding of master templates, fabricated from silicon or metal MNs, while some methods have reported direct fabrication, without a master template, which is advantageous, as this avoids the silicon or metal MNs in the first place. Nevertheless, polymeric MNs hold huge promise in the field of transdermal drug delivery of various APIs, because of the various advantages discussed here.

2.3.3 Microneedle Design

The various fabrication methods described earlier use different materials; however, the prime purpose of MNs is to demonstrate sufficient strength to penetrate into the skin or any other biological tissue without breaking or bending during application. Additionally, the elasticity of the skin remains a major challenge to the reproducibility of MN penetration. It has been shown that the skin can fold around the MN and may result in either partial or incomplete piercing of the skin, which counteracts the penetration, depending upon the MN height [203]. Furthermore, MNs should increase the blood concentration of the molecule delivered transdermally. However, some MNs that demonstrated good strength have failed to enhance the transdermal drug delivery, because not all of them have the ability to increase the blood concentration. Major factors accountable for MN performance are the type of material, needle height, tip radius, base diameter, needle geometry, needle thickness and needle density, which, in turn, determine the overall insertion and fracture force of the MN [204]. Therefore, understanding these relationships between the MNs and the skin will allow one to design an "intelligent" or "optimised" MN for widespread clinical applications, with low insertion force and high fracture force. For all these reasons, a wide range of MN designs can be found in the literature (Figure 2.12). On the other hand, before designing an optimised MN, it is necessary to understand the structure of the skin. Human skin consists of three layers: epidermis, dermis and hypodermis and the thickness of these layers varies with body parts, sex, age or skin condition. Therefore, it is imperative to understand the key mechanical properties of the skin, especially that of the *stratum corneum* and viable epidermis (VE) in MN applications.

Kendall *et al.* (2007) [206] studied the mechanical properties of skin by penetrating micro-nano-projections on a patch through the intact *stratum corneum* and VE for the efficient targeting of molecules to the immunologically sensitive cells in the viable skin. This study determined the storage modulus, the Young's modulus and the breaking strength of skin, through different layers of the *stratum corneum* and VE in intact and freshly excised murine skin, by using a micro-indentation probe fitted to a NANO-indenter. It was observed that the mechanical properties described decreased with depth through the *stratum corneum*. The authors demonstrated that variation in skin properties, such as thickness variation at different body sites, and variation due to age, sex, race and body mass index (BMI) require consideration in order to ensure consistent performance of MNs. In contrast, a clinical study evaluated the efficiency of

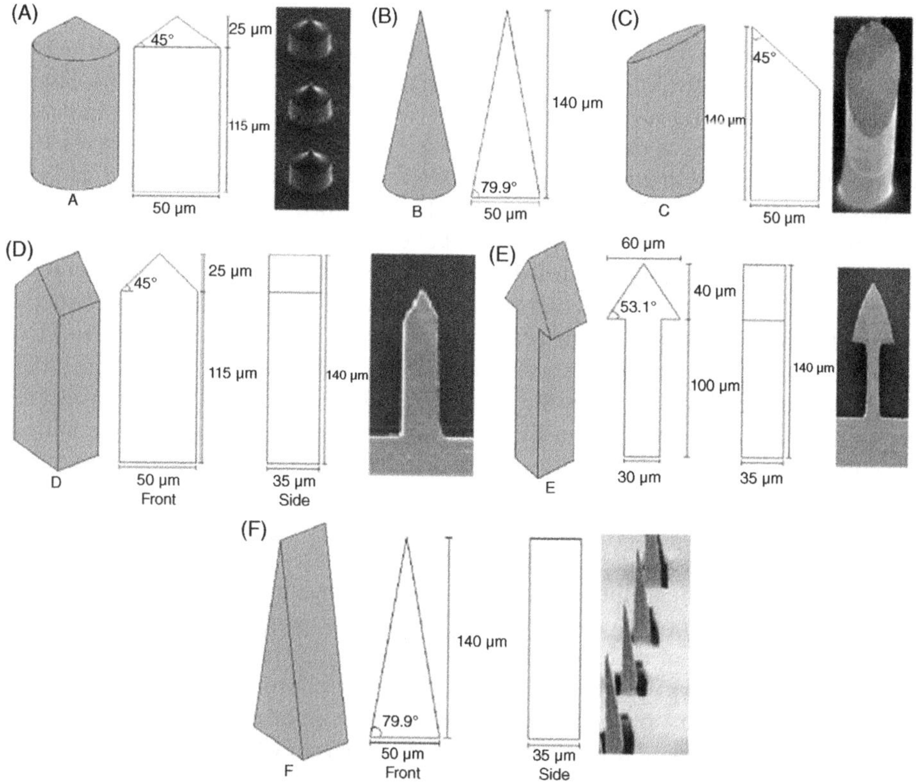

Figure 2.12 Different MN designs. Reproduced with permission from: [205] A. Davidson, B. Al-Qallaf and D.B. Das (2008) Transdermal drug delivery by coated microneedles: Geometry effects on effective skin thickness and drug permeability. *Chem. Eng. Res. Des.* 86: 1196–1206.

intradermal (ID) injection by 1.5 mm long 30G MNs on the age, sex, race, BMI and skin thickness variation within the body among 342 adult subjects (205 women, 137 men, 101 Caucasian, 118 Asian and 123 Black Africans). It was observed that the average skin thickness (epidermis–dermis) was 2.02 mm at deltoid, 2.54 mm at suprascapular, 1.91 mm at the waist and 1.55 mm at the thigh regions. Irrespective of the age, race, BMI or gender the 1.5 mm long MNs, when inserted perpendicularly, could be used efficiently for ID vaccine delivery [207]. Although the results highlight some significant differences in skin thickness, the objective of the study was to assess the appropriate MN length for ID vaccination. Laurent *et al.* (2007) demonstrated that the skin thickness varied less between people of different BMI, age, sex and ethnic origin than it did between different body sites on people with the same demographic characteristics [208]. Variation in *stratum corneum* thickness with age, sex and skin thickness variation within the body was also reported elsewhere [209]. These differences in the opinion of optimum MN height for transdermal drug delivery are particularly dependent upon the study objective, especially if the objective is to target the immune cells for vaccine in the ID region, or the systemic delivery of drug molecules. In either case, consideration should be given to the variables discussed, that is, inter- or intra-individual variability.

Although the variation in skin thickness affects the MN-based transdermal drug delivery, the other major factor to be taken into consideration is the MN geometry. The first study to demonstrate the effect of MN geometry was reported by Davis *et al.* (2004) [204], as far as the authors know. The MN insertion forces (0.1–3.0 N) varied linearly with the interfacial area of the needle tip and fracture force increased with increase in the wall thickness, wall angle and tip radius. However, the safety margins (i.e. ratio of fracture force to insertion force) between the fracture force and insertion force were high. Another study aimed to identify the most effective MN geometry for effective skin permeation [205]. This study quantified the influence of different geometrical parameters such as MN thickness, MN diameter, coating depth of drug on MN (i.e. the distance from the tip that is coated by the drug film), penetration depth, spacing between MNs and array pattern, associated with six different solid MN shapes coated with drug [205]. A 3D model, using FEMLAB scientific modelling software was used for this investigation and the MN geometry effect on skin permeation was studied. Depth of penetration and centre-to-centre spacing of MNs were found to significantly affect the effective skin permeability (P_{eff}). On the contrary, the other MN dimensions, such as the diameter of cylindrical MNs, the thickness of flat MNs and the coating depths were less significant for P_{eff}. Overall, larger, longer and more densely packed MNs resulted in greater P_{eff}.

Following MN treatment, the increase in the skin permeability and drug concentrations in the blood are not always guaranteed. Various strategies have been proposed to assess the influence of different parameters of MNs on the course of transdermal drug delivery. For example, a theoretical model was proposed to determine the influence of injection velocity, blood perfusion rate and tissue porosity on the transdermal drug delivery process using MNs [210]. Teo *et al.* (2006) showed that the key parameters of MNs, such as sharpness and type of material may affect the design of MNs [207]. A number of different MN geometries have been studied, such as straight-walled solid MNs, straight-walled hollow MNs and solid MNs with sharp tips. Despite the sharpness of the tips, it was suggested that insertion of needles by hand is difficult and not very consistent. In another study, it was proposed that the MN length should be longer than 50 μm but shorter than 200 μm [211], whereas Stoeber and Liepmann (2005) [212] anticipated that the length of the MN must be longer than 100 μm. In contrast to these two propositions, Pastorin *et al.* (2009) [213] used vertically aligned zinc oxide pyramidal nanorods (nanoneedles), where the tip size was 60 nm and the length less than 50 μm, with a base diameter of 150 nm. It was reported that these nanoneedles were sufficient for vaccine delivery. However, these differences in the MN length show the importance of, and the influence on, the transdermal drug delivery with respect to obtaining the optimum concentration of drugs in blood or to achieving optimum local delivery (e.g. vaccine). Additionally, the selection of the MNs should be such that they are not too short to be ineffective or too long to cause pain by contacting the nerve endings in the deeper layers of the dermis.

The MN length plays a crucial role in drug diffusion from the MN into the blood present in the dermal microcirculation. Drug diffusion occurs along the distance between the tip of the MN to that of the blood in dermal microcirculation. Therefore, increase in the MN length will reduce the diffusion path length and thereby increase the uptake of the drug by the dermal microcirculation to achieve higher drug concentrations. A study by Al-Qallaf's group investigated mathematical models to study the influences of a variety of variables related to MNs and their impact on the drug transport through the skin. These factors were, for example, MN length, duration of application and size of the patch, application to different anatomical regions, and were investigated with the aim of determining the influence of these factors on the blood drug concentration. In general, it was shown that an increase in the MN length or increase in patch surface area or decrease in skin thickness increased the blood concentration of the drug molecule [214, 215]. The group also proposed an optimisation model for improving drug permeability in the skin. Optimised designs for both solid and hollow MNs were proposed, in which higher skin permeability was observed when the aspect ratio of needle height over needle radius was decreased and the number of MNs was increased. Skin thickness was found to strongly affect the skin's permeability for different skin types (e.g. different race, sex, age and anatomical regions). However, this theoretical model needs to be validated experimentally. Ideally, the utility of a particular type of MN needs to be clearly demonstrated in *in vitro* experiments, followed by a sufficient number of *in vivo* experiments.

Finally, in order to achieve painless penetration into the skin, MNs should be only a few microns in height (less than 300 µm) and should penetrate the skin in a reproducible manner. However, elasticity of the skin remains a major challenge to the reproducibility of MN penetration, as discussed earlier. Therefore, it is necessary to take into consideration the flexibility of the skin in the fabrication of MN devices for an efficient and reproducible MN penetration. Furthermore, a flexible design of the MN device will allow the patients to apply it over any skin surface, specifically in the limb areas where a degree of flexibility in the design is necessary to overcome the skin contours and to avoid stress on the devices, which otherwise could damage the device during application. Hence, preliminary studies, both *in vitro* and *in vivo* are essential in order to design an optimised type of MN for its effective transdermal drug delivery. Overall, an optimised MN design should be long enough for transdermal drug delivery applications without causing pain, it should have low insertion force and high break force, its density should be optimised to deliver desired amounts of APIs, should be applied on a specific anatomical location within the body to avoid inter-individual variability, it should allow a certain degree of flexibility in the MN device to overcome the skin contours and skin deformation should be at a minimum to allow complete MN penetration.

2.4 Microneedle Mechanical Characterisation

Mechanical characterisation is a key step in the development of successful MN products. During insertion and removal, MNs are normally exposed to a wide range of stresses. The possible failure scenarios include MN buckling/bending and baseplate fracturing [23]. Accordingly, these devices should have a defined standard inherent strength to avoid failures [16]. Moreover, mechanical characterisation should be performed to evaluate whether the designed MNs are safe. One of the first MN mechanical testing methods described in the literature was developed by Zahn *et al.* [23], which entails an increasing range of forces being exerted at the MN tip until fracture. Nowadays, there are a wide range of MN testing methods reported in the literature.

2.4.1 Axial Force Microneedle Mechanical Tests

The axial force mechanical testing for MNs consists of the application of forces to the needle tips perpendicular to the baseplate [111]. For this purpose, typically the MNs are compressed against a metal block at a defined speed using a mechanical test station that records the force and the displacement throughout the process [116, 163] (Figure 2.13A). The force–displacement plots obtained show a sudden decrease when MN fracture occurs. The MN failure force can be established as the maximum force exerted directly before the fracture [163]. In addition to the purely mechanical measurements, images of the MNs can be taken before and after failure to try to determine the failure mechanism [216]. These experiments can be performed for single MNs or for MN arrays. When testing a single MN, the results should be evaluated with caution

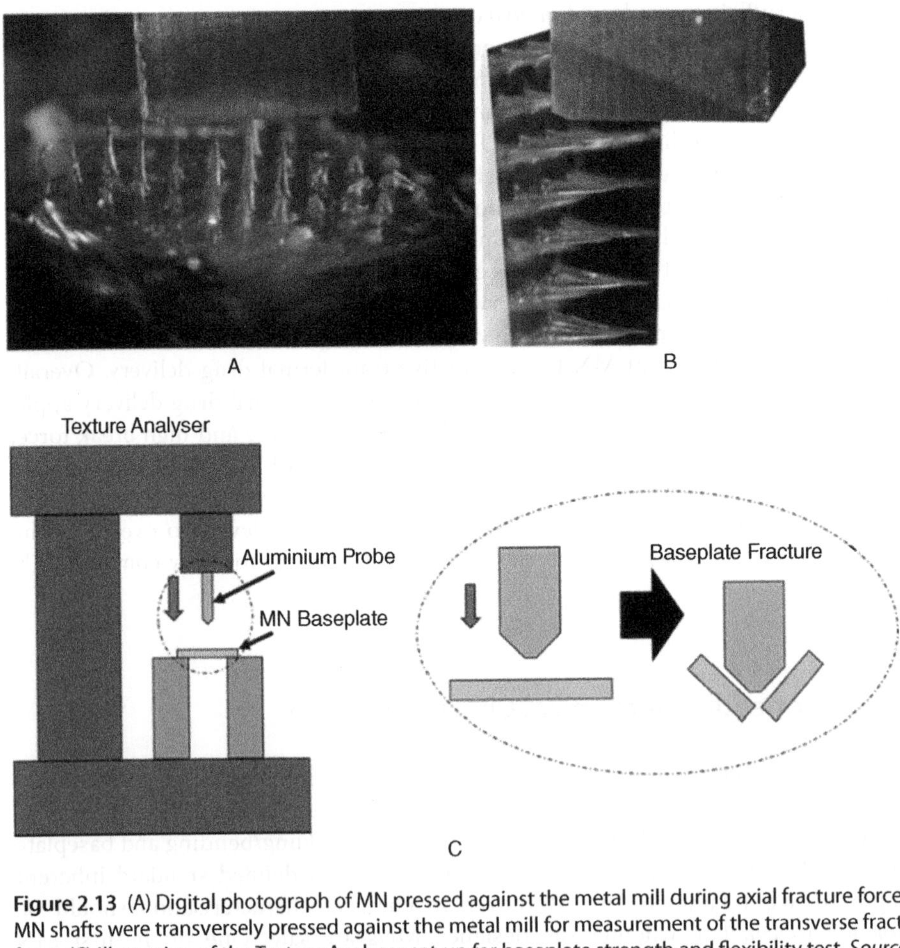

Figure 2.13 (A) Digital photograph of MN pressed against the metal mill during axial fracture force. (B) MN shafts were transversely pressed against the metal mill for measurement of the transverse fracture force. (C) Illustration of the Texture Analyser set-up for baseplate strength and flexibility test. *Source*: adapted from [9] E. Larrañeta, R.E.M. Lutton, A.D. Woolfson and R.F. Donnelly (2016). Microneedle arrays as transdermal and intradermal drug delivery systems: Materials science, manufacture and commercial development. *Mater. Sci. Eng. R-Rep.* 104: 1–32 and [116] Y.K. Demir, Z. Akan and O. Kerimoglu (2013). Characterization of polymeric microneedle arrays for transdermal drug delivery. *PLoS ONE*. 8: e77289.

[23, 217]. During the testing of MN arrays there is a density effect on the failure force that cannot be found for single MNs. Finally, it is important to note that this test is not an accurate simulation of the forces experienced by MNs during skin insertion, as during the test they are pressed against a hard metallic surface. In this situation, the forces are concentrated at the MN tip contacting the surface, while in the skin the flexible skin wraps around the needle tips generating a force distribution over a greater MN area [216].

2.4.2 Transverse Force and Shear Strength Microneedle Mechanical Tests

During insertion and due to irregularities on the skin surface MN arrays may be partially inserted. In some cases, this can produce transverse bending of the needles. Consequently, a transverse fracture force test should be developed to determine the behaviour of MNs during insertion [116, 163]. As described for the axial mechanical testing, a mechanical test station is required. In this case, a metallic prove applies a transverse force (normal to the microneedle *y*-axis) force at a defined point of the MN until the needle fractures (Figure 2.13B). The fracture force can be determined, as described previously, from the force–displacement curve [111, 116]. If the test is applied to a row of MNs, to obtain the transverse fracture force, the observed fracture force should be divided by the number of MNs in the array [111]. This test presents an important limitation, namely, that the alignment of the metal probe at a defined needle height is difficult to reproduce as in practice this is done manually [111]. Therefore, the variability of the results can be high, even in the case of alignments performed with the assistance of a microscope camera [116].

2.4.3 Baseplate Strength and Flexibility Tests

The tests described previously focus on mechanical testing of the needles. However, the majority of MN devices are MN arrays. Accordingly, the MN baseplate strength is a key parameter that needs to be evaluated. Fracture of the baseplate during MN application is not acceptable. To prevent this, the baseplate should be flexible enough to adapt to the topography of the skin without fracturing [111]. The easiest way to test the mechanical strength of MN baseplates is a three-point bending test [111]. A Texture Analyser can be used to apply a defined force to a baseplate placed between two aluminium blocks (Figure 2.13C) [111]. In this case the force–displacement graph will show a maximum that represents the force required to break the material [9]. In addition to the fracture force, the flexibility of the baseplate can be evaluated by measuring the bending of the baseplate upon fracture (Figure 2.13C).

2.4.4 Microneedle Insertion Measurements

2.4.4.1 Staining of Microneedle-treated Skin

A wide range of techniques to ascertain MN insertion can be found in the literature. However, they require the use of biological tissue and/or complex procedures [9]. The easiest way to ascertain MN insertion is to stain MN treated skin with a coloured dye such as trypan blue or methylene blue [87, 197, 218]. These dyes are able to stain the cells of the viable epidermis and not the *stratum corneum*. In this way they will only stain the micropores created after MN application. This test has been used mainly to demonstrate MN piercing as a quick qualitative test and is performed prior to measurement with other techniques, which will be described in the following sections [7]. The method can be used to evaluate the effect of higher insertion forces in MN insertion. In this case, higher insertion will cause greater skin disruption and consequently greater permeation of the dye [218]. However, this method has some limitations. The lateral diffusion of the dyes can lead to overestimation of the diameter of micropores [7]. Alternatively, in some cases the *stratum corneum* is not fully pierced, and dye pools collecting in the indentations created after MN insertion can yield false positives [7]. When using hollow or dissolving MN arrays, the dyes can be injected or administered inside the skin [7, 94]. After the delivery of the dye, microscopy images can be obtained to evaluate the successful insertion of the MNs [94].

2.4.4.2 Transepidermal Water Loss Measurements

A good alternative to staining is the measurement of the transepidermal water loss (TEWL) [197, 219]. This technique is based on the measurement of water loss from the skin's surface. When the skin barrier is disrupted the water loss increases. This technique can be used easily *in vitro* and *in vivo* [7]. After MN applications, the values of TEWL increase rapidly [7]. The difference in TEWL values from before and after MN applications have been reported to be a 10–25-fold increase [220]. Additionally, this technique can be used to evaluate the skin recovery time after MN insertion [7]. Despite being a quick and easy parameter to measure, the evaluation of TEWL presents some limitations. It is difficult to establish a correlation between TEWL and transdermal delivery of the drugs [7]. Furthermore, this technique does not provide information about MN insertion depths.

2.4.4.3 Electrical Impedance Measurements

Measurement of the electrical impedance of the skin after MN insertion has been used by several authors to ascertain skin piercing [217, 221]. In the same way as TEWL measurement, this technique is useful to evaluate if the barrier function of the *stratum corneum* has been compromised. The impedance of the skin to the flow of alternating current changes when the *stratum corneum* is breached [7]. Researchers have demonstrated a strong inverse correlation between skin electrical impedance and skin permeability [7]. The main limitation of this technique is that it does not provide information about MN insertion depths.

2.4.4.4 Histological Tissue Staining and Sectioning

Histological cryosectioning involves the treatment and analysis of biological tissue following MN insertion [7]. After the insertion, the tissue should be fixed in a suitable medium, frozen using liquid nitrogen and subsequently stored at −80 °C. The next step requires a cryostat to cryosection the skin sample. Finally, the skin sections should be stained with haematoxylin and eosin to observe the channels created by the MN [221, 222]. This technique provides information about the MN insertion depth and the dimensions of the microchannels produced [7]. However, it is not ideal as it is a destructive technique [7]. In addition, during the processing protocols the dimensions of the microchannels created by the MN array may be distorted [7, 223].

2.4.4.5 Confocal Microscopy

Confocal microscopy can be used to measure the dimensions of the MN-created pores [7, 216]. In order to obtain information using this technique, a solution containing specific fluorescent microparticles should be applied to the skin pre-treated with MNs [7]. The fluorescent compound diffuses through the microchannels created and microscopy can be used to visualise this compound inside the pores [7]. Unlike the previously described technique, this is a non-invasive and non-destructive technique [7]. The major limitation to this method is the fact that confocal microscopy is limited in application to short MN arrays, as the technique can only provide information to depths of around 200–250 μm from the skin surface [7].

2.4.4.6 Optical Coherence Tomography

The ideal technique to evaluate MN insertion in real time is optical coherence tomography (OCT) [7]. This technique is non-invasive and can be used *in vivo* [224] and *ex vivo* [121]. OCT records changes to reflected light from a tissue as a function of the tissue depth [121]. Therefore, OCT provides transverse imaging of the *stratum corneum*, epidermis and upper dermis in real time [121, 223]. This technique can be used to visualise depths of up to 2000 μm from the skin surface [7], and measurements can be made with the MNs inserted into the skin [121, 223]. However, the main limitation of this technique is that the MNs should be transparent for *in situ* imaging [7].

2.4.5 Significance of Microneedle Test Results

The results obtained from the tests just described should be evaluated carefully. These results can only be properly evaluated when compared with the corresponding insertion forces [163]. After measuring the fracture force and the insertion force, the "margin of safety" can be established. This parameter is a ratio of MN fracture force to insertion force [163, 217, 225]. Consequently, human or animal skin tissue or a skin simulant is required to measure the insertion forces. The majority of MN research only evaluates whether the MN are inserted and/or the insertion depths, but not the actual insertion forces.

However, there are several reports of research where the insertion forces for MNs have been evaluated. Davis *et al.* measured the insertion forces, in human volunteers, for MNs made of metal by means of a device that evaluates force–displacement during insertion [217]. The needles had radii varying between 30 and 80 μm and the obtained insertion forces ranged between 0.08 and 3.04 N per needle. These results suggested that this type of MN can be inserted manually. In a similar way, Khanna *et al.* measured the insertion forces of hollow 4×4 silicon MN arrays using excised cadaver skin [226]. The obtained insertion forces were dependent on the tip sharpness and they ranged between 4.75 and 0.1 N.

As an alternative methodology to evaluate the insertion forces, Loeters *et al.* described a procedure that uses devices containing MNs integrated with electrodes [222]. In this work 9×9 MN arrays were tested showing that 2.6 N were required to insert the arrays. Likewise, Roxhed *et al.* used electrical impedance measurement in the skin to evaluate the insertion of ultrasharp MNs in human skin [221]. The results showed that in this case the required forces were significantly smaller than those measured in the previous cases (less than 10 mN).

The insertion and fracture forces are defined by several parameters, such as the material used to manufacture the MN array and the design of the MN [40]. For all the MN designs, the fracture forces should be higher than the force required for insertion into the skin. These parameters can be tailored during the design and manufacturing stages, optimising their "margin of safety" values [163, 225].

Another crucial parameter that should be taken into account to define the margins of safety is the nature of the insertion process [40]. MNs can be applied manually or using an applicator [1, 9]. When using manual MN application, the insertion presents more variability than when using an applicator. If the MNs are going to be designed to be inserted manually, the range of forces that different patients can apply should be taken into account to define the margins of safety. For this purpose, Larrañeta *et al.* developed a study concluding that the range of manual forces applied by 20 volunteers, following instructions for MN insertion, presented an average value of 20 N and ranged between 10 and 50 N [123].

After evaluating all the testing methods along with their limitations, it is important to note that due to the wide range of MN designs, materials, diversity of methods used to characterise them and the use of biological tissues, which present variability, direct comparisons cannot be made. In order to overcome these limitations standardised mechanical tests should be developed. To this aim, several authors have developed artificial membranes for MN testing [123, 227].

The interpretation and relevance of MN insertion tests may be limited by a number of different of factors, including: MN geometry and materials; use of varying experimental procedures and different skin models. All these limitations could lead to divergence of results after using the different testing methods. Consequently, there is a need for the development of standardised MN testing protocols.

References

1 R.F. Donnelly, T.R.R. Singh, D.I.J. Morrow and A.D. Woolfson (2012). *Microneedle-Mediated Transdermal and Intradermal Drug Delivery*. Wiley.

2 M.R. Prausnitz (2004). Microneedles for transdermal drug delivery. *Adv. Drug Deliv. Rev.* 56: 581–587.

3 E. Larrañeta, M.T.C. McCrudden, A.J. Courtenay and R.F. Donnelly (2016). Microneedles: a new frontier in nanomedicine delivery. *Pharm. Res.* 1–19.

4 R.F. Donnelly, T.R.R. Singh, M.J. Garland, *et al.* (2012). Hydrogel-forming microneedle arrays for enhanced transdermal drug delivery. *Adv. Funct. Mater.* 22: 4879–4890.

5 S. Henry, D.V. McAllister, M.G. Allen and M.R. Prausnitz (1998). Microfabricated microneedles: a novel approach to transdermal drug delivery. *J. Pharm. Sci.* 87: 922–925.

6 S.J. Moon, S.S. Lee, H.S. Lee and T.H. Kwon (2005). Fabrication of microneedle array using LIGA and hot embossing process. *Microsyst. Technol.* 11: 311–318.

7 R.E.M. Lutton, J. Moore, E. Larrañeta, *et al.* (2015). Microneedle characterisation: the need for universal acceptance criteria and GMP specifications when moving towards commercialisation. *Drug Deliv. Transl. Res.* 1–19.

8 E. Larrañeta, S. Stewart, S.J. Fallows, *et al.* (2016). A facile system to evaluate in vitrodrug release from dissolving microneedle arrays. *Int. J. Pharm.* 497: 62–69.
9 E. Larrañeta, R.E.M. Lutton, A.D. Woolfson and R.F. Donnelly (2016). Microneedle arrays as transdermal and intradermal drug delivery systems: Materials science, manufacture and commercial development. *Mater. Sci. Eng. R-Rep.* 104: 1–32.
10 S. Badilescu and M. Packirisamy (2012). BioMEMS: Science and Engineering Perspectives. Boca Raton: CRC Press.
11 M. Gad-el-Hak (2010). *The MEMS Handbook*. Boca Raton: CRC Press.
12 O. Paul, J. Gaspar and P. Ruther (2007). Advanced silicon microstructures, sensors, and systems. *IEEJ Trans. Electr. Electron. Eng.* 2: 199–215.
13 H.J.G.E. Gardeniers, R. Luttge, E.J.W. Berenschot, *et al.* (2003). Silicon micromachined hollow microneedles for transdermal liquid transport. *J. Microelectromech. Syst.* 12: 855–862.
14 S. Indermun, R. Luttge, Y.E. Choonara, *et al.* (2014). Current advances in the fabrication of microneedles for transdermal delivery. *J. Control. Release* 185: 130–138.
15 M.A. Hopcroft, W.D. Nix and T.W. Kenny (2010). What is the Young's Modulus of silicon? *J. Microelectromech. Syst.* 19: 229–238.
16 P. Khanna, K. Luongo, J.A. Strom and S. Bhansali (2010). Sharpening of hollow silicon microneedles to reduce skin penetration force. *J. Micromech. Microeng.* 20: 045011.
17 P. Khanna, K. Luongo, J. Strom and S. Bhansali (2010). Axial and shear fracture strength evaluation of silicon microneedles. *Microsyst. Technol.* 16: 973–978.
18 R.F. Donnelly, D.I. Morrow, P.A. McCarron, *et al.* (2009). Microneedle arrays permit enhanced intradermal delivery of a preformed photosensitizer. *Photochem. Photobiol.* 85: 195–204.
19 S.J. Moon and S.S. Lee (2005). A novel fabrication method of a microneedle array using inclined deep x-ray exposure. *J. Micromech. Microeng.* 15: 903–911.
20 N. Wilke, A. Mulcahy, S.-R. Ye and A. Morrissey (2005). Process optimization and characterization of silicon microneedles fabricated by wet etch technology. *Microelectron. J.* 36: 650–656.
21 Y. Xie, B. Xu and Y. Gao (2005). Controlled transdermal delivery of model drug compounds by MEMS microneedle array. *Nanomedicine* 1: 184–190.
22 W.Z. Li, M.R. Huo, J.P. Zhou, *et al.* (2010). Super-short solid silicon microneedles for transdermal drug delivery applications. *Int. J. Pharm.* 389: 122–129.
23 J.D. Zahn, N.H. Talbot, D. Liepmann and A.P. Pisano (2000). Microfabricated polysilicon microneedles for minimally invasive biomedical devices. *Biomed. Microdevices* 2: 295–303.
24 N. Roxhed, P. Griss and G. Stemme (2008). Membrane-sealed hollow microneedles and related administration schemes for transdermal drug delivery. *Biomed. Microdevices* 10: 271–279.
25 M.G. McGrath, A. Vrdoljak, C. O'Mahony, *et al.* (2011). Determination of parameters for successful spray coating of silicon microneedle arrays. *Int. J. Pharm.* 415: 140–149.
26 A.K. Banga, Microporation applications for enhancing drug delivery (2009). *Expert Opin. Drug Deliv.* 6: 343–354.
27 J.H. Braybrook (1997). *Biocompatiblity: Assessment of Medical Devices and Materials*. New York: Wiley.

28 W.R. Runyan and K.E. Bean (1990). *Semiconductor Integrated Circuit Processing Technology*. New York: Addison-Wesley.

29 C. O'Mahony (2014). Structural characterization and in-vivo reliability evaluation of silicon microneedles. *Biomed. Microdevices* 16: 333–343.

30 R. Baran and H. Maibach (2010). *Textbook of Cosmetic Dermatology*, 2nd edn. Boca Raton: CRC Press.

31 D.R.J. Millard and D.O. Maisels (1966). Silicon granuloma of the skin and subcutaneous tissues. *Am. J. Surg.* 112: 119–123.

32 G. Kotzar, M. Freas, P. Abel, *et al.* (2002). Evaluation of MEMS materials of construction for implantable medical devices. *Biomaterials* 23: 2737–2750.

33 S.S. Stensaas and L.J. Stensaas (1978). Histopathological evaluation of materials implanted in the cerebral cortex. *Acta Neuropathol.* 41: 145–155.

34 S. Schmidt, K. Horch, R. Normann (1993). Biocompatibility of silicon-based electrode arrays implanted in feline cortical tissue. *J. Biomed. Mater. Res.* 27: 1393–1399.

35 G. Voskerician, M.S. Shive, R.S. Shawgo, *et al.* (2003). Biocompatibility and biofouling of MEMS drug delivery devices. *Biomaterials* 24: 1959–1967.

36 S.C. Bayliss, P.J. Harris, L.D. Buckberry and C. Rousseau (1997). Phosphate and cell growth on nanostructured semiconductors, *J. Mater. Sci. Lett.* 16: 737–740.

37 K. Kubo, N. Tsukasa, M. Uehara, *et al.* (1997). Calcium and silicon from bioactive glass concerned with formation of nodules in periodontal-ligament fibroblasts in vitro. *J. Oral Rehabil.* 24: 70–75.

38 T.M. Tuan-Mahmood, M.T. McCrudden, B.M. Torrisi, *et al.* (2013). Microneedles for intradermal and transdermal drug delivery. *Eur. J. Pharm. Sci.* 50: 623–637.

39 R.F. Donnelly, T.R. Raj Singh and A.D. Woolfson (2010). Microneedle-based drug delivery systems: microfabrication, drug delivery, and safety. *Drug Deliv.* 17: 187–207.

40 M. Niinomi and M. Nakai (2011). Titanium-based biomaterials for preventing stress shielding between implant devices and bone. *Int. J. Biomater.* 2011: 836587.

41 N.A. Monteiro-Riviere (2010). *Toxicology of the Skin*. Boca Raton: CRC Press.

42 F.J. Verbaan, S.M. Bal, D.J. van den Berg, *et al.* (2007). Assembled microneedle arrays enhance the transport of compounds varying over a large range of molecular weight across human dermatomed skin. *J. Control. Release* 117: 238–245.

43 D.P. Wermeling, S.L. Banks, D.A. Hudson, *et al.* (2008). Microneedles permit transdermal delivery of a skin-impermeant medication to humans. *Proc. Natl. Acad. Sci. U. S. A.* 105: 2058–2063.

44 S. Kaushik, A.H. Hord, D.D. Denson, *et al.* (2001). Lack of pain associated with microfabricated microneedles. *Anesth. Analg.* 92: 502–504.

45 F.J. Verbaan, S.M. Bal, D.J. van den Berg, *et al.* (2008). Improved piercing of microneedle arrays in dermatomed human skin by an impact insertion method. *J. Control. Release* 128: 80–88.

46 Z. Ding, F.J. Verbaan, M. Bivas-Benita, *et al.* (2009). Microneedle arrays for the transcutaneous immunization of diphtheria and influenza in BALB/c mice. *J. Control. Release.* 136: 71–78.

47 H.S. Gill and M.R. Prausnitz (2007). Coating formulations for microneedles. *Pharm. Res.* 24: 1369–1380.

48 E.M. Saurer, R.M. Flessner, S.P. Sullivan, *et al.* (2010). Layer-by-layer assembly of DNA- and protein-containing films on microneedles for drug delivery to the skin. *Biomacromolecules* 11: 3136–3143.

49 M. Cormier, B. Johnson, M. Ameri, *et al.* (2004). Transdermal delivery of desmopressin using a coated microneedle array patch system. *J. Control. Release* 97: 503–511.

50 E.R. Parker, M.P. Rao, K.L. Turner, *et al.* (2007). Bulk micromachined titanium microneedles. *J. Microelectromech. Syst.* 16: 289–295.

51 Y. Tanahashi, E. Makino, S. Toh and T. Kawashima (2008). Thermal-type blood flow sensor on titanium microneedle. *Sensor Mater.* 20: 341–349.

52 S. Chandrasekaran, J.D. Brazzle and A.B. Frazier (2003). Surface micromachined metallic microneedles. *J. Microelectromech. Syst.* 12: 281–288.

53 S. Chandrasekaran and A.B. Frazier (2003). Characterization of surface micromachined metallic microneedles. *J. Microelectromech. Syst.* 12: 289–295.

54 K. Kim, D.S. Park, H.M. Lu, *et al.* (2004). A tapered hollow metallic microneedle array using backside exposure of SU-8. *J. Micromech. Microeng.* 14: 597–603.

55 H. Suzuki, T. Tokuda and K. Kobayashi (2002). A disposable "intelligent mosquito" with a reversible sampling mechanism using the volume-phase transition of a gel. *Sens. Actuator B-Chem.* 83: 53–59.

56 K. Kobayashi, H. Suzuki, K. Kobayashi and H. Suzuki (2001). A sampling mechanism employing the phase transition of a gel and its application to a micro analysis system imitating a mosquito. *Sens. Actuator B-Chem.* 80: 1–8.

57 M. Verhoeven, S. Bystrova, L. Winnubst, *et al.* (2012). Applying ceramic nanoporous microneedle arrays as a transport interface in egg plants and an ex-vivo human skin model. *Microelectron. Eng.* 98: 659–662.

58 B. Cai, W. Xia, S. Bredenberg and H. Engqvist (2014). Self-setting bioceramic microscopic protrusions for transdermal drug delivery. *J. Mater. Chem. B.* 2: 5992–5998.

59 Q. Chen, G.A. Thouas, Q. Chen and G.A. Thouas (2015). Metallic implant biomaterials. *Mater. Sci. Eng. R Rep.* 87: 1–57.

60 S.C. Gad and M.G. McCord (2008). *Safety Evaluation in the Development of Medical Devices and Combination Products*, 3rd edn. Boca Raton: CRC Press.

61 M. Niinomi (2008). Biologically and mechanically biocompatible titanium alloys. *Mater. Trans.* 49: 2170–2178.

62 J. Black (1999). *Biological Performance of Materials: Fundamentals of Biocompatibility*, 3rd edn. Boca Raton: CRC Press.

63 A. Cowley and B. Woodward (2011). A healthy future: platinum in medical applications. *Platin. Met. Rev.* 55: 98–107.

64 M. Assad, N. Lemieux, C.H. Rivard and L.H. Yahia (1999). Comparative in vitro biocompatibility of nickel-titanium, pure nickel, pure titanium, and stainless steel: genotoxicity and atomic absorption evaluation. *Bio-Med. Mater. Eng.* 9: 1–12.

65 S. Bystrova, R. Luttge, S. Bystrova, R. Luttge (2011). Micromolding for ceramic microneedle arrays. *Microelectron. Eng.* 88: 1681–1684.

66 R. Pignatello (2011). *Biomaterials Applications for Nanomedicine.* Croatia: InTech Open Science.

67 M.P. Ginebra, T. Traykova and J.A. Planell (2006). Calcium phosphate cements as bone drug delivery systems: a review. *J. Control. Release* 113: 102–110.

68 S. Hesaraki, F. Moztarzadeh, R. Nemati and N. Nezafati (2009). Preparation and characterization of calcium sulfate-biomimetic apatite nanocomposites for controlled release of antibiotics. *J. Biomed. Mater. Res. Part B Appl. Biomater.* 91: 651–661.

69 S.D. Gittard, R.J. Narayan, C. Jin, *et al.* (2009). Pulsed laser deposition of antimicrobial silver coating on Ormocer microneedles. *Biofabrication.* 1: 041001.

70 A. Doraiswamy, A. Ovsianikov, S.D. Gittard, *et al.* (2010). Fabrication of microneedles using two photon polymerization for transdermal delivery of nanomaterials. *J. Nanosci. Nanotechnol.* 10: 6305–6312.

71 K.H. Haas and H. Wolter (1999). Synthesis, properties and applications of inorganic–organic copolymers (ORMOCER®s). *Curr. Opin. Solid State Mater. Sci.* 4: 571–580.

72 M. Navarro, A. Michiardi, O. Castaño and J.A. Planell (2008). Biomaterials in orthopaedics. *J. R. Soc. Interface* 5: 1137–1158.

73 M. Andreiotelli, H.J. Wenz and R.J. Kohal (2009). Are ceramic implants a viable alternative to titanium implants? A systematic literature review. *Clin. Oral Implant. Res.* 20: 32–47.

74 P.S. Christel (1992). Biocompatibility of surgical-grade dense polycrystalline alumina. *Clin. Orthop. Relat. Res.* 10–18.

75 M. Lewandowska-Szumieł and J. Komender (2000). Interaction between tissues and implantable materials. *Front. Med. Biol. Eng.* 10: 79–82.

76 M. Lewandowska-Szumiel and J. Komender (1990). Aluminium release as a new factor in the estimation of alumina bioceramic implants. *Clin. Mater.* 5: 167–175.

77 F. Theiss, D. Apelt, B. Brand, *et al.* (2005). Biocompatibility and resorption of a brushite calcium phosphate cement. *Biomaterials* 26: 4383–4394.

78 R.Z. LeGeros (2002). Properties of osteoconductive biomaterials: calcium phosphates. *Clin. Orthop. Relat. Res.* 81–98.

79 M. Nilsson, E. Fernández, S. Sarda, *et al.* (2002). Characterization of a novel calcium phosphate/sulphate bone cement. *J. Biomed. Mater. Res.* 61: 600–607.

80 M.V. Thomas, D.A. Puleo and M. Al-Sabbagh (2005). Calcium sulfate: a review. *J. Long Term Eff. Med. Implants* 15: 599–607.

81 A. Ovsianikov, B. Chichkov, P. Mente, *et al.* (2007). Two photon polymerization of polymer-ceramic hybrid materials for transdermal drug delivery. *Int. J. Appl. Ceram. Tech.* 4: 22–29.

82 A. Doraiswamy, C. Jin, R.J. Narayan, *et al.* (2006). Two photon induced polymerization of organic-inorganic hybrid biomaterials for microstructured medical devices. *Acta Biomater.* 2: 267–275.

83 S. Schlie, A. Ngezahayo, A. Ovsianikov, *et al.* (2007). Three-dimensional cell growth on structures fabricated from ORMOCER by two-photon polymerization technique. *J. Biomater. Appl.* 22: 275–287.

84 P.M. Wang, M. Cornwell, J. Hill and M.R. Prausnitz (2006). Precise microinjection into skin using hollow microneedles. *J. Invest. Dermatol.* 126: 1080–1087.

85 W. Martanto, J.S. Moore, O. Kashlan, *et al.* (2006). Microinfusion using hollow microneedles. *Pharm. Res.* 23: 104–113.

86 J. Gupta, E.I. Felner and M.R. Prausnitz (2009). Minimally invasive insulin delivery in subjects with type 1 diabetes using hollow microneedles. *Diabetes Technol. Ther.* 11: 329–337.
87 D.V. McAllister, P.M. Wang, S.P. Davis, *et al.* (2003). Microfabricated needles for transdermal delivery of macromolecules and nanoparticles: fabrication methods and transport studies. *Proc. Natl. Acad. Sci. U. S. A.* 100: 13755–13760.
88 C.S. Kolli and A.K. Banga (2008). Characterization of solid maltose microneedles and their use for transdermal delivery. *Pharm. Res.* 25: 104–113.
89 R.F. Donnelly, D.I. Morrow, T.R. Singh, *et al.* (2009). Processing difficulties and instability of carbohydrate microneedle arrays. *Drug Dev. Ind. Pharm.* 35: 1242–1254.
90 J. Finley and J. Knabb (1982). Cutaneous silica granuloma. *Plast. Reconstr. Surg.* 69: 340–343.
91 K.S. Parthasarathy, Y.C. Cheng, J.P. McAllister, *et al.* (2007). Biocompatibilities of sapphire and borosilicate glass as cortical neuroprostheses. *Magn. Reson. Imaging* 25: 1333–1340.
92 G. Li, A. Badkar, S. Nema, *et al.* (2009). In vitro transdermal delivery of therapeutic antibodies using maltose microneedles. *Int. J. Pharm.* 368: 109–115.
93 K. Lee, C.Y. Lee and H. Jung (2011). Dissolving microneedles for transdermal drug administration prepared by stepwise controlled drawing of maltose. *Biomaterials* 32: 3134–3140.
94 J.W. Lee, J.H. Park and M.R. Prausnitz (2008). Dissolving microneedles for transdermal drug delivery. *Biomaterials* 29: 2113–2124.
95 T. Miyano, Y. Tobinaga, T. Kanno, *et al.* (2005). Sugar micro needles as transdermic drug delivery system. *Biomed. Microdevices* 7: 185–188.
96 C.J. Martin, C.J. Allender, K.R. Brain, *et al.* (2012). Low temperature fabrication of biodegradable sugar glass microneedles for transdermal drug delivery applications. *J. Control. Release* 158: 93–101.
97 A.R. Gaines, L.R. Pierce and P.A. Bernhardt (2008) Fatal iatrogenic hypoglycemia: falsely elevated blood glucose readings with a point-of-care meter due to a maltose-containing intravenous immune globulin product. http://www.fda.gov/cber/safety/glucfalse.htm.
98 M.J. Akers (2002). Excipient-drug interactions in parenteral formulations. *J. Pharm. Sci.* 91: 2283–2300.
99 A.S. Schneider, A. Schettler, A. Markowski, *et al.* (2014). Assessment of xylitol serum levels during the course of parenteral nutrition including xylitol in intensive care patients: a case control study. *Clin. Nutr.* 33: 483–488.
100 X. Hong, L. Wei, F. Wu, *et al.* (2013). Dissolving and biodegradable microneedle technologies for transdermal sustained delivery of drug and vaccine. *Drug Des. Devel. Ther.* 7: 945–952.
101 S. Aoyagi, H. Izumi, Y. Isono, *et al.* (2007). Laser fabrication of high aspect ratio thin holes on biodegradable polymer and its application to a microneedle. *Sens. Actuator A-Phys.* 139: 293–302.
102 S.O. Choi, Y.C. Kim, J.H. Park, *et al.* (2010). An electrically active microneedle array for electroporation. *Biomed. Microdevices* 12: 263–273.

103 F. Pérennès, B. Marmiroli, M. Matteucci, *et al.* (2006). Sharp beveled tip hollow microneedle arrays fabricated by LIGA and 3D soft lithography with polyvinyl alcohol. *J. Micromech. Microeng.* 16: 473–479.
104 J.H. Park, M.G. Allen and M.R. Prausnitz (2006). Polymer microneedles for controlled-release drug delivery. *Pharm. Res.* 23: 1008–1019.
105 M. Han, D.H. Hyun, H.H. Park, *et al.* (2007). A novel fabrication process for out-of-plane microneedle sheets of biocompatible polymer. *J. Micromech. Microeng.* 17: 1184–1191.
106 J.M. Lippmann, E.J. Geiger and A.P. Pisano (2007). Polymer investment molding: Method for fabricating hollow, microscale parts. *Sens. Actuator A-Phys.* 134: 2–10.
107 F. Sammoura, J.J. Kang, Y.M. Heo, *et al.* (2007). Polymeric microneedle fabrication using a microinjection molding technique. *Microsyst. Technol.* 13: 517–522.
108 S. Sullivan, N. Murthy, M. Prausnitz, *et al.* (2008). Minimally invasive protein delivery with rapidly dissolving polymer microneedles. *Adv. Mater.* 20: 933.
109 L.Y. Chu, S.O. Choi and M.R. Prausnitz (2010). Fabrication of dissolving polymer microneedles for controlled drug encapsulation and delivery: Bubble and pedestal microneedle designs. *J. Pharm. Sci.* 99: 4228–4238.
110 Y.C. Kim, F.S. Quan, R.W. Compans, *et al.* (2010). Formulation of microneedles coated with influenza virus-like particle vaccine. *AAPS PharmSciTech.* 11: 1193–1201.
111 R.F. Donnelly, R. Majithiya, T.R. Singh, *et al.* (2011). Design, optimization and characterisation of polymeric microneedle arrays prepared by a novel laser-based micromoulding technique. *Pharm. Res.* 28: 41–57.
112 R.F. Donnelly, M.T.C. McCrudden, A. Zaid Alkilani, *et al.* (2014). Hydrogel-forming microneedles prepared from "super swelling" polymers combined with lyophilised wafers for transdermal drug delivery, *PLoS ONE* 9: e111547.
113 H. Huang, C. Fu, H. Huang and C. Fu (2007). Different fabrication methods of out-of-plane polymer hollow needle arrays and their variations. *J. Micromech. Microeng.* 17: 393–402.
114 Y. Ito, E. Hagiwara, A. Saeki, *et al.* (2006). Feasibility of microneedles for percutaneous absorption of insulin. *Eur. J. Pharm. Sci.* 29: 82–88.
115 S.G. Lee, J.H. Jeong, K.M. Lee, *et al.* (2014). Nanostructured lipid carrier-loaded hyaluronic acid microneedles for controlled dermal delivery of a lipophilic molecule. *Int. J. Nanomedicine* 9: 289–299.
116 Y.K. Demir, Z. Akan and O. Kerimoglu (2013). Characterization of polymeric microneedle arrays for transdermal drug delivery. *PLoS ONE.* 8: e77289.
117 J.H. Oh, H.H. Park, K.Y. Do, *et al.* (2008). Influence of the delivery systems using a microneedle array on the permeation of a hydrophilic molecule, calcein. *Eur. J. Pharm. Biopharm.* 69: 1040–1045.
118 P.C. DeMuth, Y. Min, B. Huang, *et al.* (2013). Polymer multilayer tattooing for enhanced DNA vaccination. *Nat. Mater.* 12: 367–376.
119 J. Kennedy, E. Larrañeta, M.T.C. McCrudden, *et al.* (2017). In vivo studies investigating biodistribution of nanoparticle-encapsulated rhodamine B delivered via dissolving microneedles. *J. Control. Release* 10 (265): 57-65. doi: 10.1016/j.jconrel.2017.04.022. Epub 2017 Apr 17.
120 M.T. McCrudden, A.Z. Alkilani, C.M. McCrudden, *et al.* (2014). Design and physicochemical characterisation of novel dissolving polymeric microneedle arrays for

transdermal delivery of high dose, low molecular weight drugs. *J. Control. Release* 180: 71–80.

121 R.F. Donnelly, M.J. Garland, D.I. Morrow, *et al.* (2010). Optical coherence tomography is a valuable tool in the study of the effects of microneedle geometry on skin penetration characteristics and in-skin dissolution. *J. Control. Release* 147: 333–341.

122 X. Hong, Z. Wu, L. Chen, *et al.* (2014). Hydrogel microneedle arrays for transdermal drug delivery. *Nano-Micro Lett.* 6: 191–199.

123 E. Larrañeta, J. Moore, E.M. Vicente-Pérez, *et al.* (2014). A proposed model membrane and test method for microneedle insertion studies. *Int. J. Pharm.* 472: 65–73.

124 E. Larrañeta, R.E.M. Lutton, A.J. Brady, *et al.* (2015). Microwave-assisted preparation of hydrogel-forming microneedle arrays for transdermal drug delivery applications. *Macromol. Mater. Eng.* 300: 586–595.

125 T.R.R. Singh, M.J. Garland, K. Migalska, *et al.* (2012). Influence of a pore-forming agent on swelling, network parameters, and permeability of poly(ethylene glycol)-crosslinked poly(methyl vinyl ether-co-maleic acid) hydrogels: Application in transdermal delivery systems. *J. Appl. Polym. Sci.* 125: 2680–2694.

126 S. Yang, Y. Feng, L. Zhang, *et al.* (2012). A scalable fabrication process of polymer microneedles. *Int. J. Nanomedicine.* 7: 1415–1422.

127 N.B. Shelke, R. James, C.T. Laurencin, *et al.* (2014). Polysaccharide biomaterials for drug delivery and regenerative engineering. *Polymer. Adv. Tech.* 25: 448–460.

128 E. Markovsky, H. Baabur-Cohen, A. Eldar-Boock, *et al.* (2012). Administration, distribution, metabolism and elimination of polymer therapeutics. *J. Control. Release* 161: 446–460.

129 N. Dhar, S.P. Akhlaghi and K.C. Tam (2012). Biodegradable and biocompatible polyampholyte microgels derived from chitosan, carboxymethyl cellulose and modified methyl cellulose. *Carbohydr. Polym.* 87: 101–109.

130 T. Miyamoto, S. Takahashi, H. Ito, *et al.* (1989). Tissue biocompatibility of cellulose and its derivatives. *J. Biomed. Mater. Res.* 23: 125–133.

131 H.D. Juneja, M. Joshi and J.P. Kanfade (2013). Synthesis and characterization of metallic gel complexes derived from carboxymethyl cellulose. *J. Chem.* 2013: 6.

132 F. Croisier and C. Jérôme (2013). Chitosan-based biomaterials for tissue engineering. *Eur. Polym. J.* 49: 780–792.

133 C.H. Goh, P.W.S. Heng and L.W. Chan (2012). Alginates as a useful natural polymer for microencapsulation and therapeutic applications. *Carbohydr. Polym.* 88: 1–12.

134 L.S. Nair, C.T. Laurencin, L.S. Nair and C.T. Laurencin (2007). Biodegradable polymers as biomaterials. *Prog. Polym. Sci.* 32: 762–798.

135 R. Lanza, R. Langer and J.P. Vacanti (2013). *Principles of Tissue Engineering*. Amsterdam: Elsevier Science.

136 J. Necas, L. Bartosikova, P. Brauner and J. Kolar (2008). Hyaluronic acid (hyaluronan): a review. *Vet. Med.* 53: 397–411.

137 D. Hreczuk-Hirst, D. Chicco, L. German and R. Duncan (2001). Dextrins as potential carriers for drug targeting: tailored rates of dextrin degradation by introduction of pendant groups. *Int. J. Pharm.* 230: 57–66.

138 S. Moreira, R.M.G. Da Costa, L. Guardão, *et al.* (2010). In vivo biocompatibility and biodegradability of dextrin-based hydrogels. *J. Bioact. Compat. Polym.* 25: 141–153.

139 A. Alrifaiy, O.A. Lindahl and K. Ramser (2012). Polymer-based microfluidic devices for pharmacy, biology and tissue engineering. *Polymers (Basel)* 4: 1349–1398.
140 R.Q. Frazer, R.T. Byron, P.B. Osborne and K.P. West (2005). PMMA: an essential material in medicine and dentistry. *J. Long Term Eff. Med. Implants* 15: 629–639.
141 K. Modjarrad and S. Ebnesajjad (2013). *Handbook of Polymer Applications in Medicine and Medical Devices*. Amsterdam: Elsevier Science.
142 T.V. Chirila, C.R. Hicks, P.D. Dalton, *et al.* (1998). Artificial cornea. *Prog. Polym. Sci.* 23: 447–473.
143 D. Kloukos, N. Pandis and T. Eliades (2013). Bisphenol-A and residual monomer leaching from orthodontic adhesive resins and polycarbonate brackets: a systematic review. *Am. J. Orthod. Dentofacial Orthop.* 143: S104.
144 D. Kloukos, E. Taoufik, T. Eliades, *et al.* (2013). Cytotoxic effects of polycarbonate-based orthodontic brackets by activation of mitochondrial apoptotic mechanisms. *Dent. Mater.* 29: E35–E44.
145 M. Watanabe (2004) Degradation and formation of bisphenol A in polycarbonate used in dentistry. *J. Med. Dent. Sci.* 51: 1–6.
146 A. Vesel, M. Mozetic, M. Jaganjac, *et al.* (2011). Biocompatibility of oxygen-plasma-treated polystyrene substrates. *Eur. Phys. J.* 56: 24024.
147 V.R. Sastri (2013). *Plastics in Medical Devices: Properties, Requirements, and Applications*. Amsterdam: Elsevier Science.
148 M.I. Baker, S.P. Walsh, Z. Schwartz and B.D. Boyan (2012). A review of polyvinyl alcohol and its uses in cartilage and orthopedic applications. *J. Biomed. Mater. Res. Part B Appl. Biomater.* 100: 1451–1457.
149 S.R. Montezuma, J. Loewenstein, C. Scholz and J.F.I. Rizzo (2006). Biocompatibility of materials implanted into the subretinal space of Yucatan pigs. *Invest. Ophthalmol. Vis. Sci.* 47: 3514–3522.
150 H. Abd El-Mohdy and S. Ghanem (2009). Biodegradability, antimicrobial activity and properties of PVA/PVP hydrogels prepared by γ-irradiation. *J. Polym. Res.* 16: 1–10.
151 K.V. Nemani, K.L. Moodie, J.B. Brennick, *et al.* (2013). In vitro and in vivo evaluation of SU-8 biocompatibility. *Mater. Sci. Eng. C* 33: 4453–4459.
152 W.D. Niles and P.J. Coassin (2008). Cyclic olefin polymers: innovative materials for high-density multiwell plates. *Assay Drug Dev. Technol.* 6: 577–590.
153 M. Polanská, H. Hulejová, M. Petrtýl, *et al.* (2010). Surface modification of cyclic olefin copolymers for osteochondral defect repair can increase pro-destructive potential of human chondrocytes in vitro. *Physiol. Res.* 59: 247–253.
154 A.C. Grayson, G. Voskerician, A. Lynn, *et al.* (2004). Differential degradation rates in vivo and in vitro of biocompatible poly(lactic acid) and poly(glycolic acid) homo- and co-polymers for a polymeric drug-delivery microchip. *J. Biomater. Sci. Polym. Ed.* 15: 1281–1304.
155 H.K. Makadia, S.J. Siegel, H.K. Makadia and S.J. Siegel (2011). Poly lactic-co-glycolic acid (PLGA) as biodegradable controlled drug delivery carrier. *Polymers (Basel)* 3: 1377–1397.
156 E.M. Mrak, G.F. Stewart and C.O. Chichester (1964). *Advances in Food Research*. Amsterdam: Elsevier Science.
157 P. Arbós, M. Wirth, M.A. Arangoa, *et al.* (2002). Gantrez AN as a new polymer for the preparation of ligand-nanoparticle conjugates. *J. Control. Release* 83: 321–330.

158 P. Ojer, A. Lopez de Cerain, P. Areses, *et al.* (2012). Toxicity studies of poly(anhydride) nanoparticles as carriers for oral drug delivery. *Pharm. Res.* 29: 2615–2627.

159 P. Ojer, L. Neutsch, F. Gabor, *et al.* (2013). Cytotoxicity and cell interaction studies of bioadhesive poly(anhydride) nanoparticles for oral antigen/drug delivery. *J. Biomed. Nanotechnol.* 9: 1891–1903.

160 E. Elizondo, A. Córdoba, S. Sala, *et al.* (2010). Preparation of biodegradable poly (methyl vinyl ether-co-maleic anhydride) nanostructured microparticles by precipitation with a compressed antisolvent. *J. Supercrit Fluids.* 53: 108–114.

161 A. Luzardo-Álvarez, J. Blanco-Méndez, P. Varela-Patiño and B. Martín Biedma (2011). Amoxicillin-loaded sponges made of collagen and poly[(methyl vinyl ether)-co-(maleic anhydride). for root canal treatment: preparation, characterization and in vitro cell compatibility. *J. Biomater. Sci. Polym. Ed.* 22: 329–342.

162 E. Moreno, J. Schwartz, E. Larrañeta, *et al.* (2014). Thermosensitive hydrogels of poly(methyl vinyl ether-co-maleic anhydride) - Pluronic(®) F127 copolymers for controlled protein release. *Int. J. Pharm.* 459: 1–9.

163 J.H. Park, M.G. Allen and M.R. Prausnitz (2005). Biodegradable polymer microneedles: fabrication, mechanics and transdermal drug delivery. *J. Control. Release* 104: 51–66.

164 S.P. Davis, W. Martanto, M.G. Allen and M.R. Prausnitz (2005). Hollow metal microneedles for insulin delivery to diabetic rats. *IEEE Trans. Biomed. Eng.* 52: 909–915.

165 R.K. Sivamani, D. Liepmann and H.I. Maibach (2007). Microneedles and transdermal applications. *Expert Opin. Drug Deliv.* 4: 19–25.

166 J. Voldman, M.L. Gray and M.A. Schmidt (1999). Microfabrication in biology and medicine. *Annu. Rev. Biomed. Eng.* 1: 401–425.

167 L. Leoni and T.A. Desai (2004). Micromachined biocapsules for cell-based sensing and delivery. *Adv. Drug Deliv. Rev.* 56: 211–229.

168 S. Katsuma and G. Tsujimoto (2001). Genome medicine promised by microarray technology. *Exp. Rev. Mol. Diagn.* 1: 377–382.

169 J.Z. Hilt and N.A. Peppas (2005). Microfabricated drug delivery devices. *Int. J. Pharm.* 306: 15–23.

170 S. Zafar Razzacki, P.K. Thwar, M. Yang, *et al.* (2004). Integrated microsystems for controlled drug delivery. *Adv. Drug Deliv. Rev.* 56: 185–198.

171 M.J. Madou (2002). *Fundamentals of Microfabrication: The Science of Miniaturization*, 2nd edn. Boca Raton: CRC Press.

172 D. Banks (2006). Microengineering, MEMS, and Interfacing: A Practical Guide. Boca Raton: CRC Press.

173 W.M. Moreau (1988). Semiconductor Lithography: Principles, Practices, and Materials. New York: Plenum Press.

174 K.F. Jensen (1989). *Chemical Vapor Deposition, Microelectronics Processing*, 199–263. American Chemical Society.

175 W. Kern (1978). Chemical etching of silicon, germanium, gallium, arsenide, and gallium phosphide. *RCA Rev.* 39: 278–308.

176 M. Shikida, K. Sato, K. Tokoro and D. Uchikawa (2000). Differences in anisotropic etching properties of KOH and TMAH solutions. *Sens. Actuator A-Phys.* 80: 179–188.

177 M.J. Arduino, L.A. Bland, L.E. Danzig, *et al.* (1997). Microbiologic evaluation of needleless and needle-access devices. *Am. J. Infect. Control* 25: 377–380.
178 E.V. Mukerjee, S.D. Collins, R.R. Isseroff and R.L. Smith (2004). Microneedle array for transdermal biological fluid extraction and in situ analysis. *Sens. Actuator A-Phys.* 114: 267–275.
179 S.J. Paik, S. Byun, J.M. Lim, *et al.* (2004). In-plane single-crystal-silicon microneedles for minimally invasive microfluid systems. *Sens. Actuator A-Phys.* 114: 276–284.
180 N. Roxhed, B. Samel, L. Nordquist, *et al.* (2008). Painless drug delivery through microneedle-based transdermal patches featuring active infusion. *IEEE Trans. Biomed. Eng.* 55: 1063–1071.
181 A. Rodriguez, D. Molinero, E. Valera, *et al.* (2005). Fabrication of silicon oxide microneedles from macroporous silicon. *Sens. Actuator B-Chem.* 109: 135–140.
182 M.K. Ramasubramanian, O.M. Barham and V. Swaminathan (2008). Mechanics of a mosquito bite with applications to microneedle design. *Bioinspiration Biomimetics* 3: 046001.
183 B. Ma, S. Liu, Z. Gan, *et al.* (2006). A PZT insulin pump integrated with a silicon microneedle array for transdermal drug delivery. *Microfluid. Nanofluid.* 2: 417–423.
184 M.W. Ashraf, S. Tayyaba, A. Nisar, *et al.* (2010). Design, fabrication and analysis of silicon hollow microneedles for transdermal drug delivery system for treatment of hemodynamic dysfunctions. *Cardiovascular Eng.* 10: 91–108.
185 U.O. Häfeli, A. Mokhtari, D. Liepmann and B. Stoeber (2009). In vivo evaluation of a microneedle-based miniature syringe for intradermal drug delivery. *Biomed. Microdevices* 11: 943–950.
186 W. Trimmer, P. Ling, C.K. Chin, *et al.* (1995). Injection of DNA into plant and animal tissues with micromechanical piercing structures. *Proc. - IEEE Micro Electro Mech. Syst.* 1995: 111.
187 C.H. Dean, J.B. Alarcon, A.M. Waterston, *et al.* (2005). Cutaneous delivery of a live, attenuated chimeric flavivirus vaccine against Japanese encephalitis (ChimeriVax)-JE) in non-human primates. *Hum. Vaccines* 1: 106–111.
188 J.B. Alarcon, A.W. Hartley, N.G. Harvey and J.A. Mikszta (2007). Preclinical evaluation of microneedle technology for intradermal delivery of influenza vaccines. *Clin. Vaccine Immunol.* 14: 375–381.
189 X.M. Wu, H. Todo and K. Sugibayashi (2007). Enhancement of skin permeation of high molecular compounds by a combination of microneedle pretreatment and iontophoresis. *J. Control. Release* 118: 189–195.
190 M.M. Badran, J. Kuntsche and A. Fahr (2009). Skin penetration enhancement by a microneedle device (Dermaroller) in vitro: dependency on needle size and applied formulation. *Eur. J. Pharm. Sci.* 36: 511–523.
191 W. Martanto, S.P. Davis, N.R. Holiday, *et al.* (2004). Transdermal delivery of insulin using microneedles in vivo. *Pharm. Res.* 21: 947–952.
192 S.K.N. Saluja, A. Badkar and A.K. Banga (2009) Optimization of fabrication of in-plane titanium microneedles. 36th Annual Meeting & Exposition of the Controlled Release Society, Copenhagen.
193 M. Cormier, B. Johnson, M. Ameri, *et al.* (2004). Transdermal delivery of desmopressin using a coated microneedle array patch system. *J. Control. Release* 97: 503–511.

194 K. Lee, H.C. Lee, D.S. Lee and H. Jung (2010) Drawing lithography: three-dimensional fabrication of an ultrahigh-aspect-ratio microneedle. *Adv. Mater.* 22: 483.

195 K. Lee and H. Jung (2012) Drawing lithography for microneedles: a review of fundamentals and biomedical applications. *Biomaterials* 33: 7309–7326.

196 M. Han, D.K. Kim, S.H. Kang, *et al.* (2009). Improvement in antigen-delivery using fabrication of a grooves-embedded microneedle array. *Sens. Actuator B-Chem.* 137: 274–280.

197 Y.A. Gomaa, D.I. Morrow, M.J. Garland, *et al.* (2010). Effects of microneedle length, density, insertion time and multiple applications on human skin barrier function: assessments by transepidermal water loss. *Toxicol. In Vitro* 24: 1971–1978.

198 S.P. Sullivan, D.G. Koutsonanos, M. Del Pilar Martin, *et al.* (2010). Dissolving polymer microneedle patches for influenza vaccination. *Nat. Med.* 16: 915–U116.

199 T. Miyano, T. Miyachi, T. Okanishi, *et al.* (2007). Hydrolyticmicroneedles as Transdermal Drug Delivery System. Proceedings of TRANSDUCERS 2007. International Solid-State Sensors, Actuators and Microsystems Conference (10–14 June 2007, Lyon, France), 355–358.

200 Y. Ito, J. Yoshimitsu, K. Shiroyama, *et al.* (2006). Self-dissolving microneedles for the percutaneous absorption of EPO in mice. *J. Drug Target.* 14: 255–261.

201 Y. Ito, A. Murakami, T. Maeda, *et al.* (2008). Evaluation of self-dissolving needles containing low molecular weight heparin (LMWH) in rats. *Int. J. Pharm.* 349: 124–129.

202 S.D. Gittard, A. Ovsianikov, N. Monteiro-Riviere, *et al.* (2009). Fabrication of polymer microneedles using a two-photon polymerization and micromolding process. *J. Diabetes Sci. Technol.* 3: 304–311.

203 F.J. Verbaan, S.M. Bal, D.J. van den Berg, *et al.* (2008). Improved piercing of microneedle arrays in dermatomed human skin by an impact insertion method. *J. Control. Release* 128: 80–88.

204 S.P. Davis, M.R. Prausnitz and M.G. Allen (2003). Fabrication and characterization of laser micromachined hollow microneedles Proceedings of TRANSDUCERS 2003. Nanotechnology Symposium, 12th International Conference on Solid-State Sensors, Actuators and Microsystems (Boston, MA, USA, 2003) vol. 2: 1435–1438.

205 A. Davidson, B. Al-Qallaf and D.B. Das (2008) Transdermal drug delivery by coated microneedles: Geometry effects on effective skin thickness and drug permeability. *Chem. Eng. Res. Des.* 86: 1196–1206.

206 M.A. Kendall, Y.F. Chong and A. Cock (2007). The mechanical properties of the skin epidermis in relation to targeted gene and drug delivery. *Biomaterials* 28: 4968–4977.

207 A.L. Teo, C. Shearwood, K.C. Ng, *et al.* (2006). Transdermal microneedles for drug delivery applications. *Mater. Sci. Eng. B* 132: 151–154.

208 A. Laurent, F. Mistretta, D. Bottigioli, *et al.* (2007). Echographic measurement of skin thickness in adults by high frequency ultrasound to assess the appropriate microneedle length for intradermal delivery of vaccines. *Vaccine* 25: 6423–6430.

209 R.H. Champion, A. Rook, J.L. Burton, *et al.* (1992). *Textbook of Dermatology.* Wiley Blackwell.

210 Y.G. Lv, J. Liu, Y.H. Gao and B. Xu (2006). Modeling of transdermal drug delivery with a microneedle array. *J. Micromech. Microeng.* 16: 2492–2501.

211 M. Shikida, T. Hasada and K. Sato (2006). Fabrication of a hollow needle structure by dicing, wet etching and metal deposition. *J. Micromech. Microeng.* 16: 2230–2239.

212 B. Stoeber and D. Liepmann (2005). Arrays of hollow out-of-plane microneedles for drug delivery. *J. Microelectromech. Syst.* 14: 472–479.

213 G. Pastorin, H. Junginger, T.R. Nayak and Z. Munrui (2009). Nanoneedles devices for transdermal vaccine delivery: in vitro and in vivo evaluation. 36th Annual Meeting & Exposition of the Controlled Release Society, Copenhagen.

214 B. Al-Qallaf and D.B. Das (2008). Optimization of square microneedle arrays for increasing drug permeability in skin. *Chem. Eng. Sci.* 63: 2523–2535.

215 B. Al-Qallaf and D.B. Das (2009). Optimizing microneedle arrays to increase skin permeability for transdermal drug delivery. *Ann. N. Y. Acad. Sci.* 1161: 83–94.

216 S.D. Gittard, C. Bo, X. Huadong, *et al.* (2013). The effects of geometry on skin penetration and failure of polymer microneedles. *J. Adhes. Sci. Technol.* 27: 227–243.

217 S.P. Davis, B.J. Landis, Z. Adams, *et al.* (2004). Insertion of microneedles into skin: measurement and prediction of insertion force and needle fracture force. *J. Biomech.* 37: 1155–1163.

218 M. Pearton, V. Saller, S.A. Coulman, *et al.* (2012). Microneedle delivery of plasmid DNA to living human skin: Formulation coating, skin insertion and gene expression. *J. Control. Release* 160: 561–569.

219 S.M. Bal, J. Caussin, S. Pavel and J.A. Bouwstra (2008). *In vivo* assessment of safety of microneedle arrays in human skin. *Eur. J. Pharm. Sci.* 35: 193–202.

220 G. Yan, K.S. Warner, J. Zhang, *et al.* (2010). Evaluation needle length and density of microneedle arrays in the pretreatment of skin for transdermal drug delivery. *Int. J. Pharm.* 391: 7–12.

221 N. Roxhed, T.C. Gasser, P. Griss, *et al.* (2007). Penetration-enhanced ultrasharp microneedles and prediction on skin interaction for efficient transdermal drug delivery. *J. Microelectromech. Syst.* 16: 1429–1440.

222 P.W.H. Loeters, R.F. Duwel, F.J. Verbaan, *et al.* (2004). Measuring the insertion of microfabricated microneedles into skin with a penetration sensor. Proceedings of μTAS 2004 8th International Conference on Miniaturized Systems in Chemistry and Life Sciences (Malmö, Sweden, 26–30 September, 2004) 497–499.

223 S.A. Coulman, J.C. Birchall, A. Alex, *et al.* (2011). In vivo, in situ imaging of microneedle insertion into the skin of human volunteers using optical coherence tomography. *Pharm. Res.* 28: 66–81.

224 R.F. Donnelly, K. Moffatt, A.Z. Alkilani, *et al.* (2014). Hydrogel-forming microneedle arrays can be effectively inserted in skin by self-application: a pilot study centred on pharmacist intervention and a patient information leaflet. *Pharm. Res.* 1–11.

225 E. Forvi, M. Soncini, M. Bedoni, *et al.* (2010). A method to determine the margin of safety for microneedles arrays. *Proceedings of the World Congress on Engineering 2010*, vol II (30 June – 2 July, 2010, London, UK).

226 P. Khanna, J.A. Strom, J.I. Malone and S. Bhansali (2008). Microneedle-based automated therapy for diabetes mellitus. *J. Diabetes Sci. Technol.* 2 1122–1129.

227 S.A. Ranamukhaarachchi, T. Schneider, S. Lehnert, *et al.* (2016). Development and validation of an artificial mechanical skin model for the study of interactions between skin and microneedles. *Macromol. Mater. Eng.* 301: 306–314.

3

Microneedle-mediated Drug Delivery

Helen L. Quinn and Ryan F. Donnelly

School of Pharmacy, Queen's University Belfast, Belfast BT9 7BL, UK

3.1 Introduction

As the largest and most accessible organ in the body, with a surface area of 1.5–2 m^2, the skin is an enticing target for drug delivery. Delivery of medication through the skin has expanded beyond the initial brief of topical delivery for local effects, to the transdermal application of drugs designed for systemic absorption and action, with the first product approved by the Food and Drug Administration (FDA) in 1979 (scopolamine). The delivery of drugs via the transdermal route offers many potential advantages in terms of drug action and also patient comfort. The primary attraction in this regard is considered to be the convenient and non-invasive nature of a transdermal patch, which can lead to increased patient adherence and satisfaction, particularly in comparison with delivery methods associated with pain and discomfort, such as injections [1]. By circumventing the gastrointestinal tract, the risk of degradation of the active is removed and hepatic first-pass metabolism is avoided, proffering the possibility for an improvement in drug bioavailability. Further to this, variables that may cause unpredictable drug absorption are minimised, such as pH, enzymatic activity and drug–food interactions [2]. Sustained delivery of the drug is possible due to the slow transport across the highly constrained skin layers, achieving a steady-state effect, instead of the typical peak–trough profile. The maintenance of drug levels within the therapeutic window for extended durations of time with the same transdermal patch is, therefore, possible. For compounds with a short biological half-life, this can often lead to a reduced dosing frequency, such as with the opioid buprenorphine, which can be administered once weekly transdermally, in contrast to a three times daily sublingual tablet [3]. The relatively constant plasma concentrations can also reduce the occurrence of adverse effects. Termination of therapy in the event of a serious adverse effect or toxicity is also easily achieved, as straightforward removal of the patch should cease drug delivery.

Any drug applied to the skin must traverse the epidermal layers in order to reach the systemic circulation. The dermis houses the rich microcirculation of the skin, so it is thought once any substance reaches this layer, it can be absorbed systemically [4]. The primary obstacle is, therefore, the epidermis, of which the *stratum corneum* has been identified as posing the greatest resistance to transdermal transport. In fact, most small, water-soluble molecules can diffuse into the systemic circulation up to one thousand

Microneedles for Drug and Vaccine Delivery and Patient Monitoring, First Edition.
Edited by Ryan F. Donnelly, Thakur Raghu Raj Singh, Eneko Larrañeta, and Maelíosa T.C. McCrudden.

times more rapidly, simply when this rate-limiting barrier is absent [5]). The number of drugs able to passively diffuse across the intact skin barrier is, therefore, inherently limited. Active pharmaceutical ingredients that display this capability tend to exhibit specific physicochemical properties. The molar mass of the drug dictates the diffusion coefficient, with more efficient transdermal transport possible for molecules of a molecular mass less than 600 Da. The lipophilicity of the molecule is also of importance for diffusion, with adequate solubility in both oil and water required. A highly hydrophilic substance cannot easily penetrate the skin because it cannot enter the lipophilic *stratum corneum* layer. In contrast, a hydrophobic substance can easily permeate into the *stratum corneum* but becomes sequestered within, due to its limited ability to partition into the hydrophilic viable epidermis. The octanol–water partition coefficient (log *P*) can be used to predict a drug's partitioning behaviour through the skin, with optimum values of log *P* deduced to lie between 1 and 3. Ideally, a low required daily dose, hence a drug of reasonably high potency, is also desirable.

This has led to a relatively small number of marketed transdermal products approved in the United States by the FDA and in Europe by the European Medicines Agency (EMA), as detailed in Table 3.1. It is pertinent to note that those molecules listed

Table 3.1 Daily dose ranges and selected physicochemical properties of currently approved transdermally delivered drugs. Adapted with permission from: [6] Wiedersberg, S. and Guy, R. (2014). Transdermal drug delivery: 30+ years of war and still fighting! *J. Control. Release* 190: 150–156.

Drug (year of FDA approval)	Dose/day (mg)	MW (Da)	Log *P* [a]
Scopolamine (1979)	0.3	303	0.98
Glyceryl trinitrate (1981)	2.4–15	227	1.62
Clonidine (1984)	0.1–0.3	230	*2.42±0.52*
Estradiol (1986)	0.025–0.100	272	4.01
Fentanyl (1990)	0.288–2.400	337	4.05
Nicotine (1991)	7–21	162	1.17
Testosterone (1993)	0.3–5	288	3.32
Estradiol and norethisterone acetate (1998)	0.025–0.050, 0.125–0.250	272, 340	4.01, 3.99
Norelgestromin and ethinyl estradiol (2001)	0.2, 0.034	327, 296	*3.90±0.47, 3.67*
Estradiol and levonorgestrel (2003)	0.050, 0.007–0.015	272, 312	4.01, *3.72±0.49*
Oxybutynin (2003)	3.9	357	*4.02±0.52*
Selegeline (2006)	6–12	187	2.90
Methylphenidate (2006)	26–80	233	*2.15±0.42*
Rotigotine (2007)	1–3	315	*4.58±0.72*
Rivastigmine (2007)	4.6–9.5	250	*2.34±0.16*
Granisetron (2008)	3.1	312	*2.55±0.28*
Buprenorphine (2010)	0.12–1.68	468	4.98

a) Log(octanol–water partition coefficient (*P*)): either experimental or calculated (mean±S.D.) values.

in Table 3.1 that may seem least suited to transdermal delivery (the steroids used for hormone replacement therapy or contraception) by way of their relatively high lipophilicity and molecular weights, are also the most potent, with daily doses of less than a milligram. Conversely, most of the molecules requiring higher systemic doses possess almost ideal physicochemical characteristics for transdermal permeation. In this manner, it can be surmised that successful passive transdermal delivery is usually a question of balancing the two criteria of physicochemical properties and required dose.

As a consequence of the outlined advantages to transdermal delivery, alongside the difficulties in achieving passive transdermal permeation, it follows that a variety of enhancement strategies have been explored. The focus of the present book is on microneedles (MNs), which are minimally invasive devices, consisting of numerous micron-sized projections, arranged on a baseplate. In order to enhance transdermal drug delivery, MNs mechanically disrupt the *stratum corneum*, circumventing the outermost skin barrier altogether. The technology was first conceptualised in the 1970s [7] and since manufacturing advances allowed the first practical realisation in 1998 [8], MNs have demonstrated great promise as a novel drug delivery platform. The removal of the rate-controlling effect of the *stratum corneum* allows transdermal delivery of a wider range of drug molecules than would be possible by passive diffusion, including macromolecules, such as proteins. This has led to a rapidly expanding MN research area, in which the boundaries of what can be delivered transdermally have been explored. This chapter aims to review some of the work performed on the transdermal delivery of low and high molecular weight compounds using various MN systems, the mechanisms of which will be described in the next section.

3.2 Microneedle Drug Delivery Strategies

MN arrays have been fabricated in a number of designs, with different drug delivery strategies, as illustrated in Figure 2.1 in Chapter 2. Solid MNs in their simplest form involve a two-step application process for drug delivery. An MN array, typically manufactured from silicon or metal, is applied to the skin and subsequently removed, temporarily increasing skin permeability. This then facilitates the passive diffusion of the drug from a reservoir, usually in the form of a patch or topical formulation, placed over the site of the MN insertion. As an advance on simple silicon or metal MN arrays, in order to create a one-step application process, solid MNs may be coated, with the drug loaded on to the surface of the needles. When applied, the drug layer dissolves, depositing the active into the skin, and the MN array can then be removed. The mechanism of action of hollow MNs resembles that of traditional injections, allowing flow of a liquid formulation through a central lumen in the MNs. Dissolving MN arrays have the drug incorporated into the MN formulation, and upon insertion into the skin the MNs dissolve, releasing the drug in the viable skin layers. Hydrogel-forming MNs are an integrated system, consisting of drug-free MNs prepared from a hydrogel-forming polymer and a separate drug reservoir attached to the supporting baseplate. Hard in the dry state, these MNs swell upon application to the skin, due to the uptake of interstitial fluid from the tissue. This triggers diffusion of the drug from the attached reservoir, through the swollen micro-projections, which act as conduits, into the dermal layers of the skin for uptake by the microcirculation. Each of these MN types has been extensively investigated for drug delivery, examples of which will be described in more detail in subsequent sections. Associated challenges with each particular methodology will also be discussed.

3.2.1 Solid Microneedles

Pre-treatment of the skin with a solid MN array, followed by application of a drug formulation for diffusion through the resulting skin pores, was the first reported strategy for MN use. Initial studies adopting this approach involved delivery of small marker compounds, with the aim of testing the delivery method, in comparison with passive permeation across the skin. A fourfold increase in permeability of calcein through human cadaver skin was demonstrated, after insertion and removal of an array of solid silicon MNs, 150 μm long [8]. Further evidence of the permeability enhancing effects of MNs for larger compounds was provided in 2003 by McAllister *et al.* [9], who showed increased permeation of insulin and bovine serum albumin (BSA) following skin pre-treatment with an array of 400 solid silicon MNs, of height 150 μm. The delivery of macromolecules transdermally, previously extremely challenging, was, therefore, established as being possible, with physiologically relevant amounts of biologically active insulin delivered *in vivo* [10]. Indeed, the transport of molecules up to at least 72 kDa was proven to be feasible, using MN arrays manufactured from commercially available 30G hypodermic needles [11]. Mohammed *et al.* [12] did, however, demonstrate a link between molecular weight and permeation, using a range of cosmeceutically relevant peptides, with the compound of the lowest molecular weight displaying the greatest transdermal enhancement.

Aside from the manufacturing of the MNs, to focus on the drug permeation enhancing effects of MNs, commercially available MN devices, which were originally designed for cosmetic applications, have been adapted for use in drug delivery. For example, the Dermaroller®, composed of 24 circular arrays of eight stainless-steel needles each (in total 192 needles) in a cylindrical assembly, designed for collagen induction therapy has been used by several researchers for drug permeation studies. In one example, the Dermaroller was used to investigate the efficiency of skin perforation and the penetration enhancement of a hydrophilic model drug [13]. Stahl *et al.* [14] also used commercially available Medik8® MN rollers composed of solid titanium needles of either 200 or 300 μm height, to enhance delivery of a number of non-steroidal anti-inflammatory drugs, with the greatest effects observed for those drugs of a more hydrophilic nature with high melting points. Zhou *et al.* [15] demonstrated that insulin delivery could also be ameliorated in this way by pre-treating rat skin with an MN roller, showing a subsequent decrease in the blood-glucose level of the animal. Other examples of topical formulations paired with MN roller devices for transdermal permeation enhancement include drugs such as ondansetron [16], sumatriptan [17], amlodipine and verapamil [18].

In combination, these early studies demonstrated the potential of MNs in transdermal drug delivery by removing the stringent physicochemical properties previously required for skin permeation. It is not to say, however, that the properties of the drug do not have an effect upon movement through MN-induced pores. To explore this concept, Banks *et al.* [19] investigated how the ionisation of naltrexone influenced its penetration across hairless guinea pig and human skin *in vitro*. Control studies revealed that the naltrexone base was significantly more permeable across guinea pig skin than the hydrochloride salt. However, the reverse was observed when the skin was pierced 20 times with an array containing five 750 μm stainless-steel MNs, before application of the drug solution. Indeed, the steady-state flux of the hydrophilic salt formed across the punctured skin was almost three times that of the free base. It was interesting to note that MN treatment made no significant improvement in the delivery of the free base ($p>0.05$) and the authors concluded that the free base was unable to utilise the alternative pathway created by MNs across the *stratum corneum*, due to the aqueous nature of the micro-channels. Continuing on from this, the permeation of the more polar active metabolite, 6-β-naltrexol, a tertiary amine with a pK_a of 7.4, was examined from donor solutions of two different pHs, 4.5 and 8.5. As noted with the parent drug, the maximum flux across treated guinea pig skin was observed with the more hydrophilic species (pH 4.5). This study, therefore, demonstrates that solid MN puncture allows enhanced permeation of more hydrophilic compounds, due to the formation of aqueous micro-channels, contrary to conventional percutaneous delivery, which tends to favour more lipophilic, un-ionised molecules. Importantly, these studies demonstrate the importance of selecting both the correct form of the drug and the optimum formulation for combination with solid MNs in order to see the greatest enhancement in transdermal permeation.

In combination with the aforementioned discussion on the properties of the drug, it is also to be expected that the design of the MN array used for skin pre-treatment will have an effect on drug permeation, with needle length and density the greatest contributors. Yan *et al.* [20] evaluated the penetration of aciclovir across human epidermis using solid silicon MN arrays with varying length (100–1100 μm) and density (400–11 900 MN/cm^2). An approximate 2–8-fold increase in aciclovir flux was reported when MNs with a length of less than 600 μm were employed and there was a 50–100-fold increase with longer needles. MN density was also shown to have an important effect. A significant enhancement in aciclovir flux was observed with human epidermis pre-treated with MN arrays with 400 μm needle length and 2000 needles/cm^2 needle density, but a lower enhancement of drug flux was observed for the MNs with same needle length but a higher needle density of 5625 needle/cm^2. The authors suggest that by applying the same application force to a higher density array, a lower force was applied to each needle, resulting in a "bed of nails" effect. Following on from this point, the force with which MN treatment occurs can also have an effect, as Cheung *et al.* [21] demonstrated, employing insulin as a model compound. Using the stainless-steel AdminPatch® for porcine skin pre-treatment, applied at two forces of 60.5 and 69.1 N, the resultant amounts of a commercially available insulin solution permeated were approximately 3 and 25 μg, respectively, over a 4-h period. In this way, it was shown that, in this case, the amount of insulin permeated was related to the initial applied force. The needle design, alongside the application process of the array, therefore needs to be carefully considered for each delivery purpose.

The first study in humans using the approach of solid MN puncture followed by application of a topical formulation was conducted using naltrexone as a model drug [22]. The skin of six healthy volunteers was pre-treated with MNs manufactured from stainless steel (620 μm length, 160 μm base width, <1 μm radius of curvature at the tip), arranged in arrays of 50 needles, followed by application of a naltrexone polymeric patch. Steady-state plasma concentrations of naltrexone were obtained within 2 h of patch application and maintained for at least 48 hours. Importantly, the dose administered (12.6 mg/day) was approximately a quarter of the daily dose administered as an oral tablet to achieve similar plasma levels, demonstrating that MN application and removal can markedly improve the transdermal delivery of low molecular weight compounds *in vivo*.

Silicon and metal are not necessarily the most appropriate choices of material for use in a transdermal drug delivery device, despite their amenable properties for MN manufacture. There are concerns that such non-biodegradable materials may present safety issues when inserted into the skin. For example, silicon remains unproven as a biocompatible material and historically there have been rare reports of silicon-related granulomas, with a latency ranging from months to years [23]. Further to this, due to the fragility of the material, silicon MNs may break upon insertion, with initial reports that the tips of several solid silicon MNs fractured when inserted into excised human skin [8]. Polymeric materials, however, are often recognised as safe, known to be biocompatible and, indeed, relatively inexpensive. Various polymers have, therefore, been investigated for the manufacture of solid MNs for skin pre-treatment, as an alternative to silicon or metal.

The fabrication of polymeric MNs from biodegradable polymers such as polylactic acid (PLA), polyglycolic acid (PGA) and their co-polymers (PLGA) and the efficacy of this type of array in improving delivery of model compounds across the skin was described by Park *et al.* [24]. It was reported that pre-treatment of human cadaver epidermis with 20 MNs led to an increase in skin permeability to BSA by two orders of magnitude, with a further increase to almost three orders of magnitude seen when 100 MNs were used. In another example, polycarbonate MNs of height 500 μm, when coupled with calcein gel, showed a 5.46-fold increase in the permeation of the hydrophilic molecule, when compared with application of the topical formulation alone [25]. Similarly, polystyrene MNs have been designed with high mechanical strength, with the intention of use as a skin pre-treatment methodology [26]. A recent study used PLA to fabricate MNs, aiming for a mechanically stable formulation, capable of inducing an improved drug permeation rate into the skin [27]. Tested *in vivo* in a rat model, using insulin as a model drug, blood glucose levels were observed to drop to 29% of the initial level at 5 h, posing a stark contrast to the subcutaneous injection of insulin, which provoked its maximum response at 1.5 h. Thus, the MN route offers an alternative for a more controlled release of active, which the authors also demonstrated can be further manipulated by optimisation of parameters such as MN height and density, drug concentration, the viscosity of drug formulation and the administration time of drug on the skin.

As research into the transdermal drug delivery field has progressed, the need for a two-step application process, consisting of pre-treatment with an MN array, followed by application of a drug formulation, has been considered impractical for patient use. Although an essentially straightforward application process, any drug delivery system that requires more than one step is unlikely to be widely acceptable by patients, with a negative impact upon patient adherence [28]. Building on the early results gathered for the permeability enhancing effects following MN insertion, other strategies for MN drug delivery have, therefore, now become more prevalent.

3.2.2 Coated Microneedles

Coated MNs, which have the drug adhered on to the surface of the needles, offer a more controlled and potentially efficient route of transdermal drug delivery than that described for skin pre-treatment with solid MNs. Typically prepared from silicon or metal, uniform and reproducible quantities of drugs, capable of inducing a therapeutic effect, must be coated onto the array microprojections. Owing to the small size of the needles, the loading capacity of such an array is inherently small, with dosing typically limited to microgram quantities, thereby dictating that a potent active must be the compound of choice. The challenge, therefore, in this approach is the coating formulation of the needles, ensuring an appropriate loading of a stable active, uniformly dispersed over the surface of the MNs. As a result, optimisation of the coating process and formulation characteristics is of paramount importance during the developmental stage of these devices. Drug formulations have been comprehensively studied, with both coating and MN parameters considered [29]. Surfactants, viscosity enhancers and peptide stabilisers, such as trehalose, are often necessary in these formulations, to ensure coating uniformity and thickness and stability of the active. Varying MN geometries have simultaneously been investigated, with one study comparing six different MN shapes to determine effective skin permeability, using insulin as a model drug [30].

Solid MNs coated with a dry-powder drug formulation that dissolves off the MNs upon insertion in the skin is a common approach for delivery of peptides, as many biomolecules have a relatively high potency and are, therefore, amenable to this strategy. Examples include salmon calcitonin [31], desmopressin [32] erythropoietin [33], human growth hormone [34] and interferon alpha [35]. Given that the majority of these biomolecules are hydrophilic, it follows that coating solutions are, for the most part, aqueous in nature. Considering further hydrophobic molecules, research has been conducted into designing formulations more suitable for compounds of this type. For example, Zhao *et al.* [36] reported a coating formulation designed for hydrophobic peptides, maintaining bioactivity despite the challenges of solvent incorporation for solubilisation.

Coated polymeric MNs have been investigated, in a bid to move away from less biocompatible materials, such as silicon and metal. In one example, poly-L-lactic acid microneedles coated in bleomycin were devised for the treatment of warts, as a convenient and painless alternative to intralesional injection [37]. The MNs were dip-coated three times in a carboxymethy cellulose (CMC) bleomycin solution, with the resulting needles being strong enough to penetrate porcine foreleg skin, representative of the thick *stratum corneum* likely to be present in warts. *In vivo* studies revealed the MNs delivered a more concentrated drug dose into the sub-epidermal layer, with levels maintained for longer than that produced by subcutaneous injection.

Lidocaine is of particular interest for MN delivery, given its role in producing local anaesthesia and the number of topical and transdermal products currently available. In a bid to achieve rapid onset of action, lidocaine was coated on non-dissolving polymer MNs, namely 3M's solid microstructured transdermal system (sMTS), and administered to pigs [38]. The resulting lidocaine levels in the skin with 1 min wear time of the patch were deemed to be sufficient to cause local analgesia, with the effect prolonged when co-administered with a vasoconstricting agent as an adjuvant. The total amount of lidocaine coated was, however, limited to 225 μg, which is unlikely to be able to provide anaesthesia for a large skin area.

In terms of commercial success, Zosano Pharma has developed a transdermal patch system composed of titanium microprojections, which is then dry-coated with a formulation of an existing drug. The aim is to achieve a faster onset of action of the chosen drug than the currently available treatment methodologies, while also exploiting the aforementioned advantages of the transdermal route. The first trial using this particular technology was for the delivery of parathyroid hormone (1-34) (PTH) [39]. In the Phase I clinical studies, MN patches coated with 30 μg PTH were applied to different sites (the abdomen, upper forearm or thigh) in healthy human subjects for 30 min to evaluate the effect of administration site, patch wear time and dose. Encouragingly, it was shown that application of MN patches, irrespective of the site, resulted in rapid peak plasma levels, indeed three times faster than that observed after administration of the subcutaneous comparator. In Phase II, 20, 30 and 40 μg PTH doses administered using MNs showed a proportional increase in plasma concentration in post-menopausal women with osteoporosis. Alongside these pharmacokinetic investigations, stability investigations of the optimised formulation revealed that over 98% of PTH coated onto the micro-projections was stable at 25 °C for two years [40]. Another drug target for the coated Zosano system is zolmitriptan for the treatment of migraine, and, indeed, this is currently the most advanced of the pipeline products, having reached Phase III studies. Zolmitriptan onset when delivered by MNs was observed to reach peak plasma concentrations within 20 min of patch application, considerably faster than when administered orally. The subsequent ZOTRIP efficacy trial demonstrated that the 3.8 mg dose met both primary endpoints in migraine patients, achieving freedom from pain and "most bothersome symptoms" at 2 h, in addition to no serious adverse events [41]. The other candidate in development is that of glucagon, for treatment of severe hypoglycaemia, with positive results from Phase II completion announced in 2015 [42].

3.2.3 Hollow Microneedles

Hollow MNs present a more complex design than solid or coated MNs, with an inherently more challenging manufacturing process, in order to prepare needles with an appropriate bore for drug infusion. In addition to needle fabrication, the array must then be integrated with some form of drug reservoir, which holds the liquid to be delivered. In its simplest form, delivery of the drug from the reservoir is by passive diffusion. However, as the interfacial area of each MN opening is small and diffusivity into the dense tissue relatively low, passive diffusion can be slow. Consequently, there have been several attempts to achieve active delivery, whereby a force is applied to the liquid reservoir, to achieve a steady drug infusion over a period of time. In comparison with solid or coated MNs, which are capable of delivering small quantities of pharmaceuticals, this approach allows for infusion of larger amounts of the drug substance. In addition, pressure-assisted injection via hollow MNs offers potential to modulate drug delivery by altering the infusion rate.

Hollow MNs can be prepared from silicon, metal or glass, with varying bore diameters, and have been used both as a single needle or an array of multiple hollow MNs. Owing to their design, hollow MNs are inherently weaker than solid MNs, and are, therefore, at greater risk of breakage [43]. The bore opening at the needle tip reduces the sharpness, making insertion into the skin more difficult. As a result, novel fabrication techniques have been investigated to produce a sharp, bevelled tip for optimal insertion, using methods capable of adaptation for mass production [44, 45]. Martanto *et al.* [46] used sulforhodamine as a model drug to optimise drug delivery protocols and identify rate-limiting barriers to flow, highlighting that resistance to the infusion of liquid is provided by compressed dermal tissue, with blockage of the opening possible. As further confirmation of this theory, the authors also demonstrated that addition of hyaluronidase, an enzyme that breaks down hyaluronic acid within skin collagen fibres, to the infusion liquid increased flow rate sevenfold. The potential for blockage may also be overcome by using an alternative design to locate the bore-opening at the side of the MN tip [47]. This ensures puncturing of the skin, without removing any tissue, thereby improving healing and, without direct pressure on the tissue, a more defined and controlled flow of liquid is permitted. Partial needle retraction following insertion has also been demonstrated to enhance fluid infusion, likely due to relaxation of the compressed tissue around the tips [48]. Despite these potential issues in administration, hollow MNs have been demonstrated to be able to deliver a variety of substances, from small molecule drugs to larger proteins and peptides, both *in vitro* and *in vivo*.

Sivamani *et al.* [49] studied injection of the vasodilator, methyl nicotinate, in 11 healthy human subjects, using an array of either pointed or symmetric hollow MNs (4×4), made from silicon (200 μm long and 40 μm lumen diameter) and attached to a syringe. It was observed that MN application provoked maximum blood flow significantly faster than topical application, with the pointed MN design the superior of the two. In addition, the same group investigated the effect of the hollow silicon MN arrays on delivery of hexyl nicotinate to five human subjects [50]. Hexyl nicotinate, a lipophilic vasodilator, which is used as a marker of *stratum corneum* penetration, was either applied topically or injected using MNs at tape-stripped and un-stripped sites of the forearms. It was demonstrated that MNs were capable of injecting the drug beyond the *stratum corneum*, as confirmed by measuring the blood flow using laser Doppler imaging. The tape-stripping did not benefit MN-mediated delivery of hexyl nicotinate, indicating that MNs bypass the *stratum corneum* barrier, and hence its prior removal, plus MN use is of no real benefit. Importantly, these relatively straightforward studies indicate that hollow MNs can be successfully used to delivery small hydrophilic and lipophilic drugs *in vivo*.

Several studies have employed insulin as a model compound for delivery through hollow MNs. Insulin is typically administered as a subcutaneous injection several times per day or, alternatively, can be given as a continuous subcutaneous infusion. Such systems consist of an insulin pump attached via tubing to a subcutaneous needle and can be used to mimic the normal basal insulin secretion, with bolus injections infused at mealtimes. Hollow MN systems potentially offer a non-invasive means of administering such infusions. McAllister *et al.* [9] reported injection of up to 32 μl of insulin into diabetic rats over 30 min at a pressure of 10 psi, using a single glass MN, producing a drop in blood glucose levels over a 5-h period. Similar results were obtained by Davis *et al.* [51] using an array of 16 hollow metal MNs of height 600 μm, paired with an insulin-filled chamber placed on top of the array to serve as a drug reservoir. It was reported that the passive diffusion driven insulin delivery resulted in the reduction of blood glucose levels over 4 h by 53% and remained constant during the 4 h post-delivery period. Continuing from these preliminary studies, the delivery of insulin via hollow MNs has been extrapolated to humans. Gupta *et al.* [52] achieved rapid insulin absorption and reduction in glucose levels by insulin infusion using a single glass MN (bevel angle of 30° and effective tip radii of between 60 and 80 mm) inserted 1 mm into the skin by a rotary drilling device. Further to this, a faster onset and offset, as well as reduced pain, were demonstrated in a group of 16 children and adolescents when insulin was delivered in this manner, compared with a subcutaneous pump catheter [53].

Hollow MNs have also been used for local treatment, with ocular delivery an area of intense interest, as discussed in more detail in Chapter 10. Another example of local delivery is that of phenylephrine delivery into the anal sphincter muscle, as a method for treating faecal incontinence [54]. In this case, a 1.5 mm hollow MN, adapted from a commercially available stainless-steel needle, was used along with a micro syringe pump to administer phenylephrine solution to the perianal region of rats, at a rate of 1 μl/s. Compared with subcutaneous, intravenous and intramuscular alternatives, the MN approach significantly increased contraction of the anal sphincter muscle over a 12-h period, with no corresponding increase in blood pressure, highlighting the potential of this treatment after further investigation.

Challenging previous reports which suggested that the volume for infusion via hollow MNs was in the microlitre range [55], 3M's hollow Microstructured Transdermal System (hMTS) is designed for infusion of volumes up to 2 ml. This fully integrated device is designed for patient self-administration and consists of twelve 1500 μm polymeric MNs, together with a conventional glass cartridge, which holds the drug solution. Demonstrated *in vivo* in a swine model, delivery of a small molecule salt, a protein and a model monoclonal antibody have all been achieved using the system, with pharmacokinetic profiles and relative bioavailabilities similar to those observed following subcutaneous injection of the same formulations [56]. This approach has attracted interest from a number of companies and clinical trials are currently underway using 3M's hMTS for delivery of two novel biotherapeutics. Radius Health Inc. are pursuing the development of a transdermal system for the delivery of abaloparatide, a parathyroid hormone-related protein analogue drug, for the treatment of post-menopausal women with osteoporosis [57]. Simultaneously, 3M and Panacea Pharmaceuticals are collaborating on an investigational therapeutic cancer vaccine, the aim of which is to deliver 1 ml of the vaccine directly to the dermis via the hMTS [58]. In addition to 3M's system, there are a number of other hollow MN licenced devices available for pairing with suitable drug formulations, the majority of which are being exploited for vaccine delivery (e.g. BD Soluvia™, MicronJet600™, Debioject™) [59–61]. In essence however, these devices reflect the nature of miniature hypodermic needles rather than the transdermal patch model of MNs typically described.

3.2.4 Dissolving Microneedles

In recent years there has been a considerable push towards developing polymeric MN arrays, using biocompatible FDA-approved materials. As a transdermal drug delivery strategy, polymeric MNs that dissolve upon insertion into the skin offer a number of advantages, in terms of both the material and, also, patient use. By incorporating the drug into the actual matrix of the device, the release rate of the active can be tailored by modification of the polymeric components. Such polymeric materials are generally inexpensive and many can be processed at ambient temperatures, thus facilitating the formulation of heat-labile active pharmaceutical ingredients, such as peptides and proteins. Moreover, as the drug is held within the needles, a one-step drug delivery system is provided, ensuring a user-friendly platform. As the MNs dissolve completely upon application, the dosage form is effectively self-disabling, meaning that not only is the re-insertion of the needles into another patient impossible, but there is also no risk of needle-stick injury to any healthcare worker who may be involved in administration. The risk of infection transmission is, therefore, reduced. The question of disposal of MNs is also simplified, as with no sharp waste remaining, there is no need for safe disposal.

One of the earliest reports of polymeric MN manufacture details the use of PLGA for MN formulation, with the drug encapsulated directly within the MN matrix or within CMC or PLA microparticles, which were then encapsulated within the needle [62]. This study reported successful controlled delivery of calcein and BSA, highlighting the feasibility of controlled drug release ranging from hours to months using polymeric MNs. The authors did, however, note a slight loss in protein activity, due to the use of elevated temperatures in processing, exposing the need for a different method of manufacture. In another study, milder processing conditions were employed to prepare MNs from amylopectin and CMC with structural and functional integrity of a model protein maintained after encapsulation and release, even after 2 months storage at room temperature [63].

The water-soluble copolymer, poly(methyl vinyl ether-co-maleic acid) (PMVE/MA), has been used to prepare dissolving MN arrays, using a micromoulding technique. Extensive work on characterisation of the needles has been conducted, demonstrating their mechanical strength and skin insertion properties [64, 65]. Delivery of a variety of hydrophilic, low molecular weight drugs has been achieved using this particular polymeric composition, including caffeine, lidocaine, theophylline and metronidazole, with successful permeation enhancement observed in each instance [66, 67]. In the case of theophylline for example, the MN array delivered 83% of the total theophylline load across the skin *in vitro* after 24 h, with levels detected after only 5 min, in contrast to the control patch formulation, which delivered only 5.5% of its loading, with no drug detected until after 4 h [65]. The same polymeric MN formulation has also been employed in the delivery of biomolecules, with *in vivo* data for insulin administration in rats showing a dose-dependent hypoglycaemic effect [68]. This particular study highlighted the difficulties of macromolecule delivery from the entirety of the polymeric array, with delivery in this instance limited to the needle drug loading only. However, a later study by the same group showed that the combination of iontophoresis with a dissolving MN array could permit delivery of biomacromolecules from the entire polymeric matrix, including the baseplate, thereby further enhancing permeation [69].

Biopolymers, such as sodium hyaluronate, offer an attractive option for dissolving MN manufacture as they are generally found abundantly within the body and are often already widely used in cosmetic and pharmaceutical products. In one example, low molecular weight sodium hyaluronate was used to formulate dissolving MN arrays containing exenatide, a peptide used in treatment of Type 2 diabetes mellitus [70]. These arrays dissolved rapidly, releasing the drug within 2 min, with an *in vivo* effect equivalent to subcutaneous injection. Moreover, the functional integrity of the peptide was preserved during preparation and storage. Sodium hyaluronate MNs have also been investigated for delivery of a number of other compounds of varying molecular weight, including fluorescein isothiocyanate-labelled dextran [71, sumatriptan [17], artemether [72] and insulin [73]. Another natural polymer, sodium chondroitin, has also been used as a polymeric base for MN manufacture, with arrays containing leuprolide acetate [74] and erythropoietin [75].

A number of other biopharmaceutical molecules have been delivered via a dissolving MN platform, including low molecular weight heparins [76] and human growth hormone [77].

One potential challenge to manufacture of dissolving MN arrays is that the incorporation of a drug may adversely influence the physical properties of the polymer. For example, reduced mechanical robustness has been associated with high drug loading into a polymeric formulation for MN preparation [62, 63]. However, it has since been demonstrated that therapeutically relevant doses of a low potency drug, ibuprofen sodium, can be delivered *in vivo* from a polymeric array composed of approximately 50% w/w drug [78], suggesting that MN strength can be preserved depending on the formulation.

As the evidence base for this particular strategy has grown, formulation approaches have become more sophisticated. A novel partially dissolving polymer MN patch has been described, using a mixture of two polymers to provide sustained release of Rhodamine 6G and fluorescein isothiocyanate-labelled BSA [79]. By modifying the ratio of poly(vinylpyrrolidone) (PVP) and PVA, the investigators were able to control the drug release rate, exploiting the different polymer release mechanisms. Similarly, two molecular weights of PVP were used to prepare an insulin-loaded MN array, which demonstrated efficient release of its loading in mice, with the resulting blood glucose profile comparable to hypodermic injection [80].

In another complex example, pH-responsive PLGA hollow microspheres were used to deliver two model drugs from a PVP MN matrix [81]. One model drug was directly incorporated into the PVP needles, along with hollow microspheres encapsulating the second model drug. Sequential release was achieved upon needle dissolution, causing rapid release of the first drug and the hollow microspheres, which then responded to the environmental pH and released the second compound. In a somewhat similar approach, hydrogel microparticles were incorporated into biodegradable PLGA MNs, in an attempt to achieve sustained delivery of both hydrophobic and hydrophilic drugs [82]. Upon insertion of the MN array into the skin, the encapsulated hydrogel particles rapidly absorb water and expand, causing cracking of the MN structures, due to the difference in volume expansion between the needle matrix polymer and the hydrogel particles. Total breakdown of the MNs then occurs, depositing the tips and loaded drugs into the skin. This was demonstrated using calcein as a model hydrophilic drug, trapped inside the hydrogel particles, and Rhodamine as a model hydrophobic compound, contained within the PLGA needle matrix. Such a system can offer sustained drug release for a predetermined period as the PLGA needle tips embedded within the skin slowly degrade by hydrolysis over a time period of a few months.

Given that the majority of polymers are water soluble, dissolving MNs are particularly amenable to drugs with more hydrophilic character. When it comes to lipophilic drugs, solvents often have to be employed for solubilisation purposes, which can result into non-homogeneity and subsequent aggregation due to the hydrophilic nature of polymers. A recent study addressed this issue, creating a dissolving MN system capable of delivering powdered lipophilic drugs [83]. The creation of a transdermal diffusible nano-sized amorphous form of two hydrophobic compounds is described and its subsequent incorporation into a polymeric array. Evidence of therapeutic efficacy in rheumatoid arthritis was presented for the resulting MN arrays containing capsaicin.

This method, however, similar to other dissolving MN delivery strategies, may lead to the deposition of polymer in the skin. Likely of little concern for occasional use such as vaccination, it would be preferable to minimise polymer delivery into the skin when considering long-term MN use [84].

3.2.5 Hydrogel-forming Microneedles

Hydrogel-forming MNs are a relatively new concept, described by Donnelly *et al.* in 2012 [85]. These first hydrogel-forming MNs were manufactured from aqueous blends of specific polymeric materials, namely PMVE/MA, cross-linked with poly(ethylene glycol) (PEG) [85]. Upon heating, these two polymers undergo an esterification reaction to produce a cross-linked material, confirmed using attenuated total reflectance (ATR)-Fourier transform infra-red (FTIR) spectroscopy and the observation of a carbonyl peak shift from 1708 to 1731 cm^{-1}, due to formation of an ester carbonyl [86, 87].

Developed in response to the challenges associated with other MN types, hydrogel-forming MNs can be can be easily sterilised, they resist hole closure while in place, thereby allowing controlled release of the active, and are removed completely intact from the skin following use. In addition, as the drug is prepared in a separate reservoir, the loading capacity is not linked to the size of the needles, removing any limitations on dosing. Since the first report in 2012, there have been other manifestations of hydrogel MNs emerging, where the drug has not been housed separately from the needles [88, 89]. Regardless of the initial location of the drug, however, the rate of drug diffusion into the lower skin layers is controlled by the degree of swelling of the cross-linked polymer, which forms the hydrogel matrix. Altering the polymer cross-link density can, therefore, control the rate of MN swelling, thus conferring the ability to govern drug release rate, which can be tailored for specific drugs. As with other MNs, hydrogel-forming MNs are painless and blood-free upon application, with the additional benefit of being removed intact from the skin following use, thereby depositing no measurable residual polymer. Importantly, however, the MNs are suitably softened by the interstitial fluid, preventing reinsertion of the array, thereby reducing the risk of infection transmission that may arise from needle reuse.

Hydrogel-forming MNs been shown to successfully deliver a range of molecules of varying molecular weight *in vivo*, from small, hydrophilic molecules, such as metronidazole and caffeine, to larger molecular weight peptide and protein molecules, such as insulin and fluorescein isothiocyanate labelled-BSA [85]. The combination of hydrogel-forming MNs with iontophoresis, with the aim of achieving pulsatile or bolus delivery, was observed to provoke a marked increase in the rate and extent of in-skin swelling of the arrays. In general, this led to a greater rate and overall extent of transdermal delivery for each of the molecules; however, the difference was only significant for the biomolecules under investigation.

Hydrogel-forming MN arrays described as "super swelling" have also been tested for their drug delivery capabilities [90]. Differing from the originally described formulation only by the addition of sodium carbonate as a modifying agent, the resultant formulation displays an increased swelling capacity. For example, after 1 h, the percentage swelling of these "super swelling hydrogels" (20% w/w acid form of PMVE/MA, 7.5% w/w PEG 10 000 and 3% w/w Na_2CO_3) was 1119%, compared with only 250% for the original formulation (15% w/w PMVE/MA, 7.5% w/w PEG 10 000). When paired with a lyophilised drug reservoir, delivery of a clinically relevant dose of a low potency, high dose drug substance (ibuprofen) was achieved, as well as rapid delivery of a model protein (ovalbumin). Hardy *et al.* [88] also investigated ibuprofen delivery, using a light-responsive 3,5-dimethoxybenzoin conjugate of the drug incorporated in a cross-linked 2-hydroxyethyl methacrylate (pHEMA) formulation, in order to test light-triggered transdermal drug delivery. *In vitro*, this hydrogel array was able to deliver up to three doses of 50 mg of ibuprofen upon application of an optical trigger, over a prolonged period of time (up to 160 h). This type of system offers great potential as a stimulus responsive delivery platform, where "on-demand" drug delivery is required, with patient- or physician-controlled analgesia an obvious example.

Yang *et al.* [89] described a polymeric MN array prepared from poly(vinyl alcohol) (PVA), dextran and CMC, which had the drug incorporated directly into the hydrogel matrix of the needle tip. Terming the system a "phase-transition MN patch," they loaded insulin into the tips of the polymeric needles, cross-linked with microcrystalline domains formed by a freeze–thaw process. Uptake of interstitial fluid by the insoluble but water-swellable needles ensured efficient transepidermal delivery of insulin, with sufficient mechanical strength retained by the hydrated state for complete withdrawal from the skin, without deposition of excess materials. Initially tested *in vitro*, this patch demonstrated rapid release of the insulin over a 6 h time period. In a diabetic pig model, application of a patch containing insulin at a dose of 2.0 IU/kg was able to achieve comparable blood insulin area under curve (AUC) to a 0.4 IU/kg injection, indicating the relative availability of insulin delivered by the MNs to be around 20% of the total insulin contained within. In fact, over a 2-month period, the PVA, dextran and CMC MN array showed significantly better control of long-term blood glucose over the subcutaneous injection, as measured by the HbA1c level, likely due to the post-peak sustained release of insulin from the array. This offers the opportunity for non-invasive insulin delivery, with future work hoping to load insulin in selected regions along the MN shafts to accommodate the different insulin-dosing regimens. As the drug is included inside the hydrogel-forming MN patch rather than in an external reservoir, it must be pointed out that the quantity of drug that can be loaded and, therefore, delivered is potentially limited. The use of a separate drug reservoir, paired with a drug-free hydrogel-forming array, provides flexibility in the formulation of the attached drug reservoir, which inherently confers greater loading capacity than an array containing the drug within itself.

In a different approach to those already discussed, Sivaraman and Banga [91] reported the use of a poloxamer-based drug formulation to form hydrogel MNs *in situ*. The skin was first punctured using a tetrahedron-shaped solid array of 81 maltose microneedles, each of 500 μm in length. Subsequently a poloxamer-based drug formulation, with a sol–gel transition temperature of 32 °C, was applied to the treated site. The solution, which flows at room temperature, filled the microcavities created by the maltose needles, before transitioning into a gel at skin temperature, thereby attaining a needle shape *in situ*. As the poloxamer formulation is then embedded within the skin, the encapsulated drug can be released over a sustained period of time. The authors tested their delivery theory *in vitro* using methotrexate, achieving a slower but more prolonged release profile, than when compared with a non-poloxamer formulation. Although the authors point out that the polymer of choice is biocompatible and water-soluble and, therefore, does not present a biohazardous risk, this would still be the first use of such a formulation intradermally, from where its removal would be uncharacterised [92].

3.3 Conclusion

Transdermal drug delivery offers a number of advantages over oral delivery: avoidance of first pass metabolism, zero-order delivery, reduced adverse effects and reduced dosing to name a few. Since the first transdermal patches were licenced in the early 1980s, only a relatively small number of patch-based systems have made their way to the market. In part, this is due to the relatively complicated processes required for their manufacture and that few pharmaceutical companies possess this expertise. However, the main stumbling block is that many promising therapeutic agents are limited by their inability to reach the systemic circulation due to the excellent barrier properties of the *stratum corneum*. It follows that, over recent years, there has been considerable interest in developing enhancement strategies in order to expand the range of drugs able to be delivered transdermally. There is, therefore, an increasing body of evidence describing the drug permeation enhancing effects of MNs. Studies vary greatly in terms of drug model, skin model, formulation and application technique. Moreover, the types of MN employed can vary in terms of their design, material, geometry and needle density. Consequently, comparing the findings of different studies can prove difficult. What is clear, however, is that MNs display huge potential in the field of transdermal drug delivery, with notable evidence for their permeation enhancing effects.

References

1 Jenkins, K. (2014). Needle phobia: A psychological perspective. *Br. J. Anaesth.* 113: 4–6.

2 Morrow, D.I.J., McCarron, P.A., Woolfson, A.D. and Donnelly, R.F. (2007). Innovative strategies for enhancing topical and transdermal drug delivery. *Open Drug Delivery J.* 1: 36–59.

3 Margetts, L. and Sawyer, R. (2007). Transdermal drug delivery: principles and opioid therapy. *Contin. Educ. Anaesthesia, Crit. Care Pain* 7: 171–176.

4 Rolland, P., Bolzinger, M.A., Cruz, C., *et al.* (2013). *Toxicol. In Vitro* 27 (1): 358–66.

5 Barry, B.W. (2001). *Eur. J. Pharm. Sci.* 14 (2): 101–114.

6 Wiedersberg, S. and Guy, R. (2014). Transdermal drug delivery: 30+ years of war and still fighting! *J. Control. Release* 190: 150–156.

7 Gerstel, M.S. and Place, V.A. (1976). Drug delivery device. US Patent 3964482 A, filed 17 May 1971 and issued 22 June 1976. https://www.google.com/patents/US3964482.

8 Henry, S., McAllister, D.V., Allen, M.G. and Prausnitz, M.R. (1998). Microfabricated microneedles: A novel approach to transdermal drug delivery. *J. Pharm. Sci.* 87: 922–925.

9 McAllister, D.V., Wang, P.M., Davis, S.P., *et al.* (2003). Microfabricated needles for transdermal delivery of macromolecules and nanoparticles: fabrication methods and transport studies. *Proc. Natl. Acad. Sci. U. S. A.* 100: 13755–13760.

10 Martanto, W., Davis, S.P., Holiday, N.R., *et al.* (2004). Transdermal delivery of insulin using microneedles in vivo. *Pharm. Res.* 21: 947–952.

11 Verbaan, F.J., Bal, S.M., van den Berg, D.J., *et al.* (2007). Assembled microneedle arrays enhance the transport of compounds varying over a large range of molecular weight across human dermatomed skin. *J. Control. Release* 117: 238–245.

12 Mohammed, Y.H., Yamada, M., Lin, L.L., *et al.* (2014). Microneedle enhanced delivery of cosmeceutically relevant peptides in human skin. *PLoS One* 9: e101956.

13 Badran, M.M., Kuntsche, J. and Fahr, A. (2009). Skin penetration enhancement by a microneedle device (Dermaroller) *in vitro*: dependency on needle size and applied formulation. *Eur. J. Pharm. Sci.* 36: 511–523.

14 Stahl, J., Wohlert, M. and Kietzmann, M. (2012). Microneedle pretreatment enhances the percutaneous permeation of hydrophilic compounds with high melting points. *BMC Pharmacol. Toxicol.* 13: 5.

15 Zhou, C.-P., Liu, Y.-L., Wang, H.-L., *et al.* (2010). Transdermal delivery of insulin using microneedle rollers in vivo. *Int. J. Pharm.* 392: 127–133.

16 Patel, D.R., Joshi, A., Patel, H.H. and Stagni, G. (2015). Development and *in-vivo* evaluation of ondansetron gels for transdermal delivery. *Drug Dev. Ind. Pharm.* 41: 1030–1036.

17 Nalluri, B.N., Anusha, S.S. V, Bramhini, S.R., *et al.* (2015). *In vitro* skin permeation enhancement of sumatriptan by microneedle application. *Curr. Drug Deliv.* 12: 761–769.

18 Kaur, M., Ita, K.B., Popova, I.E., et al. (2013). Microneedle-assisted delivery of verapamil hydrochloride and amlodipine besylate. *Eur. J. Pharm. Biopharm.* 86: 284–291.

19 Banks, S.L., Pinninti, R.R., Gill, H.S., *et al.* (2008). Flux across microneedle-treated skin is increased by increasing charge of naltrexone and naltrexol *in vitro*. *Pharm. Res.* 25: 1677–1685.

20 Yan, G., Warner, K.S., Zhang, J., *et al.* (2010). Evaluation needle length and density of microneedle arrays in the pretreatment of skin for transdermal drug delivery. *Int. J. Pharm.* 391: 7–12.

21 Cheung, K., Han, T. and Das, D.B. (2014). Effect of force of microneedle insertion on the permeability of insulin in skin. *J. Diabetes Sci. Technol.* 8: 444–452.

22 Wermeling, D.P., Banks, S.L., Hudson, D.A. *et al.* (2008). Microneedles permit transdermal delivery of a skin-impermeant medication to humans. *Proc. Natl. Acad. Sci. U. S. A.* 105: 2058–2063.

23 Millard, D.R. and Maisels, D.O. (1966). Silicon granuloma of the skin and subcutaneous tissues. *Am. J. Surg.* 112: 119–123.

24 Park, J., Allen, M.G. and Prausnitz, M.R. (2005). Biodegradable polymer microneedles : Fabrication, mechanics and transdermal drug delivery. *J. Control. Release* 104: 51–66.

25 Oh, J., Park, H., Do, K., *et al.* (2008). Influence of the delivery systems using a microneedle array on the permeation of a hydrophilic molecule, calcein. *Eur. J. Pharm. Biopharm.* 69: 1040–1045.

26 Luangveera, W., Jiruedee, S., Mama, W., and Master, A. (2015). Fabrication and characterization of novel microneedles made of a polystyrene solution. *J. Mech. Behav. Biomed. Mater.* 50: 77–81.
27 Yu Li, Q., Nan Zhang, J., Zhi Chen, B., *et al.* (2017). A solid polymer microneedle patch pretreatment enhances the permeation of drug molecules into the skin. *RSC Adv.* 7: 15408–15415.
28 Osterberg, L. and Blaschke, T. (2005). Adherence to medication. *N. Engl. J. Med.* 353: 487–497.
29 Gill, H.S. and Prausnitz, M.R. (2007). Coating formulations for microneedles. *Pharm. Res.* 24: 1369–1380.
30 Davidson, A., Al-Qallaf, B. and Das, D.B. (2008). Transdermal drug delivery by coated microneedles: Geometry effects on effective skin thickness and drug permeability. *Chem. Eng. Res. Des.* 86: 1196–1206.
31 Tas, C., Mansoor, S., Kalluri, H., *et al.* (2012). Delivery of salmon calcitonin using a microneedle patch. *Int. J. Pharm.* 423: 257–263.
32 Cormier, M., Johnson, B., Ameri, M., *et al.* (2004). Transdermal delivery of desmopressin using a coated microneedle array patch system. *J. Control. Release* 97: 503–511.
33 Peters, E.E., Ameri, M., Wang, X., *et al.* (2012). Erythropoietin-coated ZP-microneedle transdermal system: Preclinical formulation, stability, and delivery. *Pharm. Res.* 29: 1618–1626.
34 Ameri, M., Kadkhodayan, M., Nguyen, J., *et al.* (2014). Human growth hormone delivery with a microneedle transdermal system: Preclinical formulation, stability, delivery and PK of therapeutically relevant doses. *Pharmaceutics* 6: 220–234.
35 Kusamori, K., Katsumi, H., Sakai, R., *et al.* (2016). Development of a drug-coated microneedle array and its application for transdermal delivery of interferon alpha. *Biofabrication* 8: 15006.
36 Zhao, X., Coulman, S.A., Hanna, S.J., *et al.* (2017). Formulation of hydrophobic peptides for skin delivery via coated microneedles. *J. Control. Release* 10 (265): 2–13.
37 Lee, H.S., Ryu, H.R., Roh, J.Y. and Park, J.-H. (2016). Bleomycin-coated microneedles for treatment of warts. *Pharm. Res.* 34: 101–112.
38 Zhang, Y., Brown, K., Siebenaler, K., *et al.* (2012). Development of lidocaine-coated microneedle product for rapid, safe, and prolonged local analgesic action. *Pharm. Res.* 29: 170–177.
39 Daddona, P.E., Matriano, J.A., Mandema, J. and Maa, Y.-F. (2011). Parathyroid hormone (1-34)-coated microneedle patch system: clinical pharmacokinetics and pharmacodynamics for treatment of osteoporosis. *Pharm. Res.* 28: 159–165.
40 Ameri, M., Daddona, P.E. and Maa, Y.-F. (2009). Demonstrated solid-state stability of parathyroid hormone PTH(1–34) coated on a novel transdermal microprojection delivery system. *Pharm. Res.* 26: 2454–2463.
41 Zosano Pharma (2015). Zosano Pharma announces positive Phase 2 results for its ZP-glucagon patch program for treatment of severe hypoglycemia. http://ir.zosanopharma.com/releasedetail.cfm?ReleaseID=936338 (accessed 7 March 2017).
42 Zosano Pharma (2017). Zosano Pharma announces 3.8mg dose of M207, its novel transdermal therapeutic, meets both co-primary endpoints in the ZOTRIP

pivotal efficacy trial in migraine. http://ir.zosanopharma.com/releasedetail.cfm?releaseid=1011563 (accessed 6. February 2017).

43 Davis, S.P., Landis, B.J., Adams, Z.H., *et al.* (2004). Insertion of microneedles into skin: measurement and prediction of insertion force and needle fracture force. *J. Biomech.* 37: 1155–1163.

44 Wang, P.-C., Paik, S.-J., Kim, S.-H. and Allen, M.G. (2014). Hypodermic-needle-like hollow polymer microneedle array: fabrication and characterization. *J. Microelectromech. Syst.* 23: 991–998.

45 Pérennès, F., Marmiroli, B., Matteucci, M., *et al.* 2006. Sharp beveled tip hollow microneedle arrays fabricated by LIGA and 3D soft lithography with polyvinyl alcohol. *J. Micromech. Microeng.* 16: 473–479.

46 Martanto, W., Moore, J., Kashlan, O., *et al.* (2006). Microinfusion using hollow microneedles. *Pharm. Res.* 23: 104–113.

47 Griss, P. and Stemme, G. (2003). Side-opened out-of-plane microneedles for microfluidic transdermal liquid transfer. *J. Microelectromech. Syst.* 12: 296–301.

48 Wang, P.M., Cornwell, M., Hill, J. and Prausnitz, M.R. (2006). Precise microinjection into skin using hollow microneedles. *J. Invest. Dermatol.* 126: 1080–1087.

49 Sivamani, R.K., Stoeber, B., Wu, G.C., *et al.* (2005). Clinical microneedle injection of methyl nicotinate: Stratum corneum penetration. *Skin Res. Technol.* 11: 152–156.

50 Sivamani, R.K., Stoeber, B., Liepmann, D. and Maibach, H.I. (2009). Microneedle penetration and injection past the stratum corneum in humans. *J. Dermatolog. Treat.* 20: 156–159.

51 Davis, S.P., Martanto, W., Allen, M.G. and Prausnitz, M.R. (2005). Hollow metal microneedles for insulin delivery to diabetic rats. *IEEE Trans. Biomed. Eng.* 52: 909–915.

52 Gupta, J., Felner, E.I. and Prausnitz, M.R. (2009). Minimally invasive insulin delivery in subjects with type 1 diabetes using hollow microneedles. *Diabetes Technol. Ther.* 11: 329–337.

53 Norman, J.J., Brown, M.R., Raviele, N.A., *et al.* (2013). Faster pharmacokinetics and increased patient acceptance of intradermal insulin delivery using a single hollow microneedle in children and adolescents with type 1 diabetes. *Pediatr. Diabetes* 14: 459–465.

54 Jun, H., Han, M.R., Kang, N.G., *et al.* (2015). Use of hollow microneedles for targeted delivery of phenylephrine to treat fecal incontinence. *J. Control. Release* 207: 1–6.

55 Gardeniers, H.J.G.E., Luttge, R., Berenschot, E.J.W., *et al.* (2003). Silicon micromachined hollow microneedles for transdermal liquid transport. *J. Microelectromech. Syst.* 12: 855–862.

56 Burton, S.A., Ng, C.-Y., Simmers, R., *et al.* (2011). Rapid intradermal delivery of liquid formulations using a hollow microstructured array. *Pharm. Res.* 28: 31–40.

57 Radius Health Inc. (2017). Pipeline. http://radiuspharm.com/pipeline/ (accessed 7 March 2017).

58 Panacea Pharmaceuticals (2017). 3M Drug Delivery Systems announces collaboration with Panacea Pharmaceuticals, Inc. on new cancer vaccine. http://panaceapharma.com/news/2017/2/21/3m-drug-delivery-systems-announces-collaboration-with-panacea-pharmaceuticals-inc-on-new-cancer-vaccine (accessed 7 March 2017).

59 Frenck, R.W., Belshe, R., Brady, R.C., *et al.* (2011). Comparison of the immunogenicity and safety of a split-virion, inactivated, trivalent influenza vaccine (Fluzone®) administered by intradermal and intramuscular route in healthy adults. *Vaccine* 29: 5666–5674.

60 Levin, Y., Kochba, E., Hung, I. and Kenney, R. (2015). Intradermal vaccination using the novel microneedle device MicronJet600: Past, present, and future. *Hum. Vaccin. Immunother.* 11: 991–997.

61 Vescovo, P., Rettby, N., Ramaniraka, N., *et al.* (2017). Safety, tolerability and efficacy of intradermal rabies immunization with DebioJect. *Vaccine* 35: 1782–1788.

62 Park, J.-H., Allen, M.G. and Prausnitz, M.R., 2006. Polymer microneedles for controlled-release drug delivery. *Pharm. Res.* 23: 1008–1019.

63 Lee, J.W., Park, J.H. and Prausnitz, M.R. (2008). Dissolving microneedles for transdermal drug delivery. *Biomaterials* 29: 2113–2124.

64 Donnelly, R.F., Garland, M.J., Morrow, D.I.J., *et al.* (2010). Optical coherence tomography is a valuable tool in the study of the effects of microneedle geometry on skin penetration characteristics and in-skin dissolution. *J. Control. Release* 147: 333–341.

65 Donnelly, R.F., Majithiya, R., Thakur, R.R.S., et al. (2011). Design, optimization and characterisation of polymeric microneedle arrays prepared by a novel laser-based micromoulding technique. *Pharm. Res.* 28: 41–57.

66 Caffarel-Salvador, E., Tuan-Mahmood, T.-M., McElnay, J.C., *et al.* (2015). Potential of hydrogel-forming and dissolving microneedles for use in paediatric populations. *Int. J. Pharm.* 489: 158–169.

67 Garland, M.J., Migalska, K., Tuan-Mahmood, T.-M., *et al.* (2012). Influence of skin model on *in vitro* performance of drug-loaded soluble microneedle arrays. *Int. J. Pharm.* 434: 80–89.

68 Migalska, K., Morrow, D.I.J., Garland, M.J., *et al.* (2011). Laser-engineered dissolving microneedle arrays for transdermal macromolecular drug delivery. *Pharm. Res.* 28: 1919–1930.

69 Garland, M.J., Caffarel-Salvador, E., Migalska, K., *et al.* (2012). Dissolving polymeric microneedle arrays for electrically assisted transdermal drug delivery. *J. Control. Release* 159: 52–59.

70 Zhu, Z., Luo, H., Lu, W., *et al.* (2014). Rapidly dissolvable microneedle patches for transdermal delivery of exenatide. *Pharm. Res.* 31: 3348–3360.

71 Liu, S., Jin, M.N., Quan, Y.S., *et al.* (2014). Transdermal delivery of relatively high molecular weight drugs using novel self-dissolving microneedle arrays fabricated from hyaluronic acid and their characteristics and safety after application to the skin. *Eur. J. Pharm. Biopharm.* 86: 267–276.

72 Qiu, Y., Li, C., Zhang, S., *et al.* (2016). Systemic delivery of artemether by dissolving microneedles. *Int. J. Pharm.* 508: 1–9.

73 Liu, S., Jin, M., Quan, Y., *et al.* (2012). The development and characteristics of novel microneedle arrays fabricated from hyaluronic acid, and their application in the transdermal delivery of insulin. *J. Control. Release* 161: 933–941.

74 Ito, Y., Murano, H., Hamasaki, N., *et al.* (2011). Incidence of low bioavailability of leuprolide acetate after percutaneous administration to rats by dissolving microneedles. *Int. J. Pharm.* 407: 126–131.

75 Ito, Y., Hasegawa, R., Fukushima, K., et al. (2010). Self-dissolving micropile array chip as percutaneous delivery system of protein drug. *Biol. Pharm. Bull.* 33: 683–690.

76 Gomaa, Y.A., Garland, M.J., McInnes, F., *et al.* (2012). Laser engineered dissolving microneedles for active transdermal delivery of nadroparin calcium. *Eur. J. Pharm. Biopharm.* 82: 299–307.

77 Lee, J.W., Choi, S.-O., Felner, E.I. and Prausnitz, M.R. (2011). Dissolving microneedle patch for transdermal delivery of human growth hormone. *Small* 7: 531–539.

78 McCrudden, M.T.C., Alkilani, A.Z., McCrudden, C.M., *et al.* (2014). Design and physicochemical characterisation of novel dissolving polymeric microneedle arrays for transdermal delivery of high dose, low molecular weight drugs. *J. Control. Release* 180: 71–80.

79 Lee, I.-C., He, J.-S., Tsai, M.-T. and Lin, K.-C. (2014). Fabrication of a novel partially dissolving polymer microneedle patch for transdermal drug delivery. *J. Mater. Chem. B* 3: 276–285.

80 Lee, I.-C., Wu, Y.-C., Tsai, S.-W., *et al.* (2017). Fabrication of two-layer dissolving polyvinylpyrrolidone microneedles with different molecular weights for *in vivo* insulin transdermal delivery. *RSC Adv.* 7: 5067–5075.

81 Ke, C.-J., Lin, Y.-J., Hu, Y.-C., *et al.* (2012). Multidrug release based on microneedle arrays filled with pH-responsive PLGA hollow microspheres. *Biomaterials* 33: 5156–5165.

82 Kim, M., Jung, B. and Park, J.H. (2012). Hydrogel swelling as a trigger to release biodegradable polymer microneedles in skin. *Biomaterials* 33: 668–678.

83 Dangol, M., Yang, H., Li, C.G., *et al.* (2016). Innovative polymeric system (IPS) for solvent-free lipophilic drug transdermal delivery via dissolving microneedles. *J. Control. Release* 223: 118–125.

84 Donnelly, R.F. and Woolfson, A.D. (2014). Patient safety and beyond: what should we expect from microneedle arrays in the transdermal delivery arena? *Ther. Deliv.* 5: 653–662.

85 Donnelly, R.F., Thakur, R.R.S., Garland, M.J., *et al.* (2012). Hydrogel-forming microneedle arrays for enhanced transdermal drug delivery. *Adv. Funct. Mater.* 22: 4879–4890.

86 Luppi, B., Cerchiara, T., Bigucci, F., *et al.* (2003). Crosslinked poly(methyl vinyl ether-co-maleic anhydride) as topical vehicles for hydrophilic and lipophilic drugs. *Drug Deliv.* 10: 239–244.

87 Thakur, R.R.S., McCarron, P.A., Woolfson, A.D. and Donnelly, R.F. (2009). Investigation of swelling and network parameters of poly(ethylene glycol)-crosslinked poly(methyl vinyl ether-co-maleic acid) hydrogels. *Eur. Polym. J.* 45: 1239–1249.

88 Hardy, J.G., Larrañeta, E., Donnelly, R.F., *et al.* (2016). Hydrogel-forming microneedle arrays made from light-responsive materials for on-demand transdermal drug delivery. *Mol. Pharm.* 13: 907–914.

89 Yang, S., Feng, Y., Zhang, L., Chen, N., Yuan, W., Jin, T. (2012). A scalable fabrication process of polymer microneedles. *Int. J. Nanomed.* 7: 1415–1422.

90 Donnelly, R.F., McCrudden, M.T.C., Zaid Alkilani, A., *et al.* (2014). Hydrogel-forming microneedles prepared from "super swelling" polymers combined with lyophilised wafers for transdermal drug delivery. *PLoS One* 9 e111547.
91 Sivaraman, A. and Banga, A.K. (2016). Novel *in situ* forming hydrogel microneedles for transdermal drug delivery. *Drug Deliv. Transl. Res.* 7: 16–26.
92 Dumortier, G., Grossiord, J.L., Agnely, F. and Chaumeil, J.C. (2006). A review of poloxamer 407 pharmaceutical and pharmacological characteristics. *Pharm. Res.* 23: 2709–2728.

4

Microneedle-mediated Vaccine Delivery

Maelíosa T.C. McCrudden, Aaron J. Courtenay and Ryan F. Donnelly

School of Pharmacy, Queen's University Belfast, Belfast BT9 7BL, UK

4.1 Introduction

The use of microneedle (MN) arrays for facilitated delivery of therapeutic substances to target cells residing in the skin layers is one of the most obvious applications of this innovative technology. For example, breaching the skin's *stratum corneum* barrier via MN arrays allows administration of vaccines into the skin where they will activate various populations of immune cells. In this chapter, the application of MN-mediated intradermal delivery of vaccines will be discussed in detail.

4.2 Vaccine Delivery

4.2.1 Vaccination

Vaccination is one of the most important, cost-effective and successful public health interventions available to healthcare systems worldwide in the prevention of infectious disease-related morbidity and mortality. Vaccination effectively prevents disease, can improve the morbidity of a population base and helps to reduce mortality rates. On a fundamental level, a vaccine is a biological preparation that improves immunity to a particular disease. Vaccines activate the immune response of the body by imitating infection by a specific disease, thus instigating the production of T-lymphocytes and antibodies but without the associated illness. The first vaccination was famously carried out by Edward Jenner in 1796 and resulted in the development of the smallpox vaccine [1]. It is now known that introduction of antigenic material into the body can improve a person's immunity against a specific disease, through production of antibodies or through cell-mediated responses [2]. Vaccines now encompass a range of different types including: live attenuated, inactivated, subunit, toxoid and even DNA vaccines, which will be discussed in considerable detail in Chapter 5. Many vaccines aim to mimic a disease-causing organism and ultimately stimulate the host's immune system. To achieve effective vaccination, the body must be able to recognise the antigen as foreign, inactivate or destroy the vaccine and ultimately remember the vaccine, such that any further provocation by the organism can be quickly and effectively combated.

Microneedles for Drug and Vaccine Delivery and Patient Monitoring, First Edition.
Edited by Ryan F. Donnelly, Thakur Raghu Raj Singh, Eneko Larrañeta, and Maelíosa T.C. McCrudden.

Vaccine composition is unique in each case and has a significant impact on the way these vaccines are administered and, ultimately, how their effects are mediated [3].

An ideal vaccine is safe, cost-effective and efficient after a single dose. The route by which a vaccine is delivered can have considerable bearing on these factors through its influence on the efficiency of the procedure, the dose required, compliance and safety. For vaccination to succeed holistically in contributing to public health, vaccine delivery systems must allow efficient delivery without compromising product stability during storage and transport and without negatively influencing patient perception [4]. To be considered safe, new delivery systems should reduce the risk of injury and infection of healthcare workers, and prevent illicit reuse. A delivery system combining all these qualities would facilitate vaccination of greater proportions of the population [4–6]. Currently-licenced vaccines are delivered via one of five main administration routes: intramuscular for the majority of vaccines including hepatitis A and B, rabies, influenza and diphtheria–tetanus–pertussis based combination vaccines; subcutaneous for vaccines such as measles, mumps and rubella (MMR) and yellow fever; intradermal for bacillus Calmette-Guérin (BCG) and rabies; intranasal for live attenuated influenza vaccine; and oral for poliomyelitis, cholera, rotavirus and typhoid fever [4, 6, 7]. With the rare exception of jet injectors (devices that use a coiled spring or compressed gas to propel liquid or powder jets through the *stratum corneum* into the skin), intramuscular, subcutaneous and intradermal delivery are all achieved using conventional needles. In fact, of the more than 5 billion human vaccine doses given each year, 3 billion are delivered using hypodermic needles [4, 8].

These techniques, whilst having proven efficacy in terms of achieving the required immune responses, have significant drawbacks relating to safety and patient compliance. The invasive nature of the parenteral injection procedure and the potential for inappropriate reuse of equipment, particularly in developing countries, exposes patients to the risk of transmission of blood-borne pathogens. A study by the World Health Organization (WHO) estimated that at least 50% of childhood vaccinations were unsafe in 14 countries located in five different developing world regions [4, 8]. Moreover, and critically, the use and disposal of equipment is associated with the risk of needle-stick injury. An estimated three million healthcare workers worldwide are injured annually with a sharp object contaminated with the hepatitis C virus, hepatitis B virus or human immunodeficiency virus (HIV) [4, 8]. Despite these statistics however, vaccine delivery, to date, has relied heavily on the use of hypodermic needles and syringes in intramuscular, intradermal and subcutaneous injection usage. The introduction of alternative delivery devices, engineered to prevent needle reuse and reduce the risk of needle-stick injuries, will reduce injury incidence and the costs of specialist sharps disposal facilities. The perceived or real pain and trauma sometimes associated with needle-based vaccination can prove to be a substantial barrier to vaccination uptake however, particularly by needle-phobic patients, who constitute at least 10% of the population [7, 9]. As a result of such considerations, and in line with the development of new types of vaccines, the pharmaceutical industry and public health organisations are driving research and innovation in the search for new delivery methods that are safe, cost-effective and efficient. While the majority of vaccines in clinical development are envisioned as hypodermic needle and syringe products, a number of research groups and vaccine manufacturers are exploring the advantages of new parenteral delivery systems, in addition to mucosal and transcutaneous delivery systems [10].

Mucosal delivery is currently only used for live attenuated vaccines against poliomyelitis [4–6], typhoid fever (oral), rotavirus and influenza (nasal). Mucosally administered vaccines have a number of benefits. They eliminate the risk of transmission of blood-borne diseases and needle-stick injury. They can potentially be given by personnel with little medical training, which provides significant practical and cost benefits, particularly in the context of large-scale immunisation programmes in the developing world. This route can also, in theory, elicit both mucosal and humoral immunity, offering advantages against diseases contracted via mucosal surfaces. However, there are also a number of drawbacks. The live attenuated viruses in oral poliomyelitis vaccine (OPV) can revert to virulence, causing vaccine-associated paralytic poliomyelitis (VAPP) in the vaccinated child or their close contacts, particularly in immuno-suppressed patients. This has resulted in a shift from the use of OPV to the use of injectable poliomyelitis vaccine containing inactivated virus, especially in countries that have eliminated naturally occurring polio [6, 10]. Oral vaccines have to overcome problems associated with poor absorption or degradation within the digestive system that may require the concomitant administration of antacids.

With all of these considerations in mind, new vaccines under development should, ideally, have additional safety features built into their development profile. They should be needle-free or, if needle-based administration is unavoidable, the needle should be retractable or otherwise self-disabling to prevent reuse and reduce the risk of needle-stick injuries. Reducing or eliminating pain during vaccination is likely to improve uptake of vaccines by both needle-phobic individuals and parents concerned about the pain endured by their child during multiple-injection vaccination schedules [7, 9]. Administering vaccines without the use of a needle and syringe would have knock-on benefits in reducing healthcare training for vaccinators. This would be especially helpful should mass vaccination be required in the case of, for example, a natural pandemic or bioterrorism attack. Significant benefits could also be seen in the developing world, where there is often a shortage of trained healthcare workers. Increasing the speed of vaccine delivery, while not compromising safety, has obvious advantages. The production, transport, storage and administration of any newly developed vaccines should not be costlier than currently available counterparts and this is especially true considering the limited resources for vaccination in the developing world. A further aspect of vaccine development, and indeed storage, which must be given due consideration is the fact that the majority of vaccines currently in use worldwide must be maintained within a specific temperature range (often 2–8 °C) to maintain potency. Some vaccines, such as oral polio vaccine, MMR, varicella and yellow fever, are sensitive to heat, while others, such as diphtheria and tetanus toxoids, pertussis vaccine and hepatitis B vaccine, are sensitive to freezing [4–10]. The "cold chain" refers to the materials, equipment and procedures required to maintain vaccines within their optimum temperature range from the point and time of manufacture until administration into patients. Owing to the expense of maintaining cold storage facilities, cold transportation and cold boxes that preserve vaccines at the proper temperature, the cold chain is estimated to cost vaccine programmes worldwide US$200–300 million annually. Accordingly, any new vaccine strategies that avoid the necessity of the cold chain would subsequently further decrease the cost of vaccine delivery.

In considering alternative methods for vaccine delivery that circumvent the use of conventional needle and syringe delivery strategies, it is imperative that the vaccine still elicits a protective response in patients. Parenteral administration of vaccines by needle and syringe has proven successful in initiating and activating immune effectors, thus protecting those vaccinated from infection. Alternative methods of vaccine delivery must therefore prove to be at least equally immunogenic and protective. There is growing interest, and indeed evidence, to promote exploitation of the intradermal route for efficacious vaccine delivery and this research field continues to attract great interest by both industrial and academic researchers. Before exploring the status of intradermal delivery using MNs, it is worth considering other strategies or routes which have, or may yet be, explored in vaccine delivery.

4.2.2 Alternative Vaccine Delivery Options

Many vaccine preparations include adjuvants, which act to potentiate the immune-stimulatory property of the antigen whilst being non-immunogenic, nontoxic and biodegradable in isolation [2]. Vaccine delivery systems including emulsions, immune-stimulating complexes, microparticles and liposomes are often used to adjuvant vaccine antigens. Similarly, aluminium hydroxide and aluminium phosphate along with polynucleotides are examples of adjuvants used in commercially available vaccine products. Polymeric delivery systems allow for targeted delivery, with particle size playing a crucial role in antigen uptake by antigen presenting cells (APC). Repeated vaccination dosing, usually required with antigenic material, may be circumvented by the use of such microparticle technology. Larger particles, which cannot be phagocytosed by macrophages, begin to biodegrade yielding smaller fragments of polymer and antigen. This has been shown to provide pulsatile dosing, which could reduce the need for repeated vaccination, as the pulsatile release of antigen imitates the conventional prime and boost vaccination schedules [11].

Many disease-causing organisms reach the body by first entering through mucosal tissue. Appropriately, a range of significantly important immune cells reside within surface mucosal tissues [12]. Figure 4.1 shows the main mucosal surfaces throughout the human body, indicating many potential vaccination sites.

Certain vaccination sites are more effective in producing immune responses in distal parts of the body. For example, intranasal immunisation is currently the preferred route for targeting respiratory, gastric and genital tracts. Similarly, oral vaccination is most effective for inducing antibodies in the mammary glands (ultimately secreted in the milk) and in the gut. Rectal vaccination elicits significant induction of colonic and rectal immunity while intravaginal immunisation is potentially the most effective with respect to the production of antibody responses and T-lymphocyte immunity in the genital tract [13].

Vaccination at any body site is not without its challenges. For example, nasal delivery of vaccines results in short antigen exposure time, which may be a key factor in ineffective antibody response in a patient. Mucoadhesive adjuvants have been used to overcome the rapid clearance of vaccines from the nasal mucosa, thus increasing residence time within the nasal cavity and prolonging antigenic exposure. A further drawback of nasal delivery of vaccine agents is due to the proteolytic barrier present at this site, which

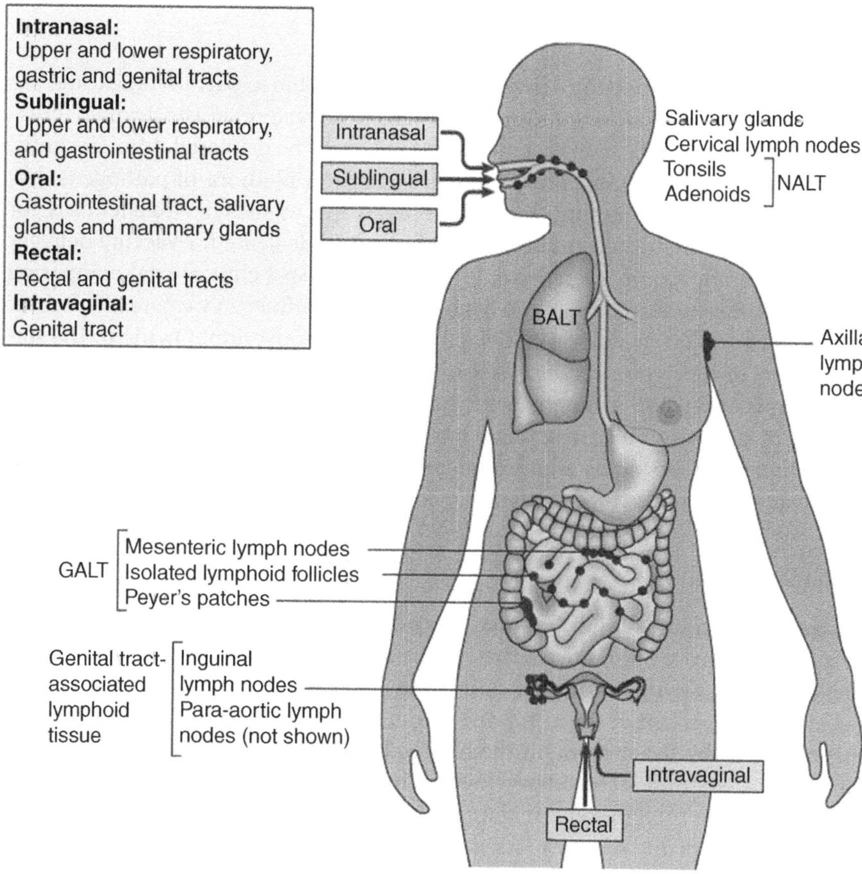

Figure 4.1 Schematic representation of established and potential mucosal immunisation routes within the human body. Reproduced with permission from [13].

can hamper vaccine delivery. A number of live attenuated flu vaccines have overcome these formulation issues and have had commercial success, namely: FluMist®, FluMist® Quadrivalent and Fluenz®. A number of other vaccines are currently under investigation for intranasal delivery, such as hepatitis B, *Mycobacterium tuberculosis, Bacillus anthacis*, HIV, *Streptococcus pneumoniae* and shigella [14].

Vaginal delivery of vaccines has shown significant promise with robust immune responses elicited at the site of pathogen entry, which could have subsequent ramifications on the spread of sexually transmitted diseases [15]. One limitation of intravaginal vaccination is the difficulty of ensuring prolonged exposure of the antigen to APCs due to physiological events such as mucosal shedding, high immunotolerance and mucus and epithelial changes according to the fluctuating hormone cycle [16]. Intravaginal delivery of DNA and small interfering ribonucleic acid (siRNA) to treat HIV is still at a very early developmental stage, however results have shown preliminary success *in vivo* in the reduction of HIV transmission [17]. Intradermal DNA vaccination strategies will be covered in significant detail in Chapter 5.

4.3 Intradermal Vaccination

The skin has been a target for drug delivery for many years due in part to its size and the associated ease of administration of topical medicines. It was previously deemed, however, to be a barrier through which vaccines could not easily be delivered. More recently, the skin, which provides the first line of defence against a plethora of pathogens, has been recognised as a highly immune-competent organ and, as such, an excellent site for vaccine administration. The acceptance of the skin as an ideal site for vaccine delivery was first supported by Combadière *et al.* [18]. In this Phase-I clinical trial comparing transdermal and intramuscular injection of an inactivated influenza vaccine, the former delivery method resulted in superior CD8 effector T-cell activation. Owing to the significant numbers of APCs present in the skin, delivery of vaccines to the skin may result in dose-sparing effects when compared with intramuscular vaccination, thus permitting the induction of strong immune responses upon exposure to lower antigen levels [19]. As a result, targeting the skin as an immunogenic organ in its own right has garnered considerable attention in the recent past.

4.3.1 Skin Structure

The skin, as discussed in detail in Chapter 1, is one of the most complex organs of the body. It is multilayered comprising of the epidermis, dermis and subcutis, which fulfil functionalities such as temperature regulation, mechanical protection and maintenance of homeostasis. Regulation of the entry and outlet of chemicals and water are integral to maintaining healthy functioning in the human body. The skin has a surface area of approximately 2–3 m^2 and is considered the largest organ of the human body, receiving up to one third of the total blood circulation [20]. The skin has been a target for drug delivery for many years due in part to its size and the associated ease of administration of topical medicines. The upper layers of the skin can be segregated into five distinct *strata* namely: *stratum corneum* (*SC*), *stratum lucidum* (*SL*), *stratum granulosum* (*SG*), *stratum spinosum* (*SS*) and the *stratum basale* (*SB*). These layers rest on top of the dermis and subcutis, as shown in Figure 4.2.

The *SC*, the outermost layer of the skin, is comprised of keratinised, dehydrated and compacted layers of dead cells. This layer is on average 10 μm in thickness, equating to approximately 15–20 cells and can swell to many times this thickness when wet. The cells within the *SC* are physiologically inactive and are continuously being shed and replaced by the outward migration of cells from the tissues underlying it. The *SC* is known as the rate-limiting membrane of the skin, thus limiting chemical impasse from the external environment and also controlling water loss from the body [22]. The viable epidermis is then composed of keratinocytes and immunocompetent cells [5, 10]. The dermis supports the epidermis with collagen fibres and also contains a dense network of capillary blood vessels and lymphatics in which dermal dendritic cells, monocytes, polymorphonuclear lymphocytes and mast cells circulate. Lymphatic vessels drain the dermis to satellite lymph nodes, while fibroblasts are the most abundant cell type. The subcutaneous layer consists of loose connective tissue and elastin located immediately beneath the dermis. The arteries and veins that drain the skin dermis, issue from the vascular plexus located in subcutaneous tissue. When entering the skin, dermal arteries form a dense network of capillary loops in the papillary dermis layer. Numerous lymphatic

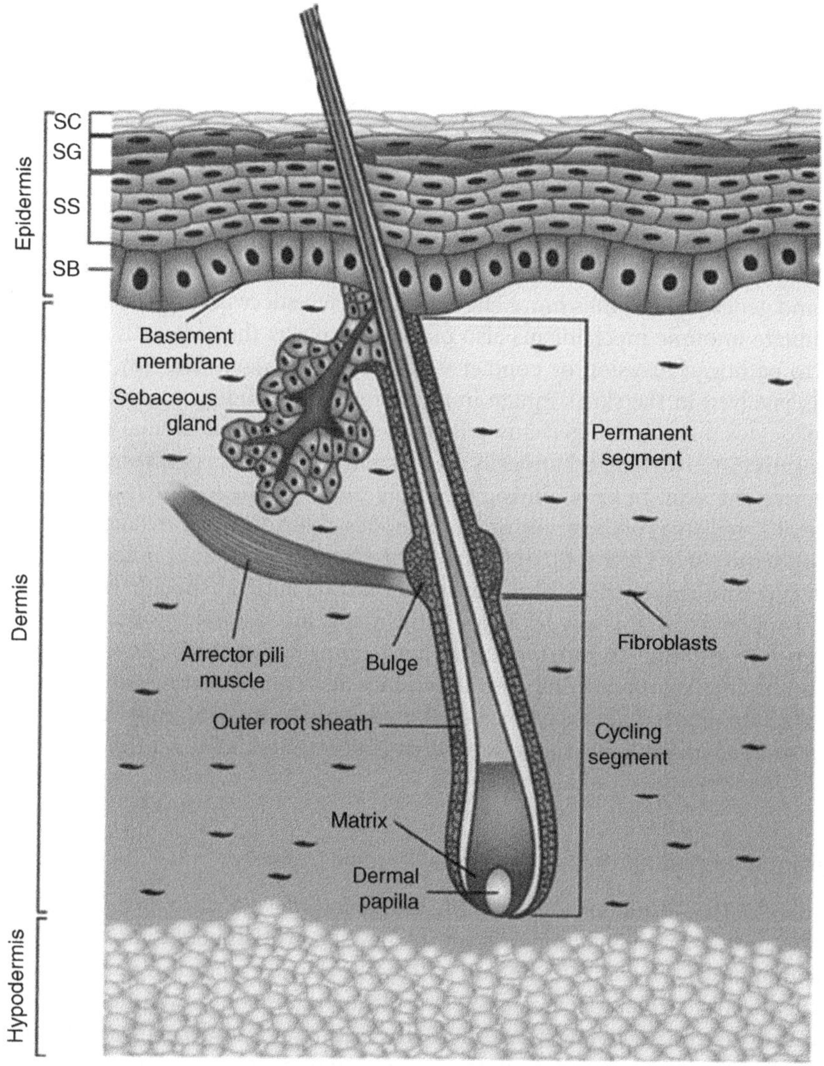

Figure 4.2 Schematic representation of skin morphology The epidermis layer is composed of the *stratum corneum* (*SC*), *stratum lucidum* (*SL*) (not shown), *stratum granulosum* (*SG*), *stratum spinosum* (*SS*), *stratum basale* (*SB*), the dermis and hypodermis layers of the skin. Reproduced with permission from [21].

vessels draining the skin dermis pass through the hypodermis before reaching draining lymph nodes. While skin thickness can vary significantly between different parts of the body, a number of recent studies have shown consistency in skin thickness across people with different ethnic backgrounds at deltoid and suprascapular body sites – likely sites for intradermal vaccine administration [23]. This consistency in skin thickness across people with different demographic profiles represents a major advantage over classic intramuscular vaccination as, to correctly perform an intramuscular vaccination, it is important to select the appropriate needle length based on considerations of the muscle mass of the injection site, the amount of subcutaneous fat and the weight of the patient.

4.3.2 Skin Immune Response

The skin generates both innate (antigen nonspecific responses without immunological memory) and adaptive immune responses (antigen specific responses with immunological memory). While the adaptive response is primordial in generating a response to vaccination and generally becomes more effective with each successive encounter with an antigen, innate immune mechanisms also play a key role as they are activated first in response to pathogen invasion or contact with foreign antigens. The key group of immune cells involved in the skin's innate immune response is bone-marrow derived dendritic leukocytes, namely Langerhans cells in the epidermis and dermal dendritic cells in the dermis [5, 6, 10]. Langerhans cells are efficient APCs that process and present antigens entering the skin. In brief, during their migration via the local lymphatics to the paracortical T-cell areas of the draining lymph nodes, these cells mature and present antigenic peptides to naïve T-cells [6]. Activated Langerhans cells produce inflammatory cytokines and co-stimulatory molecules that induce strong antigen-specific responses by B- and T-lymphocytes. This occurs in an antigen-specific fashion, resulting in the expansion of potent immune-stimulatory cells, thus helping to control the development of adaptive immunity. Worthy of due consideration when contemplating intradermal vaccine delivery, are reports that excessive UVB exposure can affect immune responses by depletion of Langerhans cells, increased dermal recruitment of macrophages and release of pro-inflammatory cytokines [5, 6, 10].

4.3.3 Conventional Strategies for Intradermal Vaccine Delivery

Reports of more potent immune responses elicited by intradermal vaccine delivery, as opposed to intramuscular or subcutaneous routes, have been attributed to the superior collection of immune cells residing in the skin [6, 22]. The skin also appears to behave as a mucosal tissue, as intradermal vaccine delivery results in immune responses at mucosal sites, such as those in the gastrointestinal tract. This feature of intradermal vaccination could prove of tremendous importance in the generation of protective vaginal mucosal immunity to sexually transmitted infections, such as HIV.

A number of methods for intradermal vaccine delivery have been developed. Over a century ago, Mantoux pioneered the standard intradermal injection technique. This methodology consists of stretching the surface of the skin and inserting the tip of a 27 gauge (G), 3/8 inch bevel needle attached to a 1-ml syringe [6]. The needle is inserted bevel upwards, almost parallel to the skin surface, and vaccine is injected slowly into the uppermost layer of the skin. If placed properly, there is considerable resistance to injection and a raised papule immediately appears, which can cause pain during injection. The correct placing of the needle tip in the dermis is critical to avoid fluid injection difficulties. This technique is not without its limitations, however, and is associated with: inconsistency in injection volume; unavoidable leakage of vaccine from the injection site; fluid wastage when filling disposable syringes and when purging the needle of air; and a large dead volume of the assembled needle and syringe. An alternative strategy for intradermal delivery is via the use of bifurcated (dual pronged) needles, which can accommodate approximately 2 μl of vaccine solution between their prongs. Only part of this volume is capable of introduction into the skin however, and precise control of dosing accuracy is not possible. The multipuncture unit is a cylinder-like device with small needles, approximately 1 mm in length, which should be pressed firmly against the skin, within the area where the vaccine dose is spread on the skin surface. In contrast with the intradermal technique, this percutaneous method does not allow a precise estimation of the dose injected into the epidermis or dermis skin layers.

Needle-free jet injectors are another alternative tool for use in the delivery of vaccine to the dermis, subcutaneous tissue and muscle, with some vaccine also deposited in the epidermis [5, 7]. As a result, precise control of intradermal dosing is unachievable using these devices. Studies have shown, however, that vaccination via jet injection can elicit an immune response comparable to or better than other delivery methods. This may be due to increased inflammation at the injection site, as a result of jet injector usage, which may thus lead to the recruitment and activation of more immune cells at the injection site. Indeed, jet injectors are known to cause at least as much pain, bruising, swelling and induration as needle-mediated vaccine delivery. Finally, despite the fact that jet injectors do not require specialist medical training for usage, they have yet to replace conventional needle-and-syringe based vaccination methods.

MN devices have the ability to enhance transdermal and intradermal delivery of vaccines. Unsurprisingly, as a result of the great potential of the intradermal route for immunisation and the largely unsatisfactory nature of conventional methods of administration, MN devices have been investigated extensively for vaccine delivery. Indeed, there have been considerable investigations into the delivery of antigens, adjuvants, inactivated and attenuated infectious agents and antigen-encoding genes, using a wide range of MN modalities and these will be explored in more detail here.

4.4 MN Delivery of Vaccine Therapeutics

As stated previously, the production of a vaccination method that is effective, user friendly, painless, minimises sharps disposal requirements and is also mass producible has been the challenge of the vaccine industry for many years [24]. The inherent ability of MN technologies to enhance transdermal and intradermal vaccine delivery has resulted in a swathe of research exploring these possibilities. As an example, the world's first MN based vaccine device was marketed as Soluvia® in 2009. This encompassed a 1.5 mm hollow MN (silicon), which was attached to a typical barrel syringe. This device has been marketed with the intradermal vaccines Fluzone Intradermal®, IDFlu® and Intanza®. This was a major step forward in the usage of MN devices in this field and highlights the progress that has been made in this area. MNs can be classified into: dissolving/biodegrading; hollow; solid and hydrogel-forming. Each of these will be considered in terms of vaccine delivery here.

4.4.1 Dissolving/Biodegrading Polymeric MNs

Based on the fact that dissolving MN designs show most promise with respect to potential future commercialisation, due mostly to their obvious advantages of targeted delivery of an antigen to a highly immunogenic region and their inherent self-disabling nature, a tremendous number of research teams have considered them as innovative devices for use in vaccination strategies. The premise here is that the MNs, consisting of fast-dissolving matrix materials such as sugars or polymers into which a drug/antigen has been mixed, rapidly dissolve in skin interstitial fluid in the viable epidermis and/or dermis, releasing their payload. If MN vaccine delivery devices can be supplied in such a self-disabling manner (as the needles on the device dissolve, they cannot then accidentally be inserted into a second person), this will have ensuing benefits in reducing the need for specialist waste disposal, and the devices could also be administered by non-specialist personnel, for example in a home setting. Additionally, their use will greatly reduce any risk of infection transmission. The solid-state nature of the contained/encapsulated vaccine should reduce the need for cold chain storage and transport. Successful optimisation and commercialisation of such systems will be of undoubted benefit to patients, especially those in the developing world.

4.4.1.1 Viral Vaccines

A wealth of published work documents the use of dissolving MN devices in vaccination against various viruses, for example influenza, polio and measles [25–27]. Two seminal papers published in 2010 described the capabilities of dissolving MN arrays in the vaccination of mice [25, 28]. The former work will be discussed now and the latter study is described in further detail in a subsequent paragraph. To begin, the Prausnitz group introduced dissolving MN patches for influenza vaccination using a simple patch-based system fabricated at room temperature by photopolymerisation of liquid vinyl pyrrolidone monomer within MN moulds to form poly(vinyl pyrrolidone) MNs 650 μm in height and encapsulating 3 μg of lyophilised inactivated influenza virus vaccine [25]. MNs were inserted into mouse skin *in vivo* by gentle hand pressure and dissolved within minutes. MN vaccination generated robust antibody and cellular immune responses that provided complete protection against lethal challenge. Compared with conventional intramuscular injection, MN vaccination resulted in more efficient lung virus clearance and enhanced cellular recall responses after challenge. The group's results suggested that dissolving MN patches could provide a new technology for simpler and safer vaccination, with improved immunogenicity that could facilitate increased vaccination coverage. To add to the studies using dissolving MNs in combination with the flu vaccine, a 2012 study documented the use of MNs fabricated from trehalose and carboxymethylcellulose (CMC), which were then utilised to deliver cell-culture-derived influenza vaccine with considerable effect [29].

Moving on considerably from those initial studies, *The Lancet* recently published the results of a Phase-I trial to deliver influenza vaccine via dissolving MN devices that was led by researchers from Emory University [30]. This work explored self and healthcare worker application of MN patches versus intramuscular administration of the same vaccine. Succinctly, the geometric mean titres determined by haemagglutination inhibition antibody assay were similar at day 28 between the healthcare worker administered MNs and the intramuscular groups for the three virus strains employed in the study (H1N1, H3N2 and B strain). Similar titres were also seen for the self-administered MN cohort. Administration of the patches was well tolerated but specific local, mild, self-limited reactions were more frequent in the MN, compared with the intramuscular patient

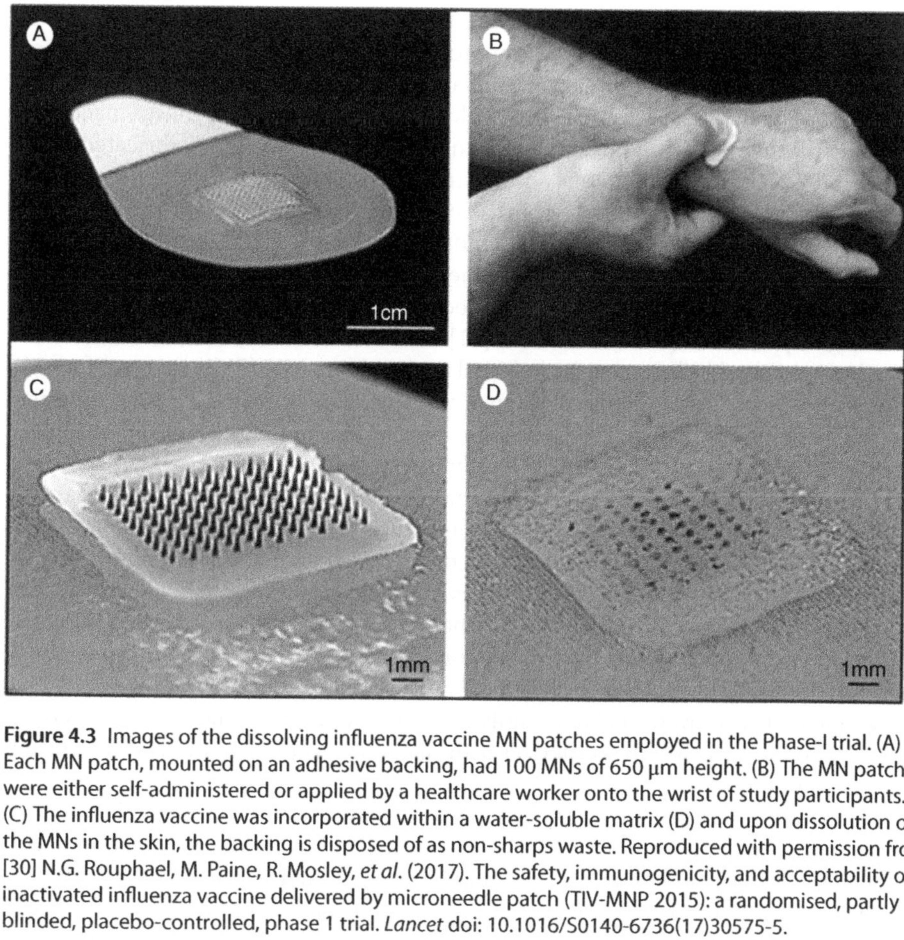

Figure 4.3 Images of the dissolving influenza vaccine MN patches employed in the Phase-I trial. (A) Each MN patch, mounted on an adhesive backing, had 100 MNs of 650 μm height. (B) The MN patches were either self-administered or applied by a healthcare worker onto the wrist of study participants. (C) The influenza vaccine was incorporated within a water-soluble matrix (D) and upon dissolution of the MNs in the skin, the backing is disposed of as non-sharps waste. Reproduced with permission from [30] N.G. Rouphael, M. Paine, R. Mosley, *et al.* (2017). The safety, immunogenicity, and acceptability of inactivated influenza vaccine delivered by microneedle patch (TIV-MNP 2015): a randomised, partly blinded, placebo-controlled, phase 1 trial. *Lancet* doi: 10.1016/S0140-6736(17)30575-5.

cohorts. The MN patches were preferred by participants over intramuscular administration of the vaccine, which could have knock-on benefits in increasing acceptability of MN devices and concurrently increasing influenza vaccination rates. The images in Figure 4.3 illustrate how this study was carried out.

Although this was a thorough study, the authors did acknowledge that it was not without its limitations, such as the fact that the subjects had high baseline, pre-vaccination titres, thus making the differentiation of the effects between administration routes difficult. Despite this, this paper will undoubtedly be one of the first in a long line of similar clinical trials that are absolutely necessary and warranted if dissolving MN vaccination platforms are to become a viable, realistic alternative to conventional vaccination methods.

In 2015, work on two different viruses emerged from the Georgia Institute of Technology [26, 27]. In the former study, rhesus macaques (*Macaca mulatta*) were vaccinated with inactivated polio vaccine (IPV) using conventional intramuscular administration of vaccine or using dissolving gelatine/sucrose MN arrays of 100 needle geometries with heights of 650 μm across a 1.27 cm diameter. The vaccination was well tolerated by the monkeys with equivalent responses measured in both intramuscular and MN cohorts, leading the authors to surmise that this approach could, in the future, be adopted as a means of combating polio outbreaks in human populations [26]. In the latter research study, live-attenuated measles virus was employed, again to vaccinate rhesus macaques. Dosing the monkeys with equivalent amounts of vaccine via MNs (fabricated using sucrose and CMC) or subcutaneous administration resulted in the production of comparable levels of antibody neutralisers that correlated with protection [27]. Interestingly, this was one of the first studies to demonstrate that a vaccine incorporated into a MN array could retain its potency when subjected to storage at elevated temperatures, and this theme was explored further in subsequent publications arising from the same research group and using either subunit [31] or inactivated [32] vaccines. In the latter study, inactivated influenza virus in liquid or dissolving MN (PVA/sucrose) formats were subjected to a range of different temperatures: 4, 25, 37 and 45 °C. Haemagglutination (HA) activity was then used to determine the potency of the vaccines post-storage. Liquid vaccine was found to have completely lost its potency within 1–2 weeks when stored without refrigeration. In contrast, although vaccine in the dissolving MN arrays lost 40–50% of its HA activity over the manufacturing period, when stored with desiccant, no further loss in HA activity was recorded, independent of the temperature at which the MN arrays were stored [32]. These findings were then validated when the vaccinated mice were challenged with a lethal dose of live, mouse-adapted A/Puerto Rico/8/34 virus. Mice vaccinated with MN arrays, regardless of the temperature at which they were stored, were completely protected against the viral challenge. Mice that had been vaccinated via intramuscular administration of the vaccine which had been stored at 45 °C were, unsurprisingly, conferred with only 50% protection and lost significant amounts of weight.

The overarching theme of this publication was to highlight the thermostability of MN arrays and their potential for usage for both seasonal and pandemic vaccination needs [32], which is complimented by other similar studies carried out in BALB/c mice [33, 34]. These studies are further bolstered by the recent publication of a very thorough study using trivalent subunit influenza vaccine incorporated into a variety of different formulations to determine which formulation was the most efficacious in maintaining immunogenicity during prolonged storage or sterilisation processes [35]. The formulations contained combinations of trehalose/sucrose, sucrose/arginine and arginine/heptagluconate, with the arginine/heptagluconate combination being deemed the most robust in *in vitro* investigations with subsequent *in vivo* studies confirming this assertion. Succinctly, the influenza vaccine formulated into the optimised MN arrays remained stable after exposure to various conditions: 25 °C, 24 months; 60 °C, 4 months; five freeze–thaw cycles; or sterilisation irradiation via electron beam. The authors correctly considered this one of the most thorough investigations of potential MN excipients for maintenance of vaccine immunogenicity [35].

The second seminal paper published in 2010 and documenting the delivery of vaccines via dissolving MN array was carried out by researchers in The University of Queensland. The paper described the micromoulding of dissolving MN arrays from master templates of the Nanopatch™ design [28]. Replica MNs were formed from CMC by multiple castings into poly(dimethylsiloxane) moulds. The MNs were 88 μm in height, with base diameters of 17 μm, showing slight shrinkage from the master templates. MNs dissolved in mouse ear skin within 5 min and successfully delivered model dyes. Importantly, enhanced diffusion of the dye payloads was observed in the dermis compared with the skin layers above. The authors suggested that this was due to the known higher hydration levels in the skin's deeper layers. The MN administration of vaccine was shown to elicit a robust systemic immune response in the mouse model employed [28].

Moving towards patient benefit, a 2015 human clinical study examined the use of the dissolving MN (hyaluronic acid) device, namely MicroHyala®, in the delivery of influenza antigen [36]. The study volunteers showed no signs of severe local or systemic reaction and showed comparable immune responses against A/H1N1 and A/H3N2 strains of the virus when compared with the subcutaneous control groups. Interestingly the study highlighted that the MicroHyala® vaccination route yielded a stronger immune response to the influenza B strain than that seen in the subcutaneous control cohort. The authors surmised that immunisation against the influenza B virus, epidemics of which occur at low frequency, would therefore be more efficacious if delivered transcutaneously [36]. According to the authors, this was due, in no small part, to the wealth of immune cells residing in the skin.

Despite the tremendous and undeniable focus that has been placed on the delivery of influenza vaccines via dissolving MN arrays, exciting studies into vaccination against other viruses such as HIV [37] and enterovirus 71 (EV71), the main causative agent of hand-foot-and-mouth disease (HFMD) [38, 39], have also undergone investigation. Additionally, dissolving MN vaccination strategies have not been limited solely to the end goal of immunisation of human subjects. Rather interesting work has proposed the use of dissolving MN devices for the simple and cost-effective vaccination of dogs as a means to combat high costs and complex schedules of post-exposure vaccinations of humans, particularly in low-resource settings. One such study carried out in 2016 involved the immunisation of beagle dogs with either 5 or 50 μg of DNA plasmid expressing rabies protein, incorporated into dissolving (PVA/sucrose) MN arrays in a prime and boost schedule [40]. In virus neutralisation tests, titres of neutralising antibodies were comparable (42 days post-prime) or higher (56 days post-prime) in the MN cohorts that received MN loaded with 50 μg DNA plasmid, compared with the intramuscular cohort of beagles. Although this work is promising, the manual application of such dissolving MN to wild dogs in rural, low resource settings undoubtedly requires careful planning and consideration if this particular platform is to move forwards.

4.4.1.2 Bacterial Vaccines

Although the majority of MN-based vaccines have focused on targeting of viral infections, most notably influenza, there is also some interest in vaccination against bacterial infections, with tetanus having attracted some interest in the research community. In 2012, Matsuo *et al.* described the use of their dissolving MN (hyaluronic acid) device, MicroHyala®, as a transcutaneous immunisation device [41]. They reported that transcutaneous immunisation of non-pregnant rats using the MicroHyala® device, containing unadjuvanted tetanus toxoid, induced effective immune responses, equivalent to those observed via intramuscular or intradermal delivery routes after five vaccine doses.

Similar work using tetanus toxoid was also published in 2016 when dissolving MN arrays (PVA/sucrose/CMC) containing unadjuvanted tetanus toxoid were prepared using a two-step micro-moulding process [42]. The authors examined the host immune responses in pregnant and non-pregnant mice vaccinated intradermally or via intramuscular injection. Both immune responses and survival rates were higher in the MN treated cohort and the passive immunity of the off-spring of MN treated mice was shown to be more robust and longer lasting than their intramuscular counterparts, with a 100% survival rate reported for tetanus challenge up to an age of 6 weeks [42].

4.4.1.3 Model and Novel Vaccines

In variations on the traditional theme of vaccine delivery, research teams have also moved towards the delivery of alternative and model vaccine agents. For example, in one of the first studies of its kind, the Banga Group at Mercer University, Atlanta, investigated the MN-mediated *in vitro* transdermal delivery of human immunoglobulin (IgG) and demonstrated the applicability of MNs in the delivery of a monoclonal antibody [43]. Microchannels created by maltose MNs inserted into full thickness hairless rat skin were visualised using methylene blue staining and the depth of penetration was determined using haematoxylin and eosin staining. *In vitro* penetration studies were conducted using freshly-excised full thickness hairless rat skin. MNs were applied and allowed to dissolve in the skin, before adding an antibody solution to the skin surface. The pathway of IgG transport across the skin was then confirmed by immunohistochemical (IHC) studies. The monoclonal antibody was then delivered under optimised conditions. Methylene blue was taken up by microchannels, indicating disruption of the *stratum corneum* and cryosections showed that MNs just reached the dermis. Human IgG delivery increased with an increase in the numbers of MNs, IgG concentration and length of MNs. IHC studies demonstrated that IgG moved through microchannels for transport across the skin. Transdermal delivery was also demonstrated for the monoclonal antibody [43].

Owing to its relatively low cost and excellent stability profile, a wide range of dissolving MN array studies have involved the delivery of ovalbumin (OVA) as a model antigen. One of the earliest such studies was published in a 2010 paper from the Kendall group [28]. Dual-layer MNs containing OVA, along with the adjuvant Quil-A, elicited post-immunisation schedule antibody levels in mice that were comparable to an intramuscular OVA/Quil-A immunisation group at day 28 and superior to the intramuscular group at day 102, despite using a lower antigen dose in the MNs. Similar results were also seen using influenza vaccine, as described previously. In 2013, Zaric *et al.* took advantage of poly-D,L-lactide-*co*-glycolide (PLGA) nanoparticles (NP) encapsulating OVA, suspended in aqueous blends of poly(methylvinylether-*co*-maleic acid) to target dendritic cells in the skin [44]. Dissolving within minutes, the dissolution strategy resulted in considerable Th1 CD4+ and potent cytotoxic CD8+ T-cell responses [44]. This work also addressed the recurring theme of antigen stability in the MN arrays. In brief, MNs were fabricated laden with OVA–NP or soluble OVA and then stored under ambient conditions for 4, 6 or 10 weeks. These MNs were then used to vaccinate mice and the levels of OVA-specific CD8+ T-cell proliferation was maintained when nanoencapsulated antigen was formulated into MN arrays, up to 10 weeks prior to vaccination. This elegant and thorough study also elucidated that the growth of an OVA-expressing melanoma tumour was suppressed in vaccinated mice via a Th1 interferon gamma (IFN-γ) mediated response and vaccination protected mice against challenge with an OVA-expressing virus. Then in 2014, the Donnelly group published one study on intradermal vaccination with dissolving MN arrays, measuring anti-OVA specific IgG responses [45]. In this instance the prime, boost and second boost vaccination schedule resulted in significantly higher antibody serum titres following vaccination with dissolving MN arrays, compared with intradermal administration. Over and above the studies discussed here, there is a considerable body of other work which has been published charting the use of OVA as a model antigen delivered via dissolving MN. Examples of these studies include the following references [46–51], all of which serve to highlight the utility of this model antigen delivered via this unique platform.

To move away from these model antigens and towards a disease with a high profile in the public psyche, vaccination against Alzheimer's disease (AD) based on the amyloid cascade hypothesis has been attempted for a number of years, initially by Schenk *et al.* in 1999 [52]. The group aimed to deliver an amyloid-beta 42-amino acid peptide (Aβ1–42) as an antigenic stimulant via the traditional parenteral method. This particular approach failed at Phase-II clinical trials due to over stimulation of the Th1 immune response, resulting in a number of cases of meningoencephalitis [53]. More recently, however, this peptide has been revisited and its potential reassessed by delivery into mice via the MicroHyala® dissolving MN patch [54]. Although, as conceded by the authors, their most recent work did not result in improved brain function, it demonstrated that efficient anti-Aβ1–42 immune responses could be attained by MN administration [54].

The selected studies discussed here serve to highlight the immense bodies of work which have focused on vaccination via dissolving MN platforms.

4.4.2 Hollow MNs

Hollow MN arrays have, by definition, a central bore or lumen, through which vaccines and antigens can pass into the dermal microcirculation. Vaccine delivery through a hollow MN could be seen to emulate the Mantoux technique (dermal needle technique based on shallow angle needle insertion), which has produced inconsistent dermal cellular responses in some studies [55, 56]. However, hollow MN studies with anthrax vaccination in rabbit models showed the potential for an up to 50-fold vaccine dose reduction when delivered by this MN platform. Subsequent lethal challenge with anthrax spores resulted in equivalent survival profiles across groups, compared with the traditional intramuscular dosing route [57, 58]. The 2006 work recognised the potential of recombinant protective antigen (rPA) of *Bacillus anthracis* as a promising anthrax vaccine [58]. This work compared serum IgG levels and toxin-neutralising antibody titres in rabbits following delivery of various doses of vaccine by MN-based intradermal delivery or intramuscular injection using conventional needles. MN intradermal delivery required less antigen to induce levels of antibody similar to those produced via intradermal injection during the first 2 weeks following primary and booster inoculation. This dose-sparing effect was less evident at the later stages of the immune response. Rabbits immunised intradermally with 10 μg of rPA displayed 100% protection from aerosol spore challenge, while intramuscular injection of the same dose provided slightly lower protection (71%). Groups immunised with lower antigen doses were partially protected (13–29%), regardless of the mode of administration. Overall, the authors suggested that rPA formulated with aluminium adjuvant and administered to the skin by a MN-based device is as efficacious as intramuscular vaccination.

The same research team, at BD Technologies, have also carried out a series of investigations using hollow MNs and influenza vaccination. They described the first preclinical use of MN technology for intradermal administration of three different types of influenza vaccines: a whole inactivated influenza virus, a trivalent split-virion human vaccine and a plasmid DNA encoding the influenza virus haemagglutinin [59]. In this study, a 34G stainless-steel MN with inner diameter of 76 μm, an outer diameter of 178 μm and an exposed length of 1 mm was used. The MN was inserted perpendicularly to the skin surface to its full exposed length in order to control the dermal penetration depth. In a rat model, intradermal delivery of the whole inactivated virus provided up to 100-fold dose-sparing compared with intramuscular injection. In addition, intradermal delivery of the trivalent human vaccine enabled at least 10-fold dose-sparing for the H1N1 strain and elicited levels of response across the dose range similar to those of intramuscular injection for the H3N2 and B strains. Furthermore, at least fivefold dose-sparing from intradermal delivery was evident in animals treated with multiple doses of DNA plasmid vaccine, although such effects were not apparent after the first immunisation. The authors concluded that their results demonstrated that MN-based intradermal delivery elicits antibody responses that are at least as strong as via intramuscular injection and that, in many cases, dose-sparing can be achieved by this new immunisation method.

A similar conclusion was drawn from their 2010 study using the rabies vaccine in healthy adult human trials, which demonstrated adequate immune-efficacy when contrasted with intramuscular injection [60]. In this pilot study, they investigated intradermal delivery using a BD MN from 1 to 3 mm in length and epidermal delivery (BD skin abrader) through an abraded skin surface relative to standard intramuscular injection [60]. The prototype skin-microabrader, known as Onvax™, consists of an array of plastic microprojections with a height of approximately 200 μm from the base to the top and spacing between projections of twice the height. Rubbing the skin surface with Onvax results in *stratum corneum* disruption, with possible antigen delivery as close as possible to Langerhans cells distributed in the germinative layer of the epidermis. The research team emphasised the fact that rabies vaccination by the intradermal route is widely promoted by the WHO. However, well-designed studies to formally demonstrate the degree of dose-sparing, and to determine whether there are real differences between the intradermal and intramuscular routes, are currently lacking and would be valuable. Rabies represents a good model vaccine for evaluating novel intradermal devices in naïve recipients. In this study, circulating neutralising antibodies were measured against the rabies virus after the Vero cells rabies vaccine was administered at Day 0, Day 7, Day 21 and Day 49. This clinical evaluation in 66 healthy volunteers showed that intradermal delivery using BD MN technology of a quarter the intramuscular antigen dose is safe, efficient and reliable, resulting in a protective seroconversion rate. In contrast, the epidermal delivery route did not produce an immune response against the rabies vaccine.

In 2009, Van Damme *et al.* investigated the safety and efficacy of a novel MN device for dose-sparing intradermal influenza vaccination in healthy adults [61]. They conducted a prospective, randomised trial in 180 intended-to-treat healthy adults. Study objectives were to evaluate the safety and immunogenicity of low-dose intradermal influenza vaccines delivered using a novel MN device (Micronjet®). The MN device used in this study comprised an array of four MNs, each 0.45 mm in length. The needles were made of silicon crystal and were bonded to the tip of a plastic adapter, which could be mounted on any standard syringe. The device could be used, in theory, in the same way as any other needle, except that the needles can only be used for intradermal delivery due to the limited length of the MNs. Subjects were randomly assigned to receive either the full-dose standard flu vaccine (containing 15 μg haemagglutinin per strain) delivered intramuscularly using a conventional needle (intramuscular, IM group), a medium-dose intradermal injection (6 μg haemagglutinin per strain) delivered with the Micronjet (ID2 group), or a low-dose intradermal injection (3 μg haemagglutinin per strain) delivered with the Micronjet (ID1 group). A marketed influenza vaccine for the 2006/2007 influenza season (α-RIX®, GSK Biologicals) was used for all injections. Adverse events were recorded over a 42-day period. Immunogenicity was evaluated by changes in haemagglutination inhibition (HAI) antibody titre, and by comparing geometric mean titres (GMTs), seroconversion and seroprotection rates between the study groups. Local reactions were significantly more frequent following intradermal vaccination, but were mild and transient in nature. At 21 days after injection, GMT fold increases were: 22.18 and 22 in the ID1, ID2 and IM groups, respectively, for the H1N1 strain; 9, 9 and 16 for the H3N2 strain; and 9, 13 and 11 for strain B. The Committee for Proprietary Medicinal Products (CPMP) criteria for re-licensure of seasonal influenza vaccines were met in full for all study groups. The authors concluded that low-dose influenza vaccines delivered intradermally (ID) using MNs elicited immunogenic responses similar to those elicited by the full-dose intramuscular vaccination. The MN injection device used in this study was found to be effective, safe and reliable [61].

Multiple vaccine components amalgamated into a single combined product have also received some attention in the MN field. A 2008 study successfully demonstrated the use of hollow MNs to deliver a combination vaccine providing protection from staphylococcal toxic shock, botulism, anthrax and plague in rhesus macaque monkeys [62].

The Bouwstra research group has also carried out considerable work on vaccine delivery using hollow MNs. In a series of linked experiments, they have charted the delivery of IPV to rats [63–65]. This work culminated in a study involving the fractional dosing of rats, over multiple days, with IPV serotype 1 (IPV 1). This resulted in the induction of higher and stronger immune responses compared with intramuscular or intradermal bolus dosing, with the inclusion of an adjuvant having no significant impact on the responses [65]. A different comparative study deciphering the efficacy of tetanus toxoid loaded chitosan NP delivery via MN was carried out in 2016. Both hollow and solid MNs were employed in this study with both eliciting similar levels of IgG and IgG1 titres in the *in vivo* mouse model tested, although the IgG levels induced following hollow MN usage were higher [66]. The authors concluded that hollow MN delivery platforms should be favoured over their solid counterparts due to the fact that hollow MN assisted vaccination was deemed to have induced both humoral and cellular immune responses with better administration accuracy and precision than solid MNs or indeed commercial vaccine.

It is evident, however, that despite the successes demonstrated by hollow MN usage in vaccine delivery, they have not generated the same wealth of published studies as their dissolving and coated MN counterparts.

4.4.3 Solid MN

Solid MNs can be delineated into "Poke and Patch" MNs and coated MNs. The former involves the application of solid MN to the skin, followed by their removal and topical application of vaccine onto the surface where the MNs have been inserted. The latter, solid MN, as their name suggests, have been coated with vaccine, which is deposited into the skin when MNs are applied.

4.4.3.1 "Poke and Patch" Methodologies

The Bouwstra group based in Leiden have carried out considerable work using "Poke and Patch" approaches. They investigated mouse immune responses after intradermal immunisation using two model antigens, diphtheria toxoid and influenza subunit vaccine [67]. Three types of MN arrays were used in this study. Firstly, "assembled" MN arrays were manufactured from commercially available 30G hypodermic needles. The needles were assembled as a 4×4 array on a polymer plate with the surface area of around 0.5 cm^2. A series of different MN arrays were fabricated with needle lengths of 300, 550, 700 and 900 μm. The second type was made of stainless-steel wire with a diameter of about 200 μm and a length of 300 μm, with a tangentially cut tip. The third one was silicon hollow MN arrays with a length of 245 μm, available as 4×4 and 9×9 arrays. A custom-designed electric applicator enabled shorter MNs (300 μm) to pierce mouse skin effectively, as shown by Trypan blue staining and trans-epidermal water loss measurements. The vaccines were topically applied with and without cholera toxin (CT) on MN-treated skin. In diphtheria toxoid intradermal immunisation, MN array pre-treatment of the skin was essential to achieve substantial IgG and toxin-neutralising antibody titres. Addition of CT further boosted the immune response to similar levels as observed after subcutaneous injection of AlPO4-adsorbed diphtheria toxoid (DT-alum). In contrast, MN array pre-treatment showed no effect on the immune response to plain influenza vaccine. This response was strongly improved by inclusion of CT, independent of MN treatment. The authors concluded that their study indicated that intradermal immunisation of diphtheria toxoid and CT following MN treatment results in comparable protection to injection of DT-alum, and intradermal immunisation of influenza vaccine adjuvanted with CT is superior to the injection of plain vaccine [67].

In another study carried out in the same year, this group looked at modulation of the immune response against diphtheria toxoid by various adjuvants in intradermal immunisation with MN array pre-treatment [68]. Intradermal immunisation was performed on BALB/c mice with or without MN array pre-treatment using diphtheria toxoid as a model antigen co-administrated with lipopolysaccharide (LPS), Quil A, CpG oligo deoxynucleotide (CpG) or CT as adjuvant. The immunogenicity was evaluated by measuring serum IgG subtype titres and neutralising antibody titres. Intradermal immunisation with MN array pre-treatment resulted in a 1000-fold increase of diphtheria toxoid-specific serum IgG levels. The immune response was further improved by co-administration of adjuvants, showing a progressive increase in serum IgG titres when adjuvanted with LPS, Quil A, CpG and CT. IgG titres of the CT-adjuvanted group reached levels comparable to those obtained after DT-alum subcutaneous injection. The authors concluded that the potency and quality of the immune response against diphtheria toxoid mediated by MN arrays can be modulated by co-administration of adjuvants.

The same group aimed to gain insight into the delivery and immunogenicity of *N*-trimethyl chitosan (TMC) adjuvanted diphtheria toxoid formulations applied transcutaneously with MNs [69]. Mice were vaccinated with diphtheria toxoid-loaded TMC nanoparticles, a solution of TMC and diphtheria toxoid (TMC/DT) or diphtheria toxoid alone. The formulations were applied onto the skin before or after MN treatment with two different 300 µm long MN arrays and also injected intradermally. As a positive control, alum-adjuvanted diphtheria toxoid (DT-alum) was injected subcutaneously. *Ex vivo* confocal microscopy studies were performed with rhodamine-labelled TMC. Independent of the MN array used and the sequence of MN treatment and vaccine application, intradermal immunisation with the TMC/DT mixture elicited eightfold higher IgG titres compared with the TMC nanoparticles or diphtheria toxoid solution. The toxin-neutralising antibody titres from this group were similar to those elicited by SC DT-alum. After intradermal immunisation, both TMC-containing formulations induced enhanced titres compared with a diphtheria toxoid solution. Confocal microscopy studies revealed that transport of the TMC nanoparticles across the MN conduits was limited compared with a TMC solution. They concluded that TMC has an adjuvant function in transcutaneous immunisation with MNs, but only if applied in solution [69].

Ding and colleagues then investigated MN pre-treatment followed by topical applications of either diphtheria toxoid or vesicle formulations containing the same antigen [70]. As with other similar studies, the pre-treatment of the skin with MN was seen to enhance immune responses with free diphtheria toxoid inducing faster IgG development than antigen-vesicle formulations. In fact, regardless of the vesicle formulation employed (cationic liposomes or anionic surfactant-based vesicles), these formulations were shown to have no influence on the immunogenicity of the antigen [70].

This team has, more recently, carried out a comparative study between the "Poke and Patch" MN delivery method and a coated pH-sensitive MN delivery platform [71]. They showed the induction of anti-OVA specific IgG in conjunction with robust activation of CD4+ and CD8+ T cells in BALB/c mice following a prime, boost and second boost vaccination schedule. Owing to superior antigen delivery from the coated MN arrays, the authors conceded that the coated MN system was better than the "Poke and Patch" methodology and as such demonstrated greater applicability for intradermal vaccination strategies [71].

4.4.3.2 Coated MNs

In 2009, a new avenue in MN delivery systems was ignited with the publication of several studies using antigen-coated MNs in vaccination strategies [72, 73]. These studies revealed that following application of an influenza vaccine coated MN device, complete protection from lethal challenge of H3N2 and H1N1 seasonal strains of influenza virus was afforded in mouse models [72, 73]. This work was capitalised upon and in 2010, the same year that the first studies using dissolving MN in vaccine delivery were published, a large body of pioneering work was carried out at Georgia Tech in Atlanta on coated MN arrays for vaccine delivery [74]. Once again, this work served to spearhead and pioneer the subsequent wealth of published studies in this field. Coating solid MNs has since proven to be a very popular choice for such strategies, one reason being that the coating process requires only minor alteration to longstanding techniques such as spray coating

[75, 76] and repeated dip coating to load stainless steel [77], titanium [78] and silicon [79] projections with antigenic material.

Focus on Influenza Vaccines In 2009, one of the first pieces of research using coated MN in a vaccination strategy emerged from Georgia Tech. This work investigated the protective immune responses elicited after a single influenza vaccination to the skin of mice via coated MNs [80]. Skin vaccination with inactivated virus-coated MNs provided superior protection against lethal challenge compared with intramuscular injection, as evidenced by effective virus clearance from the lungs. Detailed immunologic analysis suggested that induction of virus neutralising antibodies, as well as enhanced anamnestic humoral and cellular responses, contributed to improved protection by MN vaccination to the skin [80].

The following year, 2010, proved to be an extremely productive year with a deluge of varied studies published in this area. The Prausnitz group used MNs fabricated by laser-cutting stainless-steel sheets and designed in monument-shaped arrays of five MNs, which were dip-coated with vaccine [74]. The MNs were approximately 700 μm in height, similar to the depth of the outer layers of skin including the *stratum corneum*, epidermis and the upper part of the dermis. The coating solution typically contained the film-forming polymer sodium CMC and the surfactant Lutrol F-68 NF. However, recognising that dip-coating MNs with vaccine-containing solutions/suspensions can reduce antigen activity, the authors sought to determine the experimental factors and mechanistic pathways by which inactivated influenza vaccine could lose activity, as well as to develop and assess improved MN coating formulations that protect the antigen from activity loss [74]. After coating the stainless-steel MNs using a standard vaccine formulation, the stability of the influenza vaccine was reduced to 2%, as measured by haemagglutination activity. The presence of CMC, which was added to increase viscosity of the coating formulation, was shown to contribute to vaccine activity loss. After screening a panel of candidate stabilisers, the addition of trehalose to the coating formulation was found to protect the antigen and retain 48–82% antigen activity for all three major strains of seasonal influenza: H1N1, H3N2 and B. Influenza vaccine coated in this way also exhibited thermal stability, such that activity loss was independent of temperature over the range of 4–37 °C for 24 h. Dynamic light scattering measurements showed that antigen activity loss was associated with virus particle aggregation, and that stabilisation using trehalose largely blocked this aggregation. Finally, MNs using an optimised vaccine coating formulation were applied to the skin to vaccinate mice. MN vaccination induced robust systemic and functional antibodies and provided complete protection against lethal challenge infection similar to conventional intramuscular injection. The concluding remarks of this work were that antigen activity loss during MN coating could be largely prevented through optimised formulation and that stabilised MN patches could be used for effective vaccination.

A host of other studies, using influenza vaccine as the archetypal vaccine, were also published in the same year using stabilised influenza vaccine [81] and influenza virus-like particles [82–85]. In the Pearton study, MNs were coated with influenza virus-like particles (VLP) [82] and reproducibly penetrated freshly excised human skin, depositing 80% of the coating within 60 s of insertion. Human skin experiments showed that H1 and H5 VLPs, delivered via MNs, stimulated Langerhans cells, resulting in changes in cell morphology and a reduction in cell number in epidermal sheets.

Langerhans cell response was significantly more pronounced in skin treated with H1 VLPs, compared with H5 VLPs. These data provided some evidence that MN-facilitated delivery of influenza VLP vaccines initiates a stimulatory response in Langerhans cells in human skin. This was one of numerous studies that validated and supported animal data, suggesting that dendritic cells targeted through deposition of vaccines in skin generated immune responses.

The same research team subsequently investigated the immunogenicity and protective efficacy of influenza H5 VLPs containing the haemagglutinin (HA) of A/Vietnam/1203/04 (H5N1) virus delivered into the skin of mice using metal MN patches [86]. The work also studied the response of Langerhans cells in a human skin model. Prime-boost MN vaccinations with H5 VLPs elicited higher levels of virus-specific IgG1 and IgG2a antibodies, virus-specific antibody-secreting cells and cytokine-producing cells up to 8 months after vaccination, compared with the same antigen delivered via IM (intramuscular). Both prime-boost MN and IM vaccinations with H5 VLPs induced similar haemagglutination inhibition titres and conferred 100% protection against lethal challenge with the wild-type A/Vietnam/1203/04 virus 16 weeks after vaccination. MN delivery of influenza VLPs to viable human skin using MNs induced the movement of CD207+ Langerhans cells toward the basement membrane, which is significant as this indicates mobilisation of the cells in the human skin environment [86].

They then demonstrated enhanced memory responses to seasonal H1N1 influenza following intradermal vaccination using their coated MNs [81]. Mice vaccinated with a single MN dose of trehalose-stabilised influenza vaccine developed strong antibody responses that were long-lived. Compared with traditional intramuscular vaccination, stabilised MN vaccination was superior in inducing protective immunity, as was evidenced by efficient clearance of virus from the lung and enhanced humoral and antibody-secreting cell immune responses after 100% survival from lethal challenge. Vaccine stabilisation was found to be important, because mice vaccinated with an unstabilised MN vaccine elicited a weaker immunoglobulin IgG2a antibody response, compared with the stabilised MN vaccine and were only partially protected against viral challenge. It was suggested that improved trafficking of dendritic cells to regional lymph nodes as a result of MN delivery to the skin may possibly play a role in contributing to improved protective immunity [81].

The same group examined the immunogenicity and protective efficacy of influenza VLPs (H1N1 A/PR/8/34) after skin vaccination using vaccine dried onto their solid MN arrays [83]. Coating of MNs with influenza VLPs using an unstabilised formulation was found to decrease haemagglutinin (HA) activity, whereas inclusion of trehalose disaccharide preserved the HA activity of influenza VLP vaccines after MNs were coated. MN vaccination of mice in the skin with a single dose of stabilised influenza VLPs induced 100% protection against challenge infection with a high lethal dose. In contrast, unstabilised influenza as well as intramuscularly injected vaccines, provided inferior immunity and only partial protection ($\leq$40%). The stabilised MN vaccination group showed IgG2a levels that were one order of magnitude higher than those of other groups and had the lowest lung viral titres after challenge. Also, levels of recall immune responses, including haemagglutination inhibition titres, neutralising antibodies and antibody-secreting plasma cells, were significantly higher after skin vaccination with stabilised formulations. The authors stated that their results indicated that HA stabilisation, combined with vaccination via the skin using a vaccine formulated as a solid MN patch, confers protection superior to that with IM injection and enables potential dose-sparing effects, which are reflected by pronounced increases in rapid recall immune responses against the influenza virus. The group stated that while intradermal immunisation using MNs can enhance immune responses and provide antigen sparing and potentially offer improved shelf-life, the approach is not fully compatible with many vaccine adjuvants. This includes alum, the most common adjuvant used in the vaccine market globally. They studied a polyphosphazene immunoadjuvant to investigate its potential as a synergistic constituent of MN-based intradermal immunisation systems. Polyphosphazenes are synthetic macromolecules with a phosphorus–nitrogen backbone and organic side groups. They can be rendered water soluble and are known to possess potent immune-stimulating properties. Poly(di(carboxylatophenoxy)phosphazene) (PCPP) is the most advanced member of this group and has a history of use in clinical trials. The researchers used it here in coating formulations for an amended version of their metal MN system, with titanium arrays composed of 50 needles of height 600 μm. PCPP was used to replace both sodium CMC and surfactant. This may be important due to potential effects on vaccine stability from polymers and surfactants typically used to enhance the efficiency of the coating process. When used as part of an intradermal delivery system for hepatitis B surface antigen, PCPP demonstrated superior activity in pigs compared with intramuscular administration and significant antigen sparing potential. It also accelerated the MN fabrication process and reduced its dependence on the use of surfactants. In this way, PCPP-coated MNs may enable effective intradermal vaccination from an adjuvanted patch delivery system.

In that same year, 2010, the Kendall group in Australia used a probability-based theoretical approach for targeting skin APCs [87]. This research team has carried out a significant body of work using miniaturised needle arrays, termed Nanopatch™. These miniaturised arrays are two orders of magnitude smaller than standard needles and are also much smaller than typical MN arrays. The Nanopatch™ is fabricated from silicon using a process of deep reactive ion etching. The projections are solid silicon, sputter coated with a thin (≈100 nm) layer of gold. In this specific piece of work, the Nanopatch™, comprised of 21 025 needles per cm^2 and 110 μm in length, tapering to tips with a sharpness of <1000 nm were dry-coated with vaccine and applied to mouse skin *in vivo* for 2 min [87]. The group demonstrated that the Nanopatch™ was capable of delivering a seasonal influenza vaccine Fluvax® 2008 to directly target thousands of APCs, in excellent agreement with theoretical predictions. By physically targeting vaccine directly to these cells they induced protective levels of functional antibody responses in mice and also protection against an influenza virus challenge that are comparable to the vaccine delivered intramuscularly with the needle and syringe, but with less than 1/100th of the delivered antigen. They concluded that this study provided a proven mathematical/engineering delivery device template for extension into human studies and speculated that successful translation of these findings into humans could assist with problems of vaccine shortages and distribution, together with alleviating fear of the needle and the need for trained practitioners to administer vaccine during an influenza pandemic.

MN-mediated vaccination has, without question, used influenza models to a tremendous extent, as exemplified by the studies discussed here. Since that prolific year of 2010, various other pieces of work using influenza vaccine coated MN arrays have also been published [77, 88–92]. Weldon and colleagues, for example, investigated the effect of the imiquimod adjuvant on responses to H1N1 influenza subunit coated MNs and deduced that adjuvant inclusion enhanced immune responses [77]. In different work, the inclusion of trehalose helped to stabilise VLPs (which resemble a virus in conformation but are non-infectious) with resultant total virus clearance in a mouse *in vivo* model, 14 months post-single MN vaccination [88]. In the same year, Kommareddy *et al.* demonstrated that coating influenza antigen onto solid MN structures could produce antibody titres comparable to traditional intramuscular injection in a guinea pig model, thus moving studies beyond mouse models [90]. In an innovative approach emerging from Emory University, a fusion protein was constructed as a vaccine tool [91]. Specifically, a broad-spectrum vaccine against influenza was developed, containing multiple repeats of the conserved extracellular domain of the influenza matrix protein 2 (M2e) and the Toll-like receptor (TLR) 5-agonist domains from flagellin, derived from the filament of a bacterial flagellum (M2e-flagellin). Once purified, this fusion protein was coated onto stainless-steel MNs by micro-precision dip coating. Vaccination with M2e-flagellin demonstrated comparative protective efficacy in mice, when compared with intranasal delivery of the same and highlighted a significant improvement over intramuscular delivery [91].

With a view to combating influenza infection in young children, a 2015 study by Koutsonanos and colleagues studied the efficacy of stainless-steel H1N1 influenza subunit vaccine coated MNs to elicit an immune response in young mice [92]. To this effect, female BALB/c mice were vaccinated once with 5 μg of the licenced subunit A/California/07/2009 vaccine via antigen-coated MN array or intramuscular injection. Those mice in the MN cohort showed improved serum IgG1 and IgG2a antibody responses compared with their intramuscular counterparts that received the same dose. Additionally, there were greater numbers of influenza-specific antibody secreting cells (ASCs) in the bone marrow of the MN treated mice and increased activation of follicular helper T cells and formation of germinal centres, formations within secondary lymphoid organs where B-cell differentiation, proliferation, maturation, clone selection and class switch occurs, in the regional lymph nodes. The authors correctly surmised that the significant differences in immune responses observed in this work, between the two immunisation routes tested, served to validate the intradermal immunisation route and to highlight the potential of MN vaccination to improve influenza-related outcomes in all age groups of the population [92]. The latter assertion, undoubtedly, requires significantly more work as MN vaccination strategies progress forward.

This vast range of studies highlights the utility of influenza as a disease model in such vaccination studies. Coated MN arrays have not, however, been limited to coating with influenza vaccines only.

Non-influenza Vaccines Moving away from influenza vaccines, in 2012 Vrdoljak *et al.* reported the delivery of a live recombinant adenovirus and the modified vaccinia virus, Ankara [76]. This was achieved by spray coating vaccine particles onto wet-etched silicon pyramidal MNs, referred to as "ImmuPatch". Their approach produced comparable antibody or CD8+ T cellular responses to those elicited by vaccine delivery via a traditional needle and syringe approach [76]. Stainless-steel MNs have been employed to deliver the Edmonston–Zagreb measles vaccine strain [93]. Trehalose, acting as a viscosity enhancer, was employed to stabilise the vaccine through a rapid drying stage post-coating. This study demonstrated that coated MNs have the ability to produce neutralising antibody levels equivalent to those levels when subcutaneous injection is employed [93]. The successful vaccination of mice against rotavirus, using coated MNs, was reported in 2013. The highlights of this published work included the induction of higher rotavirus-specific IgG and neutralising antibody titres in MN treated mice, compared with intramuscular delivery and the memory response of dendritic cells in the spleen, although this was seen in both MN and intramuscular treatment groups [94].

Hiraishi's 2011 work focused on vaccination against a bacterial disease, namely tuberculosis (TB) caused by *Mycobacterium tuberculosis* [95]. They coated stainless-steel MNs with the licenced human vaccine, BCG, and vaccinated female guinea pigs once with the MNs. The study outcomes demonstrated that the use of the MNs, coated with a low dose of BCG (5×10^4 viable bacilli), resulted in robust antigen-specific cellular immune responses after a single application. The responses were comparable to those induced by traditional intradermal BCG vaccination using a 26G hypodermic needle. The authors concluded that this work provided proof of concept that a live attenuated BCG-coated MN array could be used in the guinea pig model without affecting the immunogenicity of the virus [95].

As discussed in previous sections of this chapter, OVA is widely used in vaccination studies, acting as a model antigen. A very early study, carried out in 2006, investigated immunisation to this model antigen using antigen dip-coated onto titanium MN arrays, which varied in height from 100 to 600 μm and in density from 280 to 1314 MNs per array [96]. The authors studied the influence of depth of vaccine delivery, dose of vaccine delivered, density of MNs on the array and area of application on the resulting immune responses. The immune response was found to be dose dependent and mostly independent of depth of delivery, density of MNs, or area of application. This paper and the conclusions drawn by the authors were the first to point towards microscopic needles, in this instance dip-coated with model antigen, as potential mediators of vaccination across the skin.

An interesting study emerged in 2014 which was designed to evaluate the potential to use coated MN for vaccine delivery in the oral cavity [97]. The theory behind this work was that vaccination in such an easily accessible mucosal area of the body may in turn elicit mucosal immune responses. As such, the subsequent effects of coated MN vaccination via this route on the systemic and mucosal immune responses were investigated. Stainless-steel MN were micro-precision dip-coated with sulforhodamine, as a tracer dye, OVA, as a model antigen, or two HIV antigens concurrently, into the inner lower lip and dorsal surface of the tongue of rabbits, see Figure 4.4.

With reference to the rationale for the use of the two HIV antigens in this work, the oral mucosa is a potential site of induction of secretory immunoglobulin A (sIgA) in the saliva. The authors declared that inducing sIgA in saliva could be a novel approach to combat HIV transmission from mothers to newborns during breastfeeding. To confirm the delivery of sulforhodamine into the tissues, rabbit tongue and lip tissues were isolated post-euthanasia and histologically characterised using fluorescence and brightfield microscopy.

Systemic immune responses against the model antigen, OVA, were measured by determining OVA-specific IgG levels in the serum. Post-8 weeks, there were no statistically significant differences in OVA-specific IgG serum levels between those rabbits vaccinated via the lip or the tongue, suggesting that both sites were equally

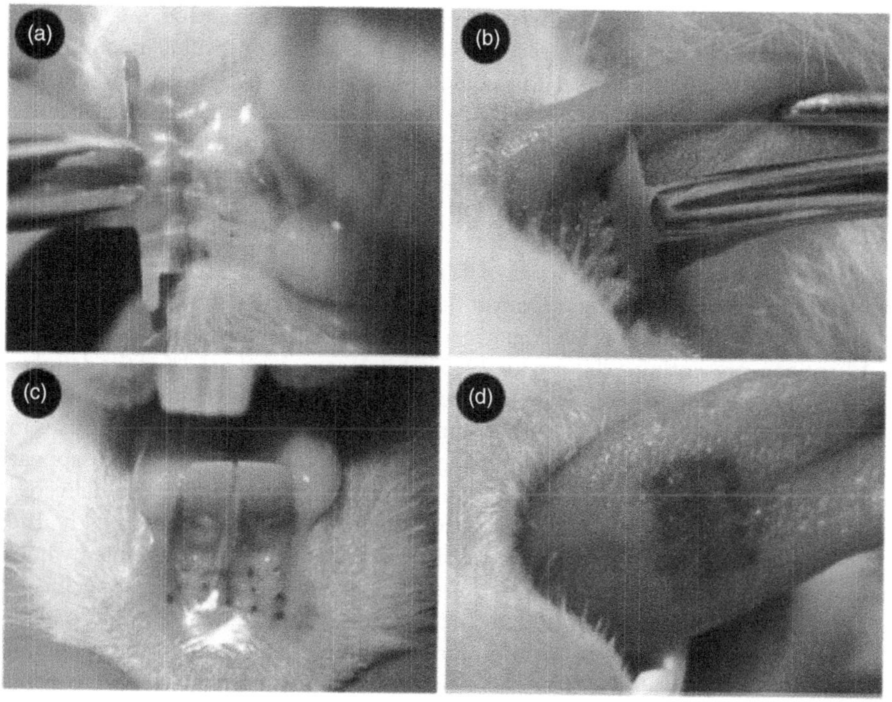

Figure 4.4 Images of the insertion of coated MN into rabbit lip or tongue. Sulforhodamine coated MN were held in Kelly locking forceps and then MN inserted into (a) the extended lower lip and (b) the tongue of rabbits. Regular arrays of dots could then be seen on (c) the lip and (d) the tongue of the rabbit upon removal of the coated MN. Reproduced with permission from [97] Y. Ma, W. Tao, S.J. Krebs, *et al.* (2014). Vaccine delivery to the oral cavity using coated microneedles induces systemic and mucosal immunity. *Pharmaceutical Research* 31 (9): 2393–2403.

immunogenic. Mucosal responses were then measured by determining OVA-specific IgA levels in the saliva. Compared with pre-exposure saliva, there was a significant increase in IgA post-vaccination, irrespective of the vaccination site. Vaccination using the two HIV antigens, E2V3, a virus-like particle expressing the V3 loop of HIV-1, and a DNA expressing gp160 protein, coated onto solid MNs were then evaluated. After two vaccine doses, the MN and intramuscular control groups elicited similar systemic responses with significantly higher specific IgG serum levels in both cohorts, compared with pre-vaccination serum. The oral and intramuscular routes were deemed equivalent in their ability to induce production of IgG in serum. In contrast, mucosal immunity, as measured via antigen-specific IgA antibodies in saliva, was superior in the MN cohort, as exemplified by significantly higher specific IgA levels in the saliva of these animals compared with pre-immunisation levels and with the intramuscular cohort. In fact, intramuscular vaccination produced only weak stimulation of salivary IgA for both antigens. The authors concluded that their hypothesis had been proven, as coated MN arrays were shown to assist in oral mucosal vaccination via the stimulation of both systemic IgG and mucosal IgA production. Mucosal vaccination by the coated MNs also generated more antigen-specific IgA in saliva than the intramuscular vaccination and the MNs also induced a systemic immune response similar to that of the intramuscular route [97]. This innovative route of MN delivery has subsequently been explored further, for example using a commercial influenza vaccine, as a test case immunotherapeutic [98] and more recently using liposome-constituted microneedle arrays (LiposoMAs) [99] and tracer dye coated MNs [100] to determine whether the presence of saliva over the insertion site led to loss of deposited drug in the oral cavity.

Branching from their work using dissolving MN arrays in vaccine delivery [28], the Kendall group used their NanoPatch™ to target intradermal vaccination against West Nile Virus and Chikungunya virus in mice [101]. An individual NanoPatch™ measured 5×5 mm in size and the central 4×4 mm area contained 3364 densely packed projections. The distance between the centres of adjacent projections was 70 mm with each micro-nanoprojection measuring 65 mm in length. The NanoPatch™ was dry-coated with antigen, adjuvant and/or DNA payloads. An OVA payload co-localised with 91.4±4.1 antigen-presenting cells/mm (or 2925 in total) representing 52% of the delivery sites within the NanoPatch™ contact area, agreeing well with a probability-based model used to guide the device design; it then substantially increases as the antigen diffuses in the skin to many more cells. APC co-localising with protein payloads rapidly disappear from the application area, suggesting APC migration. The NanoPatch™ also delivered DNA payloads, leading to cutaneous expression of encoded proteins within 24 h. The efficiency of NanoPatch™ immunisation was demonstrated using an inactivated whole chikungunya virus vaccine and a DNA-delivered attenuated West Nile virus vaccine.

In another study, the group investigated dry-coated NanoPatch™ as an alternative delivery system to intramuscular injection for delivering the alum-adjuvanted human papillomavirus (HPV) vaccine Gardasil®, now commonly used as a prophylactic vaccine against cervical cancer [102]. NanoPatch™ delivered vaccine to mouse ear skin within 5 min. To assess vaccine immunogenicity, doses corresponding to the HPV-16 component of the vaccine between 0.43±0.084 and 300±120 ng were administered to mice at day 0 and day 14. A dose of 55±6.0 ng delivered intracutaneously by NanoPatch™ was sufficient to produce a maximal virus neutralising serum antibody response at day 28 post-vaccination. Neutralising antibody titres were sustained up to 16 weeks post-vaccination and, for comparable doses of vaccine, somewhat higher titres were observed with intracutaneous patch delivery than with intramuscular delivery with the needle and syringe at this time point.

This research team continues to focus energies on NanoPatch™ vaccination strategies, having recently published work coating simian adenovirus serotype 63 and the poxvirus modified vaccinia Ankara – two vectors under evaluation for the delivery of malaria antigens to humans, onto their NanoPatch™ [103] and also inactivated polio virus [104].

The depth and breadth of the MN field, as it applies to vaccine delivery specifically, is not confined or limited to those studies explored here, rather, as the area of MN device development continues to undergo constant change and improvement, there is a growing number of published review articles appraising the status of MN devices in vaccine delivery systems. A selection of the most recent reviews covering this area which are worthy of contemplation are listed in the following references [105–111].

4.5 Future Perspectives

MN-based systems for vaccine delivery have undoubted potential. The majority of studies appear to demonstrate dose-sparing with respect to conventional routes of administration. The ability to formulate vaccines in the dry state is a significant advantage in attempts to circumvent the cold chain, while the lack of medical training required for MN application could prove to be of tremendous benefit, particularly in developing countries. With reference to dissolving MN specifically, the emerging science behind their use in vaccination certainly holds a considerable amount of promise. Formulating MNs from vaccine-loaded polymeric gels is straightforward and avoids complex and time-consuming coating processes, materials are cheap and can be processed at room temperature. Importantly, the MNs are self-disabling, dissolving or biodegrading rapidly in skin to release their payload. This means that these MNs cannot be reused following removal from a patient and also require no special disposal arrangements. Further clinical vaccination studies in human volunteers, such as that recently reported by Emory University [30], are now required to demonstrate the safety and efficacy of this vaccination approach. However, it is not difficult to foresee the major impact vaccine-loaded polymeric MNs could have on the health of human beings worldwide.

Undeniably, key aspects of vaccine formulation lie in the maintenance of vaccine component stability, crucially throughout manufacture and storage. Encapsulation into dissolving MN or, conversely, coating onto solid MN, remains the major challenge to vaccine stability. As with other biological agents, drying stages of manufacture, associated with this MN loading and formulation, tend to have the potential to cause destruction of APIs and with reference to vaccines specifically, may lead to loss of antigenicity. Long-term storage of injectable vaccines continues to present challenges with cold chain storage often required. Therefore, MN approaches may hold the key to overcoming these challenges as many of the strategies discussed here result in vaccine material formulated in the dry state and, as mentioned previously, the possibility for dose-sparing.

It is therefore undeniable that MN-mediated vaccination has gained considerable headway within the academic community and may soon be a viable alternative to traditional vaccine delivery methods, having moved past proof-of-concept, developmental stages of research and into recent clinical trials. It has been suggested that scale-up of animal-model testing should be explored. For example, Rattanapak *et al.* suggested moving from mice, rat and miniature pig models to pre-clinical and clinical studies in man to drive MN vaccination forward so that the full potential of this novel delivery methodology can be achieved [113]. The recent work from Emory University [30] goes some way to addressing these assertions. Scale-up of the manufacturing process of MN vaccines will be required to allow mass production, alongside further endeavours to improve vaccine stability.

With reference to the varying ages of potential MN-users, Phase-II and Phase-III clinical trials conducted with microinjection of a trivalent influenza vaccination observed a difference in response between healthy adult patients aged between 18 and 57 years [114] and healthy elderly patients aged >60 years [115]. Ageing is known to affect the cutaneous immune response, with an age-associated reduction in the numbers of Langerhans cells. However, recent work has highlighted no change in phenotype or function of these monocyte-derived Langerhans cells, suggesting epidermal environment may be more important [116]. Ultimately, the efficacy of MN vaccine delivery will need to be demonstrated as the various dermal components alter with age, ensuring vaccine protection across a breadth of ages.

To conclude, vaccination is considered to be one of the most significant health interventions ever developed and the skin, the largest organ of the human body, has become an attractive site for facilitated intradermal delivery of vaccines via MN arrays. Recent strides forwards in this area led to the recent publication of a National Institutes of Health funded Phase-I clinical trial using dissolving MN arrays for influenza vaccine delivery. This publication and the breadth of ongoing work, being carried out around the globe, serve to highlight the tremendous potential of MN-mediated vaccine delivery strategies.

References

1 E.E. Kris, G. Winter and J. Myschik (2012). Devices for intradermal vaccination. *Vaccine* 30 (1): 523–538.
2 C. Saroja, P. Lakshmi and S. Bhaskaran (2011). Recent trends in vaccine delivery systems: A review. *International Journal of Pharmaceuitical Investigations* 1 (2): 64–74.
3 D. Baxter (2007). Active and passive immunty, vaccine types, excipients and licensing. *Occupational Medicine* 57: 552–556.
4 G. Kersten and H. Hirschberg (2007). Needle-free vaccine delivery. *Expert Opinion on Drug Delivery* 4 (5): 459–474.
5 M. Kendall (2006). Engineering of needle-free physical methods to target epidermal cells for DNA vaccination. *Vaccine* 24 (21): 4651–4656.
6 P.H. Lambertand and P.E. Laurent (2008). Intradermal vaccine delivery: Will new delivery systems transform vaccine administration? *Vaccine* 26 (26): 3197–3208.
7 E.L. Giudice and J.D. Campbell (2006). Needle-free vaccine delivery. *Advanced Drug Delivery Reviews* 58 (1): 68–89.
8 L. Simonsen, A. Kane, J. Lloyd, *et al.* (1999). Unsafe injections in the developing world and transmission of bloodborne pathogens: a review. *Bulletin of the World Health Organization* 77 (10): 789–800.
9 J.G. Hamilton (1995). Needle phobia – a neglected diagnosis. *Journal of Family Practice* 41 (2): 169–175.
10 C.M. Huang (2007). Topical vaccination: the skin as a unique portal to adaptive immune responses. *Seminars in Immunopathology* 29 (1): 71–80.
11 M.O. Oyewumi, A. Kumar and Z. Cui (2010). Nano-microparticles as immune adjuvants: correlating particle size and the resultant immune responses. *Expert Review of Vaccines* 9 (3): 1095–1107.
12 S. Chadwick, C. Kriegel and M. Amiji (2010). Nanotechnology solutions for mucosal immunization. *Advanced Drug Delivery Reviews* 62: 394–407.
13 N. Lycke (2012). Recent progress in mucosal vaccine development: potential and limitations. *Nature Reviews Immunology* 12 (8): 592–605.
14 A. Fortuna, A. Brass, G. Karystianis, *et al.* (2014). Intranasal delivery of systemic-acting drugs: Small-molecules and biomacromolecules. *European Journal of Pharmaceutics and Biopharmaceutics* 88 (1): 8–27.
15 J.M. Steinbach (2015). Protein and oligonucleotide delivery systems for vaginal microbicides against viral STIs. *Cellular and Molecular Life Sciences* 72 (3): 469–503.

16 P. Kuo-Haller, Y. Cu, J. Blum, *et al.* (2010). Vaccine delivery by polymeric vehulces in the mouse reproductive tract induces sustained local and systemic immunity. *Molecular Pharmaceutics* 7 (5): 1585–1595.

17 S. Yang, Y. Chen. R. Ahmadie and E.A. Ho (2013). Advancements in the field of intravaginal siRNA delivery. *Journal of Controlled Release* 167 (1): 29–39.

18 B. Combadière, A. Vogt, B. Mahé, *et al.* (2010). Preferential amplification of CD8 effector-T cells after transcutaneous application of an inactivated influenza vaccine: a randomized phase I trial. *PLoS One* 5 (5): e10818.

19 S. Al-Zahrani, M. Zaric, M.T.C. McCrudden, *et al.* (2012). Microneedle-mediated vaccine delivery: harnessing cutaneous immunobiology to improve efficacy. *Expert Opinion in Drug Delivery* 9 (5): 541–550.

20 A.D. Woolfsonand and D.E. McCaffrey (1993). *Percutaneous Local Anasthesia*. London: Ellis Horwood.

21 D.J. Wong and H.Y. Chang (2009). Skin tissue engineering, StemBook, ed. *The Stem Cell Research Community*, StemBook, doi/10.3824/stembook.1.44.1. www.stembook.org (accessed 25 July 2017).

22 R.J. Scheuplein and I.H. Blank (1971). Permeability of the skin. *Physiology Review* 51 (4): 702–747.

23 A. Laurent, F. Mistretta, D. Bottigioli, *et al.* (2007). Echographic measurement of skin thickness in adults by high frequency ultrasound to assess the appropriate microneedle length for intradermal delivery of vaccines. *Vaccine* 25 (34): 6423–6430.

24 T. Gratieri, I. Alberti, M. Lapteva and Y.N. Kalia (2013). Next generation intra- and transdermal therapeutic systems: using non- and minimally-invasive technologies to increase drug delivery into and across the skin. *European Journal of Pharmaceutical Sciences* 50 (5): 609–622.

25 S.P. Sullivan, D.G. Koutsonanos, M. D. Martin, *et al.* (2010). Dissolving polymer microneedle patches for influenza vaccination. *Nature Medicine* 16 (8): 915–U116.

26 C. Edens, N.C. D ybdahl-Sissoko, W.C. Weldon, *et al.* (2015). Inactivated polio vaccination using a microneedle patch is immunogenic in the rhesus macaque. *Vaccine* 33 (37): 4683–4690.

27 C. Edens, M.L. Collins, J.L. Goodson, *et al.* (2015). A microneedle patch containing measles vaccine is immunogenic in non-human primates. *Vaccine* 33 (37): 4712–4718.

28 A.P. Raphael, T.W. Prow, M.L. Crichton, *et al.* (2010). Targeted, needle-free vaccinations in skin using multilayered, densely-packed dissolving microprojection arrays. *Small* 6 (16): 1785–1793.

29 S. Kommareddy, B. Baudner and S. Oh (2012). Dissolvable microneedle patches for the delivery of cell-culture-derived influenzae vaccine antigens. *Journal of Pharmaceutical Sciences* 101 (3): 1021–1027.

30 N.G. Rouphael, M. Paine, R. Mosley, *et al.* (2017). The safety, immunogenicity, and acceptability of inactivated influenza vaccine delivered by microneedle patch (TIV-MNP 2015): a randomised, partly blinded, placebo-controlled, phase 1 trial. *Lancet* doi: 10.1016/S0140-6736(17)30575-5.

31 M.J. Mistilis, A.S. Bommarius and M.R. Prausnitz (2015). Development of a thermostable microneedle patch for influenza vaccination. *Journal of Pharmaceutical Sciences* 104 (2): 740–749.

32 L.Y. Chu, L.K.R.W. Compans, C. Yang and M.R. Prausnitz (2016). Enhanced stability of inactivated influenza vaccine encapsulated in dissolving microneedle patches. *Pharmaceutical Research* 33 (4): 868–878.
33 E.V. Vassilieva, H. Kalluri, D. McAllister, *et al.* (2015). Improved immunogenicity of individual influenza vaccine components delivered with a novel dissolving microneedle patch stable at room temperature. *Drug Delivery and Translational Research* 5 (4): 360–371.
34 A. Vrdoljak, E.A. Allen, F. Ferrara *et al.* (2016). Induction of broad immunity by thermostabilised vaccines incorporated in dissolvable microneedles using novel fabrication methods. *Journal of Controlled Release* 225: 192–204.
35 M.J. Mistilis, J.C. Joyce, E.S. Esser, *et al.* (2017). Long-term stability of influenza vaccine in a dissolving microneedle patch. *Drug Delivery and Translational Research* 7 (2): 195–205.
36 S. Hirobe, H. Azukizawa, T. Hanafusa, *et al.* (2015). Clinical study and stability assessment of a novel transcutaneous influenza vaccination using a dissolving microneedle patch. *Biomaterials* 57 (1): 50–58.
37 A. Pattani, P.F. McKay, M.J. Garland, *et al.* (2012). Microneedle mediated intradermal delivery of adjuvanted recombinant HIV-1 CN54gp140 effectively primes mucosal boost inoculations. *Journal of Controlled Release* 162 (3): 529–537.
38 Z. Zhu, X. Ye, Z. Ku, *et al.* (2016). Transcutaneous immunization via rapidly dissolvable microneedles protects against hand-foot-and-mouth disease caused by enterovirus 71. *Journal of Controlled Release* 243: 291–302.
39 Q.Y. Mao, Y. Wang, L. Bian, *et al.* (2016). EV71 vaccine, a new tool to control outbreaks of hand, foot and mouth disease (HFMD). *Expert Review of Vaccines* 15 (5): 599–606.
40 J.M. Arya, K. Dewitt, M. Scott-Garrard, *et al.* (2016). Rabies vaccination in dogs using a dissolving microneedle patch. *Journal of Controlled Release* 239: 19–26.
41 K. Matsuo, S. Hirobe, Y. Yokota, *et al.* (2012). Transcutaneous immunization using a dissolving microneedle array protects against tetanus, diphtheria, malaria, and influenza. *Journal of Controlled Release* 160 (3): 495–501.
42 E.S. Esser, A. Romanyuk, E.V. Vassilieva, *et al.* (2016). Tetanus vaccination with a dissolving microneedle patch confers protective immune responses in pregnancy. *Journal of Controlled Release* 236: 47–56.
43 G.H. Li, A. Badkar, S. Nema, *et al.* (2009). In vitro transdermal delivery of therapeutic antibodies using maltose microneedles. *International Journal of Pharmaceutics* 368 (1-2): 109–115.
44 M. Zaric, M., O. Lyubomska, O. Touzelet, *et al.* (2013). Skin dendritic cell targeting via microneedle arrays laden with co-glycolide nanoparticles induces efficient antitumor and antiviral immune responses. *ACS Nano* 7 (3): 2042–2055.
45 M.T.C. McCrudden, A.Z. Alkilani, C.M. McCrudden, *et al.* (2014). Design and physicochemical characterisation of novel dissolving polymeric microneedle arrays for transdermal delivery of high dose, low molecular weight drugs. *Journal of Controlled Release* 180 (1): 71–80.
46 K. Matsuo, Y. Yokota, Y. Zhai, *et al.* (2012). A low-invasive and effective transcutaneous immunization system using a novel dissolving microneedle array for soluble and particulate antigens. *Journal of Controlled Release* 161 (1): 10–17.

47 S. Naito, Y. Ito, T. Kiyohara, *et al.* (2012). Antigen-loaded dissolving microneedle array as a novel tool for percutaneous vaccination. *Vaccine* 30 (6): 1191–1197.
48 P.C. Demuth, W.F. Garcia-Beltran, M.L. Ai-Ling, *et al.* (2013). Composite dissolving microneedles for coordinated control of antigen and adjuvant delivery kinetics in transcutaneous vaccination. *Advanced Functional Materials* 23 (2): 161–172.
49 M.C. Chen, S.F. Huang, K.Y. Lai and M.H. Ling (2013). Fully embeddable chitosan microneedles as a sustained release depot for intradermal vaccination. *Biomaterials* 34 (12): 3077–3086.
50 L. Guo, J. Chen, Y. Qiu, *et al.* (2013). Enhanced transcutaneous immunization via dissolving microneedle array loaded with liposome encapsulated antigen and adjuvant. *International Journal of Pharmaceutics* 447 (1-2): 22–30.
51 P.C. DeMuth, Y. Min, D.J. Irvine and P.T. Hammond (2014). Implantable silk composite microneedles for programmable vaccine release kinetics and enhanced immunogenicity in transcutaneous immunization. *Advanced Healthcare Materials* 3 (1): 47–58.
52 D. Schenk, R. Barbour, W. Dunn, *et al.* (1999). Immunization with amyloid-beta attenuates Alzheimer-disease-like pathology in the PDAPP mouse. *Nature* 400 (6740): 173–177.
53 J.M. Orgogozo, S. Gilman, J.F. Dartigues, *et al.* (2003). Subacute meningoencephalitis in a subset of patients with AD after Abeta42 immunization. *Neurology* 61 (1): 46–54.
54 K. Mutsuo, H. Okamoto and Y. Kawai (2014). Vaccine efficacy of transcutaneous immunisation with amloid beta using a dissolving microneedle array in a mouse model of Alzheimer's disease. *Journal of Neuroimminology* 122 (1-2): 1–11.
55 P.M. Flynn, J.L. Shenep, *et al.* (1994). Influence of needle gauge in Mantoux skin testing. *Chest* 106 (5): 1463–1465.
56 P.E. Laurent, S. Bonnet and P. Alchas (2007). Evaluation of the clinical performance of a new intradermal vaccine administration technique and associated delivery system, *Vaccine* 25 (52): 8833–8842.
57 J.A. Mikszta, V.J. Sullivan, C. Dean, *et al.* (2005). Protective immunization against inhalational anthrax: a comparison of minimally invasive delivery platforms. *Journal of Infectious Diseases* 191 (2): 278–288.
58 J.A. Mikszta, J.P. Dekker III,, N.G. Harvey, *et al.* (2006). Microneedle-based intradermal delivery of the anthrax recombinant protective antigen vaccine. *Infection and Immunity* 74 (12): 6806–6810.
59 J.B. Alarcon, A.W. Hartley, N.G. Harvey and J.A. Mikszta (2007). Preclinical evaluation of microneedle technology for intradermal delivery of influenza vaccines. *Clinical and Vaccine Immunology* 14 (4): 375–381.
60 P.E. Laurent, H. Bourhy, M. Fantino, *et al.* (2010). Safety and efficacy of novel dermal and epidermal microneedle delivery systems for rabies vaccination in healthy adults. *Vaccine* 28 (36): 5850–5856.
61 P. Van Damme, F. Oosterhuis-Kafeja, M. Van der Wielen, *et al.* (2009). Safety and efficacy of a novel microneedle device for dose sparing intradermal influenza vaccination in healthy adults. *Vaccine* 27 (3): 454–459.

62 G.L. Morefield, R.F. Tammariello and P.K. Purcel (2008). An alternative approach to combination vaccines: intradermal administration of isolated components for control of anthrax, botulism, plague and staphylococcal toxic shock, *Journal of Immune Based Therapy Vaccines* 6 (1): 5–16.
63 K. Van der Maaden, S.J. Trietsch, H. Kraan, *et al.* (2014). Novel hollow microneedle technology for depth-controlled microinjection-mediated dermal vaccination: a study with polio vaccine in rats. *Pharmaceutical Research* 31: 1846–1854
64 P. Schipper, K. van der Maaden, S. Romeijn, *et al.* (2016). Determination of depth-dependent intradermal immunogenicity of adjuvanted inactivated polio vaccine delivered by microinjections via hollow microneedles. *Pharmaceutical Research* 33: 2269–2279.
65 P. Schipper, K. van der Maaden, S. Romeijn, *et al.* (2016). Repeated fractional intradermal dosing of an inactivated polio vaccine by a single hollow microneedle leads to superior immune responses. *Journal of Controlled Release* 242: 141–147.
66 K. Siddhapura, H. Harde and S. Jain (2016). Immunostimulatory effect of tetanus toxoid loaded chitosan nanoparticles following microneedles assisted immunization. *Nanomedicine* 12 (1): 213–222.
67 Z. Ding, F.J. Verbaan, M. Bivas-Benita, *et al.* (2009). Microneedle arrays for the transcutaneous immunization of diphtheria and influenza in BALB/c mice. *Journal of Controlled Release* 136 (1): 71–78.
68 Z. Ding, E. Van Riet, S. Romeijn, *et al.* (2009). Immune modulation by adjuvants combined with diphtheria toxoid administered topically in BALB/c mice after microneedle array pretreatment. *Pharmaceutical Research (Dordrecht)* 26 (7): 1635–1643.
69 S.M. Bal, Z. Ding, G.F.A. Kersten, *et al.* (2010). Microneedle-based transcutaneous immunisation in mice with *N*-trimethyl chitosan adjuvanted diphtheria toxoid formulations. *Pharmaceutical Research* 27 (9): 1837–1847.
70 Z. Ding, S.M. Bal, S. Romeijn, *et al.* (2011). Transcutaneous immunization studies in mice using diphtheria toxoid-loaded vesicle formulations and a microneedle array. *Pharmaceutical Research* 28 (1): 145–158.
71 K. Van der Maaden, E.M. Varypataki, S. Romeijn, *et al.* (2014). Ovalbumin-coated pH-sensitive microneedle arrays effectively induce ovalbumin-specific antibody and T-cell responses in mice. *European Journal of Pharmaceutics and Biopharmaceutics* 88 (2): 310–315.
72 Q. Zhu, V.G. Zarnitsyn, L. Ye, *et al.* (2009). Immunization by vaccine-coated microneedle arrays protects against lethal influenza virus challenge. *Proceedings of the National Academy of Sciences USA* 106 (19): 7968–7973.
73 D.G. Koutsonanos, M.M. del Pilar, V.G. Zarnitsyn, *et al.* (2009). Transdermal influenza immunization with vaccine-coated microneedle arrays. *PLoS ONE* 4 (3): e4773.
74 Y.C. Kim, F.S. Quan, R.W. Compans, *et al.* (2010). Formulation and coating of microneedles with inactivated influenza virus to improve vaccine stability and immunogenicity. *Journal of Controlled Release* 142 (2): 187–195.
75 M.G. McGrath, A. Vrdoljak, C. O'Mahony, *et al.* (2011). Determination of parameters for successful spray coating of silicon microneedle arrays. *International Journal of Pharmaceutics* 415 (1-2): 140–149.

76 A. Vrdoljak, M.G. McGrath, J.B. Carey, *et al.* (2012). Coated microneedle arrays for transcutaneous delivery of live virus vaccines. *Journal of Controlled Release* 159 (1): 34–42.
77 W.C. Weldon, V.G. Zarnitsyn, E.S. Esser, *et al.* (2012). Effect of adjuvants on responses to skin immunization by microneedles coated with influenza subunit vaccine. *PLoS One* 7 (7): e41501.
78 H.J. Choi, B.J. Bondy, D.G. Yoo, *et al.* (2013). Stability of whole inactivated influenza virus vaccine during coating onto metal microneedles. *Journal of Controlled Release* 166 (2): 159–171.
79 J. Chen, Y. Qiu, S. Zhang and Y. Gao (2016). Dissolving microneedle-based intradermal delivery of interferon-α-2b. *Drug Development and Industrial Pharmacy* 42 (6): 890–896.
80 Y.C. Kim, F.S. Quan, D.G. Yoo, *et al.* (2009). Improved influenza vaccination in the skin using vaccine coated microneedles. *Vaccine* 27 (49): 6932–6938.
81 Y.C. Kim, F.S. Quan, D.G. Yoo, *et al.* (2010). Enhanced memory responses to seasonal H1N1 influenza vaccination of the skin with the use of vaccine-coated microneedles. *Journal of Infectious Diseases* 201 (2): 190–198.
82 M. Pearton, S.M. Kang, J.M. Song, *et al.* (2010). Influenza virus-like particles coated onto microneedles can elicit stimulatory effects on Langerhans cells in human skin. *Vaccine* 28 (37): 6104–6113.
83 F.S. Quan, Y.C. Kim, A. Vunnava, *et al.* (2010). Intradermal vaccination with influenza virus-like particles by using microneedles induces protection superior to that with intramuscular immunization. *Journal of Virology* 84 (15): 7760–7769.
84 F.S. Quan, Y.C. Kim, R.W. Compans, *et al.* (2010). Dose sparing enabled by skin immunization with influenza virus-like particle vaccine using microneedles. *Journal of Controlled Release* 147 (3): 326–332.
85 Y.C. Kim, F.S. Quan, R.W. Compans, *et al.* (2010). Formulation of microneedles coated with influenza virus-like particle vaccine. *AAPS PharmSciTech* 11 (3): 1193–1201.
86 J.M. Song, Y.C. Kim, A.S. Lipatov, *et al.* (2010). Microneedle delivery of H5N1 influenza virus-like particles to the skin induces long-lasting B- and T-cell responses in mice. *Clinical and Vaccine Immunology* 17 (9): 1381–1389.
87 G.J.P. Fernando, X.F. Chen, T.W. Prow, *et al.* (2010). Potent immunity to low doses of influenza vaccine by probabilistic guided micro-targeted skin delivery in a mouse model. *Plos One* 5 (4): e10266.
88 F.S. Quan, Y.C. Kim, J.M. Song, *et al.* (2013). Long-term protective immunity from an influenza virus-like particle vaccine administered with a microneedle patch. *Clinical and Vaccine Immunology*, 20 (9): 1433–1439.
89 M.C. Kim, J.W. Lee, H.J. Choi, *et al.* (2015). Microneedle patch delivery to the skin of virus-like particles containing heterologous M2e extracellular domains of influenza virus induces broad heterosubtypic cross-protection. *Journal of Controlled Release* 210: 208–216.
90 S. Kommareddy, B.C. Baudner, A. Bonificio, *et al.* (2013). Influenza subunit vaccine coated microneedle patches elicit comparable immune responses to intramuscular injection in guinea pigs. *Vaccine* 31 (34): 3435–3441.

91 B.Z. Wang, H.S. Gill, C. He, *et al.* (2014). Microneedle delivery of an M2e-TLR5 ligand fusion protein to skin confers broadly cross-protective influenza immunity. *Journal of Controlled Release* 178: 1–7.
92 D.G. Koutsonanos E.S. Esser, S.R. McMaster, *et al.* (2015). Enhanced immune responses by skin vaccination with influenza subunit vaccine in young hosts. *Vaccine* 33 (37):4675–4682.
93 C. Edens, M.L. Collins, J. Ayers, *et al.* (2013). Measles vaccination using a microneedle patch. *Vaccine* 31 (34): 3403–3409.
94 S. Moon, Y. Wang, C. Edens, *et al.* (2013). Dose sparing and enhanced immunogenicity of inactivated rotavirus vaccine administered by skin vaccination using a microneedle patch. *Vaccine* 31 (34): 3396–3402.
95 Y. Hiraishi, S. Nandakumar, S.O. Choi, *et al.* (2011). Bacillus Calmette-Guérin vaccination using a microneedle patch. *Vaccine* 29 (14): 2626–2636.
96 G. Widera, J. Johnson, L. Kim, *et al.* (2006). Effect of delivery parameters on immunization to ovalbumin following intracutaneous administration by a coated microneedle array patch system. *Vaccine* 24 (10): 1653–1664.
97 Y. Ma, W. Tao, S.J. Krebs, *et al.* (2014). Vaccine delivery to the oral cavity using coated microneedles induces systemic and mucosal immunity. *Pharmaceutical Research* 31 (9): 2393–2403.
98 C.L. McNeilly, M.L. Crichton, C.A. Primiero, *et al.* (2014). Microprojection arrays to immunise at mucosal surfaces. *Journal of Controlled Release* 196: 252–260.
99 T. Wang and N. Wang (2016). Preparation of the multifunctional liposome-containing microneedle arrays as an oral cavity mucosal vaccine adjuvant-delivery system. *Methods in Molecular Biology* 1404: 651–667.
100 L. Serpe, A. Jain, C.G. de Macedo, *et al.* (2016). Influence of salivary washout on drug delivery to the oral cavity using coated microneedles: An in vitro evaluation. *European Journal of Pharmaceutical Sciences* 93: 215–223.
101 T.W. Prow, X.F. Chen, N.A. Prow, *et al.* (2010). Nanopatch-targeted skin vaccination against West Nile virus and Chikungunya virus in mice. *Small* 6 (16): 1776–1784.
102 H.J. Corbett, G.J.P. Fernando, X.F. Chen, *et al.* (2010). Skin vaccination against cervical cancer associated human papillomavirus with a novel micro-projection array in a mouse model. *Plos One* 5 (10): e13460.
103 F.E. Pearson, C.L. McNeilly, M.L. Crichton, C.A. Primiero, *et al.* (2013). Dry-coated live viral vector vaccines delivered by nanopatch microprojections retain long-term thermostability and induce transgene-specific T cell responses in mice. *PLoS One* 8 (7): e67888.
104 D.A. Muller, F.E. Pearson, G.J. Fernando, *et al.* (2016). Inactivated poliovirus type 2 vaccine delivered to rat skin via high density microprojection array elicits potent neutralising antibody responses. *Scientific Reports* 6: 22094.
105 S. Marshall, L.J. Sahm and A.C. Moore (2016). The success of microneedle-mediated vaccine delivery into skin. *Human Vaccines and Immunotherapeutics* 12 (11): 2975–2983.
106 E. Larrañeta, M.T.C. McCrudden, A.J. Courtenay and R.F. Donnelly (2016). Microneedles: A new frontier in nanomedicine delivery. *Pharmaceutical Research* 33 (5): 1055–1073.

107 L. Yang, W. Li, M. Kirberger, *et al.* (2016). Design of nanomaterial based systems for novel vaccine development. *Biomaterials Science* 4 (5): 785–802.

108 J. Arya and M.R. Prausnitz (2016). Microneedle patches for vaccination in developing countries. *Journal of Controlled Release* 240: 135–141.

109 J. Li, M. Zeng, H. Shan and C. Tong (2017). Microneedle patches as drugs and vaccine delivery platform. Current Medicinal Chemistry doi: 10.2174/0929867324666170526124053.

110 C.I. Shin, S.D. Jeong, N.S. Rejinold and Y.C. Kim (2017). Microneedles for vaccine delivery: challenges and future perspectives. *Therapeutic Delivery* 8 (6): 447–460.

111 M.R. Prausnitz (2017). Engineering microneedle patches for vaccination and drug delivery to skin. *Annual Review of Chemical and Biomolecular Engineering* 8: 177–200.

112 A.P. Raphael, M.L. Crichton, R.J. Falconer and M.A. Kendall (2016). Formulations for microprojection/microneedle vaccine delivery: Structure, strength and release profiles. *Journal of Controlled Release* 225 (1): 40–52.

113 T. Rattanapak, J.C. Birchall, K. Young, *et al.* (2014). Dynamic visualization of dendritic cell-antigen interactions in the skin following transcutaneous immunization. *PLoS One* 9 (2): e89503.

114 J. Beran, A. Ambrozaitis, A. Laiskonis, *et al.* (2009). Intradermal influenza vaccination of healthy adults using a new microinjection system: a 3-year randomised controlled safety and immunogenicity trial. *BMC Medicine* 7: 13.

115 R. Arnou, G. Icardi, M. De Decker, et al. (2009). Intradermal influenza vaccine for older adults: a randomized controlled multicenter phase III study. *Vaccine* 27 (52): 7304–7312.

116 S. Ogden, R.J. Dearman, I. Kimber and C.E.M. Griffiths (2011). The effect of ageing on phenotype and function of monocyte-derived Langerhans cells. *The British Journal of Dermatology*, 165 (1): 184–188.

5

Microneedles for Gene Therapy: Overcoming Extracellular and Intracellular Barriers

Grace Cole[1], *Nicholas J. Dunne*[2] *and Helen O. McCarthy*[1]

[1] *School of Pharmacy, Queen's University Belfast, Belfast BT9 7BL, Northern Ireland, UK*
[2] *School of Mechanical & Manufacturing Engineering, Dublin City University, Dublin 9, Ireland*

5.1 Gene Therapy

Gene therapy may be defined as the use of genetic material to modify cells *in vitro* or *in vivo* for corrective or therapeutic purposes [1]. Central to the concept of gene therapy is that the introduction of DNA to the nucleus or of mRNA (messenger RNA) to the cytoplasm of the host cell results in the translation of the encoded protein. Nucleic acids may therefore be employed as a means to replace or repair the mutated or disease-causing genes (gene transfer therapy) [2, 3]. On the other hand, new genetic material may be introduced to cause expression of a desired therapeutic or immunogenic protein. Alternatively, abnormal genes may be deleted or "knocked down" to restore normal function via interfering RNAs [4, 5]. As such gene therapy is a broad field that holds the potential to treat any number of heritable or acquired disease states, which are typically treated with recombinant proteins [6].

Gene therapy utilises the host's cellular machinery to synthesise the desired protein thus ensuring correct folding in the translation process. Indeed, one or more proteins may be delivered within the sequence of the nucleic acid. The advent of plasmid vector refinements, such as mini-circle technology, and the scaffold–matrix attachment regions (S/MAR) facilitate persistence of introduced genes within host cells, without the need for chromosomal insertion, thereby negating regular dosing [7, 8]. Additionally, expression of a therapeutic/cytotoxic protein within desired target cells and/or tissues may be controlled by delivery vehicles, or specific promoters, thereby reducing off-target side-effects. Gene therapy is being investigated for a broad range of diseases, from cancer to conditions characterised by abnormal protein expression such as cystic fibrosis [6]. The majority of these disease states require expression of the therapeutic protein at the target tissue and delivery to the skin would constitute an additional barrier to efficacy. Therefore, the use of microneedles to deliver nucleic acids has two distinct applications: firstly, for vaccination, where the dermis and epidermis make an attractive immunogenic target [9–11]; and secondly, for the treatment of diseases where local transfection within the skin is desirable, such as proliferative disorders, genetic mutations, local cancerous lesions or wound healing.

Microneedles for Drug and Vaccine Delivery and Patient Monitoring, First Edition.
Edited by Ryan F. Donnelly, Thakur Raghu Raj Singh, Eneko Larrañeta, and Maelíosa T.C. McCrudden.

5.2 DNA Vaccination

DNA vaccination is a type of immunotherapy that aims to harness the host's immune system to mount an antigen-specific immune response against the encoded protein [12]. Genetic immunisation has garnered much interest since the first experiments in the early 1990s where Wolff *et al.* demonstrated that intra-muscular (IM) injection of reporter gene cargo could induce long term protein expression *in vivo* [13]. To date four commercially available DNA vaccines have been licenced for use in animals, including West Nile Innovator DNA (Pfizer), Apex-IHN (Novartis Aqua Health), Lifetide SW 5 (VGX Animal Health) and ONCEPT (Vical), which was licenced in 2007 for the treatment of canine malignant melanoma [14]. According to *The Journal of Gene Medicine*, of the 2409 gene therapy clinical trials ongoing, 19.3% of these deliver DNA coding for an antigen (www.abedia.com/wiley/genes.php) and the first licensure of DNA vaccines for human use is eagerly anticipated.

5.2.1 Advantages of DNA Vaccination

Endogenous production of antigenic peptides leads to several advantages over conventional vaccination strategies. (1) There is no need for handling potentially hazardous pathogens or the processing of live viruses, resulting in reduced biosafety requirements and hence cost. (2) DNA vaccines have no risk of reversion to virulence, raise a highly specific immune response against the encoded antigen and have proven to be safe and are highly tolerable in multiple clinical trials [15, 16]. (3) DNA vaccines may contain genes coding the entire length of an antigenic peptide, offering multiple potential epitopes for immune activation [17]. (4) DNA vaccines have the potential to encode multiple antigenic peptides and/or molecular adjuvants, and hence may be useful in immunising against multiple strains or diseases at one time, or targeting multiple antigens synergistically [18–20]. (5) DNA is a relatively stable molecule, which can be easily produced via replication of transformed bacteria and so production and storage should be relatively inexpensive [21, 22]. (6) Lastly, DNA vaccines are potent activators of both cellular and humoral immune responses and hence can be used for prophylactic and therapeutic purposes [17, 23].

5.2.2 Mechanism of Action of DNA Vaccines

Central to the induction of an adaptive immune response is the capture and presentation of antigenic peptides by antigen presenting cells (APCs) and the subsequent expansion of an antigen specific T-cell population. DNA vaccines delivered to the host transfect local non-APCs (e.g. myocytes or keratinocytes) or "professional" APCs (e.g. dendritic cells (DCs)), leading to the endogenous production of the antigenic peptide. These host-derived antigens are then able to be displayed by APCs to naïve CD8+ or CD4+ T cells in the form of major histocompatibility class I (MHC-I) or class II (MHC-II) complexes, respectively, via several mechanisms [12]. Figure 5.1 provides a schematic representation of DNA vaccination.

Direct transfection of APCs leads to the production of the antigenic protein/peptide in the cell cytosol and subsequent proteolysis via a proteasome. Proteolytic products are translocated to the endoplasmic reticulum (ER) lumen via the transporter associated

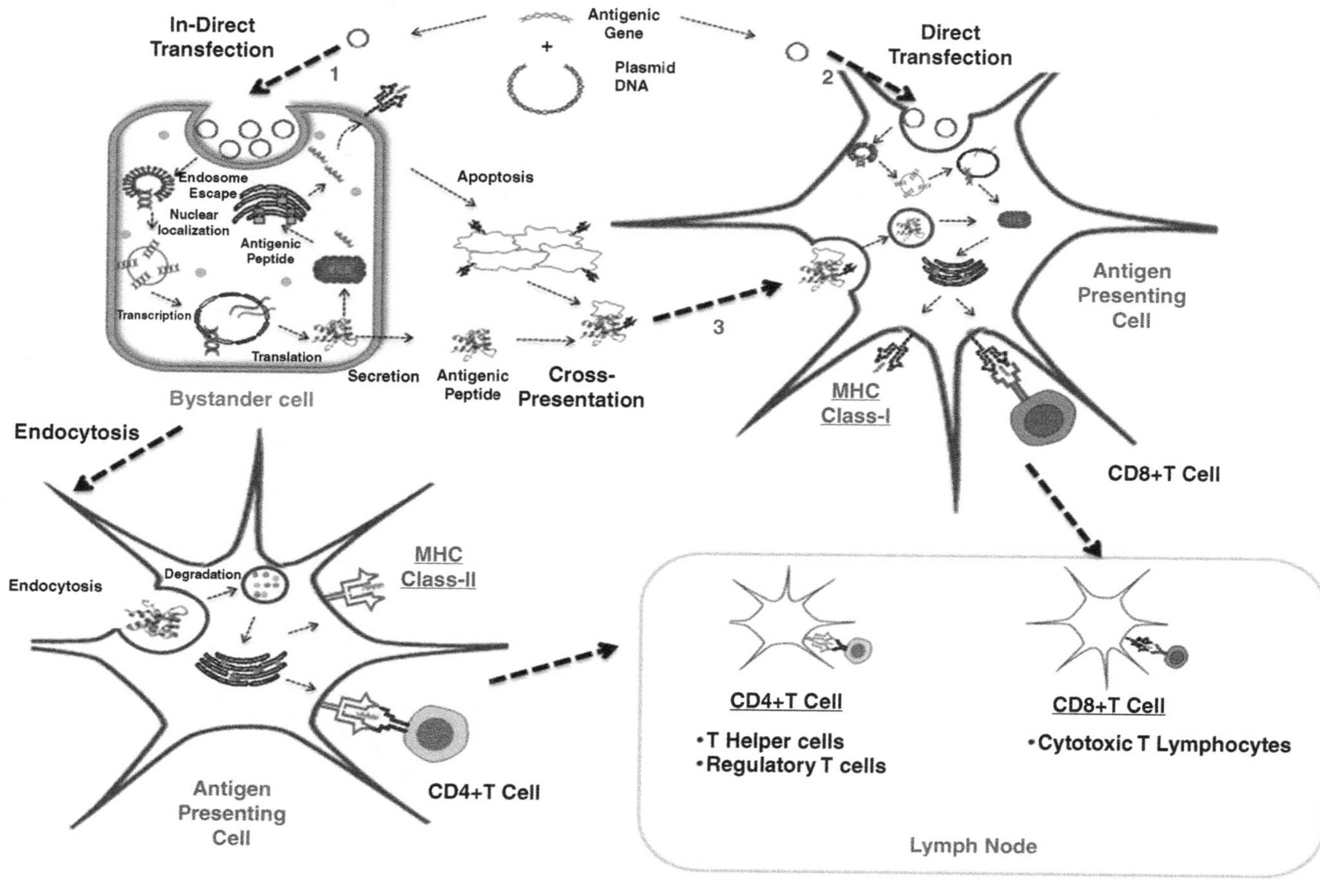

Figure 5.1 Schematic representation of DNA vaccination. (1) In order to be effective, DNA encoding antigenic peptide(s) must transfect local antigen-presenting cells (APCs) and non-antigen presenting cells (non-APCs), resulting in the endogenous production of the antigenic peptide(s). Transfection of non-APCs results in secretion of the antigenic peptide into circulation, subsequent capture by APCs and presentation to naïve CD4+ T cells via complexation with major histocompatibility (MHC) class II molecules. (2) Direct transfection of APCs results in antigen presentation via complexation with MHC class I molecules and the subsequent presentation to naïve CD8+ T cells. (3) Additionally, APCs are able to present some exogenous antigen on MHC class I molecules leading to the subsequent presentation to CD8+ T cells via a process known as cross-presentation.

with antigen processing (TAP), and bind to MHC-I molecules (which have been assembled in the ER) before being transported to the cell membrane via the golgi [24].

Alternatively, transfection of non-APCs leads to the secretion of antigenic peptide into the host circulation where it can be captured by APCs through several complementary mechanisms. Particulate antigen or apoptotic cells expressing the antigenic peptide(s) may be phagocytosed by APCs into a phagosome [25], while soluble antigens may be introduced into an endosome via receptor-mediated endocytosis or pinocytosis [26]. Both organelles eventually merge with lysosomes and the contents are subjected to lysosomal proteolysis. MHC-II molecules that have been assembled within the ER of APCs are transported to the phagocytic and endocytic pathways, by Ii, a chaperone associated with the MHC class II dimer [27]. In the phagolysosome/endosome Ii is degraded and the intermediate products of proteolysis bind to MHC-II molecules to form complexes that migrate to the cell surface for presentation [24, 26].

MHC-I molecules may also access peptides from these pathways through the process of cross-presentation [24, 28, 29]. Following phagosome formation within DCs it has been shown that inhibition of protease activity by the nicotinamide adenine dinucleotide phosphate (NAPDH) oxidase, NOX2, enhances cross-presentation [30–32]. As opposed to other antigen-presenting cells, such as macrophages, which rapidly acidify and degrade phagosome contents, DCs have lower levels of protease activity within phagosomes, allowing increased time for peptide processing and cross-presentation [31, 33]. At this stage there are three main models for how the loading of exogenous antigen onto MHC-I molecules may occur, although it is probable that all models could contribute synergistically. Firstly, it is proposed that inhibiting protease activity allows time for transport of the antigen to the cytosol where the peptide can be processed via proteasomes, and bound to MHC-I peptides, as described earlier [34]. Alternatively, ER contents have been found to fuse with phagosomes, suggesting that the MHC-I loading process could occur exclusively within the phagosome, known as the vacuolar pathway [29, 35, 36]. Lastly, certain studies support an ER–phagosome model where the ER fuses with phagosomes containing antigenic peptide(s). Antigen(s) are subsequently transported out of the phagosome to the cytosol, degraded by a proteasome and then transported back into the same phagosome for MHC-I loading within the ER–phagosome compartment [37, 38].

Cross-presentation from within the endocytic pathway is also heavily dependent on delaying the degradation of endosomal contents, and is thought to be similar to the mechanism of phagosomal cross-presentation [26]. Certain receptors involved in receptor-mediated endocytosis are associated with sorting soluble antigens into endosomes with distinct maturation characteristics, and hence target antigens towards MHC class I and/or II presentation [26, 39–41]. Examples of receptors that have been demonstrated to sort antigens into stable early endosomes, and hence promote MHC-I loading, include the mannose receptor (MR) [39, 42] and Langerin [43]. Conversely, certain receptors target antigens exclusively to lysosomal degradation, promoting MHC-II loading [39]. Once within stabilised early endosomes the mechanisms of MHC-I loading are not fully understood, but are proposed to be similar to phagosomal loading, where MHC-I molecule binding may occur following transport to the cytosol or within the endosome itself [24, 44]. Importantly, different populations of APCs display varied patterns of these receptors, making some particularly efficient for cross-presentation. For example, Langerhan cells (LCs), a type of DC found in high abundance within the epidermis, and Langerin+ DCs, display high levels of the Langerin receptor, which almost exclusively directs the antigen to the early endosome and hence CD8+ T cell priming [9, 43].

Following presentation of peptide–MHC complexes on the cell surface, DCs must undergo phenotypic changes, a process known as maturation, from immature/sentient DCs to DCs that can stimulate the expansion of an effector T-cell population. This essential process of maturation is influenced by changes in the microenvironment as well as the antigens encountered [45]. DCs express receptors for a large number of cytokines and chemokines, including, but not limited to, interleukin-1 (IL-1), granulocyte-macrophage colony-stimulating factor (GM-CSF) and tumour necrosis factor-α (TNF-α) and IFN-γ [45], which promote the maturation of DCs [10, 46–48]. Conversely, other receptors, for example IL-10, which is commonly produced within the tumour microenvironment, can suppress maturation, leading to the expansion of a regulatory T-cell population [49]. Importantly, damage to local cells may cause the release of cell membrane polysaccharides and preformed cytokines, leading to the activation of DC surface receptors and an increase of these "danger" maturation signals within the local environment [50]. DCs also express a variety of pattern recognition receptors (PRRs), which recognise highly conserved molecules produced by microbes such as CpG oligodeoxynucleotides (ODNs), LPS, carbohydrates, lipids and proteins, leading to the activation of DCs. With the exception of CpG ODNs, which are expressed in high amounts on plasmid DNA vaccines and hence may promote DC activation, this method of maturation induction is more relevant for physiological infection, rather than DNA vaccination [51].

Stimulation from these sources leads to the initiation of several events: DC migration to lymph nodes, that is promoted by the expression of CCR7 which recognises chemokines expressed within the lymph node [52]. The production of ceramide decreases the antigen-capturing capabilities of the cell, limiting the presentation of antigens to those captured outside of the lymph node [53]. The MHC presentation of captured antigen is upregulated, resulting in hundreds of MHC–peptide complexes being displayed at the cell surface where they can interact with naïve T cells [54]. The expression of a trio of molecules necessary for T-cell interaction (CD40) and co-stimulation (CD80 (B7-1) and CD86 (B7-2)) is upregulated [55]. Lastly, DCs begin to secrete cytokines and chemokines, for example IL-12, IL-15 and IFN-α, which promote the expansion of the innate (NK cells) and adaptive (T cells and B cells) branches of immunity [56].

Within the lymph node, multiple cell surface receptors mediate the interaction between mature antigen-loaded DCs and naïve T lymphocytes. The most critical for T-cell activation is the interaction between the T-cell receptor (TCR) of the T lymphocyte and the MHC–peptide complex of the DC. All T lymphocytes are equipped with hundreds of surface TCRs, composed of an α and a β chain, which are responsible for recognising non-self antigens bound to MHC molecules [57]. Unlike other receptors that are highly conserved between cells, the structure of the TCR varies between T cells due to deliberate genetic shuffling during T-cell production within the thymus [57]. Thus, there are millions of possible TCR antigen-recognition surfaces, each capable of binding to peptide–MHC ligands to differing degrees. At the cell membrane, the TCR forms a complex with CD3, a signal transduction complex composed of subunits whose cytoplasmic domains contain immunoreceptor tyrosine-based action motifs (ITAMs) [58]. High-affinity engagement of an MHC–peptide ligand into the TCR leads to biochemical changes in the CD3 cytoplasmic ITAMs, known as T-cell "triggering," which leads to the recruitment of intracellular molecules to initiate signalling cascades. In addition to this TCR–MHC complex interaction other receptor and co-receptor/ligand interactions are critical for CD3 ITAM phosphorylation and subsequent T-cell activation.

TCRs are accompanied by CD8 or CD4 co-receptors, which enable binding to MHC I or MHC II molecules, respectively. The cytoplasmic regions of these CD8 or CD4 co-receptors are linked to the cytoplasmic tyrosine kinase, LCK. Upon MHC–peptide engagement of the TCR and CD8/CD4 co-receptor simultaneously, LCK is brought into closer proximity with, and phosphorylates, the cytoplasmic CD3 ITAMs [59]. This phosphorylation is necessary for T-cell activation, but can also occur in the absence of co-receptors, although this leads to less efficient TCR triggering [59]. Following phosphorylation, CD3 ITAMs recruit the cytoplasmic tyrosine kinase, zeta chain associated protein 70 (ZAP70), that in turn phosphorylates linker of activation of T cells (LAT), CD6 and SLP-76, which form signalosomes in the cytoplasm [58, 59]. These signalosomes are responsible for the re-organisation of the T cell actin cytoskeleton and nearly all of the subsequent downstream signalling including T-cell differentiation [60–62].

T-cell activation is also dependent on co-stimulatory and co-inhibitory integrin and cytokine receptor–ligand interactions, which are separate from those produced by the MHC–peptide/TCR interaction. These play a key role in determining the functional outcome of TCR triggering based on APC and cytokine signalling [63]. The mature APC expresses the co-stimulatory ligands CD58 (LFA-3), ICAM-1 and CD80 and CD86 (also known as B7-1 and B7-2), which bind to the co-stimulatory CD2 [64], LFA-1 [65] and CD28 [66, 67] receptors, respectively, to aid in T-cell adhesion, kinase recruitment, actin re-structuring and increased sensitivity towards secreted cytokines [66, 68]. If T cells do not receive adequate co-stimulation this results in T-cell deletion or the production of regulatory T cells [69, 70].

Together the combination of the affinity of the TCR for the MHC–peptide ligand, as well as additional APC and cytokine stimulation/inhibition determine the phenotypic changes of a given CD8+ or CD4+ T cell to an effector, memory or regulatory cell [71]. The goal of DNA vaccination in cancer therapy is the generation of effector TAA-specific CD8+ cytolytic T cells (CTLs), which is driven through the action of IL-12 and IL-15 [72]. TAA-specific CTLs are able to recognise and interact with TAA-MHC I complexes expressed by tumour cells through the TCR and CD8 co-receptor. Unlike naïve CD8+ T cells, following this interaction CTLs do not require any additional stimulation to be activated, and kill tumour cells through expression of Fas and release of granulosomes (containing perforin and granzyme) into the immunological synapse (the space formed between the T cell and tumour cell upon TCR/TAA-MHC engagement) [71, 73].

CD4+ T cells may differentiate into effector T Helper-1 (T_{H1}) or T Helper-2 (T_{H2}) cells under the influence of IL-12 or IL-4, respectively [71]. T_{H1} cells promote the cell-mediated immune response by contributing to CD8+ T-cell differentiation and expansion via the release of IL-2 [71, 74]. Additionally, T_{H1} cells secrete IFN-γ and tumour necrosis factor-α (TNF-α) [75], which contribute to anti-tumour immunity via recruitment and activation of macrophages, natural killer (NK) and DCs, as well as up-regulating tumour MHC expression, allowing greater CTL recognition [75, 76]. There is also some evidence that CD4+ T_{H1} cells may directly cause apoptosis of

MHC-II expressing target cells [77, 78]. Conversely, T_{H2} cells polarise the immune response towards a humoral-based response by secreting suppressive cytokines such as IL-4, -5, -6 and -13 [79]. The development of a T_{H1} versus T_{H2} response is antagonistic, as the cytokines secreted by one population suppress the development of the other [80].

The vaccination site plays a key role in determining the efficacy of a DNA vaccine. The majority of clinical trials have delivered DNA vaccines via IM injection, where there are relatively few APCs, contributing to a low clinical efficacy. The efficacy of IM DNA vaccination is heavily dependent on the injection volume, which causes an increased local pressure, augmenting cell uptake and causing slight tissue damage, which aids in the recruitment and maturation of APCs. Therefore IM injections in a small model such as a mouse are likely to result in a more robust immune response than in a human patient, creating a need to investigate alternative delivery routes [81]. As such, the skin may be an ideal site for vaccination, owing to the abundance of "professional" APCs in the epidermis and dermis, and several studies have demonstrated that intradermal (ID) vaccine administration elicits a more potent immune response compared with the IM route and may enable dose-sparing [82–84].

5.3 Treatment of Local Skin Diseases

Nucleic acids may be delivered to the skin for the purpose of introducing therapeutic or antigenic peptides, or replacing proteins produced by defective genes, for the treatment of numerous conditions such as cancer [85], wounds [86] and immunogenic diseases [81]. Aberrant gene expression caused by a genetic disorder or activation of tumourigenic pathways can only be curtailed with gene therapy. One of the most significant advances in gene therapy came just over two decades ago with the discovery of small non-coding RNA molecules that could regulate gene expression at the post-transcriptional level [87, 88]. These small molecules generally inhibit gene expression, known as gene silencing, by a mechanism called RNA interference (RNAi) [89]. Subsequently, several classes of small RNAs have emerged as the most clinically relevant, these being small interfering RNAs (siRNAs) and microRNAs (miRNAs) [90]. Both siRNA and miRNA are short (~20–30 nucleotides) double-stranded RNA molecules that unwind under the action of RNA helicases to incorporate into the Arganoute proteins to form RNA induced silencing complexes (RISC). In this complex the RNA is able to bind to the 3′-UTR end of mRNA through complementary Watson–Crick base pairing, and in this way mRNA can no longer be translated [5, 91]. miRNAs are endogenous genome regulators residing in intergenic regions or the introns or exons of non-coding genes. Primary miRNA transcripts form imperfectly matched hairpin loops prior to cleavage and export to the cytoplasm [92]. In contrast, siRNA precursors are long double-stranded RNAs that are processed by Dicer-2. The main difference between siRNA and miRNA is specificity. To achieve gene silencing siRNA must fully complement a target and hence has only one target mRNA. In contrast, miRNA complementarity is more complicated and does not require perfect base-matching, and hence an miRNA may have many targets [93]. siRNA therapy may therefore be used to silence specific disease-causing genes, and so far more than 20 siRNA therapies have been, or are being, evaluated in clinical trials at various disease states [4, 94, 95].

Delivery of siRNA to the skin is of particular relevance to a number of cutaneous diseases that are characterised by inordinate or atypical protein expression, such as allergic disorders [96], inflammatory diseases (e.g. psoriasis) [96–98], hyperpigmentation [99] and inherited genetic disorders [100–102], which currently lack effective treatment options. In fact one of the first siRNA therapies to complete clinical trials was TD101, for the treatment of the inherited skin disorder pachyonychia congenita, which is caused by heterozygous mutations of the genes coding the keratin K6a, K6b, K16 or K17 proteins and is characterised by the development of debilitating calluses on the plantar surfaces [103, 104]. The TD101 therapy, which inhibits the N171K K6a cytosine-to-adenine mutation, completed a randomised, double-blind, 17-week Phase Ib dose-escalation trial in a single patient. Through the course of treatment, the patient received twice weekly intradermal injections of either TD101 or vehicle control, into plantar calluses on opposite feet, beginning at a dose of 0.1 mg TD101 in week one, and finishing with a dose of 17 mg in week 17. At week 10 of treatment (dose of 6 mg siRNA per injection) the treated calluses began to become statistically smaller, and by week 14 (dose of 10 mg) they began to fall away. On completion of treatment (week 17) there was an area, at the point of injection, within the treated calluses that was clinically free from disease. This relief from symptoms was a dramatic and positive result for siRNA therapy. Of note is the high (2 ml) injection volume and siRNA dose (at least 6 mg) that was necessary to observe a clinical improvement, resulting in a high pain level necessitating pre-treatment with pain killers and nerve blockers for the majority of injections [103]. This high volume is likely to have been at least partly responsible for treatment efficacy by improving cargo uptake through "pressure-fection" [105], coupled with the observation that improvement was only noted in a small area of the calluses immediately where the injection point was. This indicates the need for new gene delivery technologies to the skin capable of targeting a larger area, more efficiently, in a patient friendly (pain-free) manner.

5.4 Limitations of Gene Therapy

The DNA structure is comprised of a phosphate backbone of deoxyribose and nitrogenous base pairs linked by phosphodiester bonds. The anionic charge of DNA does not facilitate translocation across the skin. The first hurdle to overcome is therefore the introduction of DNA into the host tissue or circulation, which is traditionally achieved by injection with a needle and syringe. Following administration, DNA is vulnerable to degradation by host nucleases via hydrolysis of the phosphodiester bonds, with the half-life of DNA varying from minutes to hours depending on the site of administration [106–109]. Additionally, numerous studies have demonstrated that naked DNA is removed rapidly from host circulation by renal and hepatic clearance [110–112]. In addition to nuclease degradation and clearance, bioavailability is also decreased via the reticuloendothelial system (RES), composed of phagocytic cells that degrade plasmid DNA [113]. Finally, the hydrophilic and anionic nature of DNA repels the negatively charged phospholipid bilayer hindering cellular uptake (Figure 5.2a).

Once inside the cell, the DNA enters an early endosome, which fuses with progressively more acidic sorting endosomes (pH ~5–6.5) before reaching the late endosome, which subsequently fuses with highly acidic lysosomes (pH ~4.5–5) containing

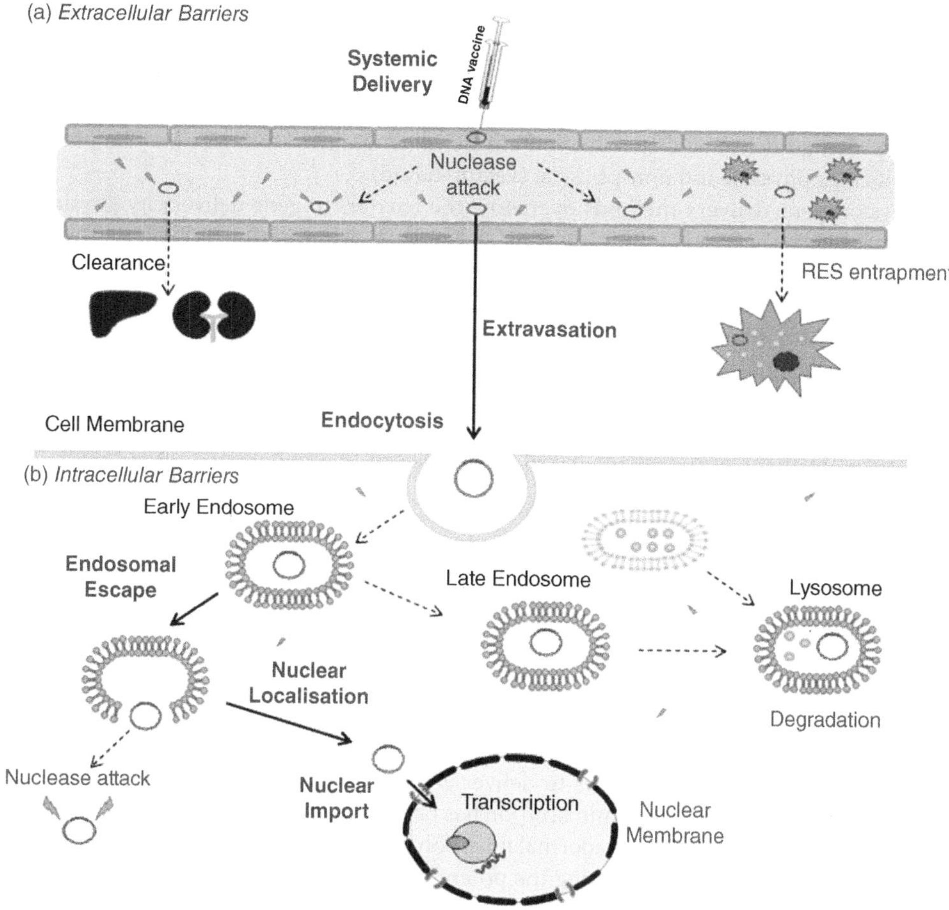

Figure 5.2 Schematic representation of the biological barriers to gene delivery. In order to be effective, the DNA must overcome a number of (a) extra- and (b) intracellular barriers to nuclear entry. (a) Following introduction to systemic circulation DNA must avoid nuclease degradation, renal and hepatic clearance and RES entrapment and localise to the target tissue. (b) DNA must then transverse the cell membrane, and, if introduced into an endosome, escape prior to lyosomal degradation. In the cytosol, DNA must avoid degradation by intracellular nucleases and overcome the steric hindrance of the cytoplasm to localise to the nuclear membrane. The DNA must then be transported into the nucleus where transcription can occur.

nucleases [114]. Thus a successful DNA delivery technology must enable endosomal escape prior to DNA degradation, or deliver DNA directly into the cytoplasm. Following escape from the endosome, the DNA must traffic through the crowded cytosol to the nucleus. The macromolecular size of the DNA prevents passive transport into the nucleus (Figure 5.2b) [115, 116]. Indeed, transport to the nucleus only occurs when the membrane dissolves during replication, which is a chance event. Hence it is not surprising that "naked" DNA vaccines are weakly immunogenic.

This inefficiency in overcoming these barriers is the rate-limiting factor of DNA vaccine immunogenicity in larger animals and humans. Similarly, limitations to the quantity of RNA reaching the cell cytoplasm results in sub-optimal knockdown to correct protein overexpression. It has been reported that doses of up to 17 mg of siRNA have been required to induce therapeutic affects in a Phase Ib clinical trial [103], and effective delivery systems are urgently needed before gene therapies can generate clinically relevant responses [23]. This need has given birth to a field dedicated to improving nucleic acid transfection efficacies, where the strategies may be broadly classified as physical and non-physical (vector-based).

Physical gene delivery methods overcome the barriers to gene delivery by physically localising genetic cargo to the desired site and/or temporarily disrupting the cell membrane allowing nucleic acid entry [12]. Physical delivery strategies that have been investigated for gene delivery within the skin include electroporation (EP) [117], the gene gun [118], jet injection [119] and tattooing [120], as well as microneedles [121]. Vector-based approaches can be viral, where the gene of interest is encoded within the viral genome [122], or non-viral, where the genetic cargo is complexed to a synthetic vector to improve the particle characteristics to favour cellular uptake [123–125]. Importantly, many of the vector-based strategies can be implemented in combination with microneedle technology.

5.5 Microneedles as a Physical Delivery Strategy for Gene Therapy

Where local expression within the skin is desired then administration of genetic cargo to the desired site of action, as opposed to systemically, may result in reduced accumulation within the major excretory organs such as the liver and kidney [126]. Previously, groups have utilised hypodermic needles to deliver genetic cargo to the epidermis or dermis and, although effective, this administration has been associated with significant pain and requires high expertise [4]. Intradermal injection may also only deliver a limited volume to a small area of the skin, reducing the potential number of cells the cargo comes into contact with and therefore there is a clear need for the development of more elegant and efficacious delivery systems. Of the current physical gene delivery systems for the skin, electroporation (EP) is the current gold standard, but use is invasive and transfection efficacy is limited by the EP intensity, which is directly related to cell damage and patient discomfort and pain [127]. In addition, EP requires a two-step procedure and the use of specialist equipment and training, limiting accessibility and increasing the expense. The gene gun is hampered by costly materials, small carrying capacity on gold particles (~2 μg/actuation) and a bias towards a T_{H2} humoral immune response, which is speculated to be related to the mode of uptake (direct insertion into cytoplasm) or the limited CpG DNA motif delivered by each immunisation [128, 129]. Furthermore, the gene gun is associated with painful administration where multiple actuations are needed, and allergic reactions to the gold carrying particles have been reported in several trials [130]. Particle-mediated epidermal delivery (PMED) is a minimally invasive needle-free delivery method that is conceptually similar to the gene gun. PMED has a high-pressure jet that propels microdroplets through the *SC* at a high velocity (100 or 200 per ms) into the epidermis, dermis or subcutaneum [131]. However, preclinical and clinical trials have indicated that other means of ID immunisation are superior in eliciting antigen-specific humoral and cellular immune responses for vaccination [119, 130]. An emerging physical delivery strategy for gene therapy is that of using tattooing (puncture mediated gene transfer). Using this method genes are introduced into the dermis via a needle or cluster of needles that repeatedly penetrate through the skin [132]. Owing to the intrinsic nature of the device, the difficulty of tattooing will mainly lie in translation to the clinic and optimising vaccination for human patients. Tattooing is an invasive technique and protein expression levels have been shown to be significantly improved by increasing the needle penetration depth and tattooing time [133]. There is, however, a trade-off between efficacy and patient acceptability. Another physical method of delivering pDNA into the dermis and epidermis via multiple punctures is that of microneedle application [134].

Microneedles have been demonstrated to be at least as equally efficacious as other well established gene delivery systems such as EP, the gene gun and tattooing in preclinical models [129]. Unlike these other physical delivery systems, microneedle application is painless and can be achieved without additional equipment via self-application [135, 136]. Therefore, microneedle delivery is likely to confer many of the benefits of other well established invasive techniques, but in a manner more easily translatable to clinical use, and may provide a much needed alternative to ID injection. Additional advantages over other types of ID delivery systems include the potential to transfect a larger number of cells (APC or non-APC) by introducing cargo at multiple points through the skin [137], and increased stability of nucleic acids stored in a solid state without the need for cold storage [138, 139]. Several distinct types of microneedle designs have been developed and utilised for gene therapy, and these can be broadly classified as solid, coated, hollow and dissolvable [134].

5.5.1 Solid Microneedles

Solid microneedles represent the first and simplest form of microneedle device, where inert microneedles fabricated from silicon or metal are used to create pores in the *SC* immediately following, or prior to, application of a drug loaded solution [140].

Mikszta *et al.* (2002) described the first use of a solid microneedle delivery system to enhance ID gene transfer *in vivo* [141]. They demonstrated that scraping solid silicon microneedles, up to 200 μm in length, across a 25 μl pDNA solution (containing 35 μg of pDNA) applied to the *SC* significantly increased reporter gene expression compared with the control (topical application without microneedle scraping). Furthermore, despite the quantity of pDNA introduced into the epidermis being significantly lower, microneedle application induced similar levels of reporter gene expression compared with simple IM or ID injection of the same 25 μl solution [141]. In a separate DNA vaccination experiment, mice immunised via the same methods with 100 μg of DNA encoding a hepatitis B surface antigen, HBsAg, showed 100% seroconversion following two immunisations with microneedle arrays, compared with 40% using IM injection and 50% using ID injection. Additionally, mice immunised via microneedle application developed similar numbers of HBsAg-specific CD8+ T cells following three injection immunisations [141].

Birchall *et al.* (2005) and Coulman *et al.* (2006) demonstrated that solid microneedles could be used to induce gene expression in excised human skin [142, 143]. Coulman *et al.* surgically removed the majority of the dermis from donated human skin, and applied a 50 μl solution containing 125 μg of pDNA coding beta-galactosidase (pCMVβ) to the *stratum corneum*. An array of 20×20 silicon microneedles (200 μm in length) was then applied to the pCMVβ solution, the skin was cultured *ex vivo* for 24 h and subsequently stained using X-Gal staining solution to determine whether microneedle treatment had resulted in gene expression. The group successfully demonstrated that there were several areas of the epidermis immediately proximal to the micro-channels that were expressing beta-galactosidase, although they noted that only a very small portion of the created channels stained positively for gene expression, and hence the device required further optimisation [143].

These proof-of-concept experiments demonstrated that microneedles could enable ID gene transfer for DNA vaccine purposes and led to a widespread interest in the use of microneedles for gene delivery and in particular vaccination. Solid microneedle devices such as the Dermaroller™, a manual roller device with microneedles ranging from 0.2 to 2 mm in length, are commercially available for cosmetic procedures [144], and pyramidal solid silicon microneedles (ImmunPatch) have been developed by researchers at University College, Cork for the purposes of vaccine delivery, including DNA vaccination [142, 145, 146]. However, despite being easy to use with encouraging preclinical results, solid microneedles represent basic devices that require two-step application and offer high variability in the quantity and efficiency of pDNA delivery into the cutaneous layers [142, 143].

One possible improvement to increase pDNA delivery and transfection efficiencies from these physical delivery systems was explored by Pearton *et al.* (2008) who utilised pDNA-loaded hydrogels to act as a pDNA reservoir in conjunction with solid silicon microneedles [147]. Two distinct hydrogels with differing functionalities were tested to determine the most efficient hydrogel–microneedle combination for pDNA-delivery. The first strategy employed pDNA-loaded polyacrylic Carbopol-940, which forms a solid mesh-like structure when rehydrated, and the second employed a pDNA-loaded thermosensitive poly(lactide-*co*-glycolide) (PLGA), poly(ethylene glycol) (PEG) copolymer, termed PLGA-PEG-PLGA, which formed a liquid at room temperature and a more rigid gel at physiological temperatures [147]. To determine the efficiency of pDNA delivery from each hydrogel across human epidermal sheets, 1 ml of each formulation, loaded with 1 μg/μl of pCMVβ, was placed into the donor compartment of a Franz type diffusion cell, and permeation of pDNA across heat-separated human epidermal skin, pre-treated with solid silicon microneedles, was determined over a five-day period. The authors found that ~10% of pDNA from PLGA-PEG-PLGA hydrogels diffused across the epidermis to the receptor compartment over a five-day period, compared with <1% of pDNA from Carbopol-940 hydrogels, possibly due to varying methods of pDNA release.

The authors next determined whether application of pDNA-loaded hydrogels to skin, followed by treatment with solid silicon microneedles over the treated area, could result in gene expression in an *ex vivo* human skin model. For gene delivery experiments the authors applied 100 μl of each hydrogel (containing 1 μg/μl pCMVβ), to a discrete area of human epidermal skin, and either rolled or scraped a silicon microneedle (composed of 4×4 projections 260 μm in length, with a 100 μm blunted tip) over the hydrogel treated area. Treated skin was subsequently cultured in complete media for 24 h, and stained in X-Gal for a further 24 h to determine points of beta-galactosidase expression. Despite the discrepancy in the quantity of pDNA that had permeated from the loaded hydrogels across the epidermal layers in the Franz-cell model, rolling of the microneedle array across both of the pCMVβ-loaded hydrogel formulations did result in positive points of gene expression within the epidermis (~3 points from Carbopol-940 gels, and ~2 points from PLGA-PEG-PLGA gels). However, this was markedly lower than the number of points of gene expression elicited by rolling silicon microneedles through a simple 1 μg/μl pCMVβ solution (~10 distinct points of gene expression), highlighting potential limitations of this hydrogel–microneedle combination, including the inefficient release of pDNA from the hydrogels, and the inability of the hydrogels to permeate through the channels created by the microneedles. Following scraping of the silicon microneedle through the pre-applied pCMVβ-loaded hydrogels the authors observed an increase in the number of points of beta-galactosidase expression (~5 from Carbopol-940 gels, and ~7 from PLGA-PEG-PLGA gels), potentially due to greater disruption of the *stratum corneum*; however, this was still lower than the levels of gene expression produced with the simple pCMVβ solution. The authors also noted that in previous studies they had observed gene expression in ~30% of channels created by microneedles when they had pre-applied a simple pDNA solution to human skin, compared with ~12% of channels expressing beta-galactosidase in the present study [142]. A reason for these limited transfection efficiencies following application of pCMVβ-loaded hydrogels may have been that for post-microneedle application the human skin was only cultured for a maximum of 24 h, and the hydrogels had released very little of the payload (<5%). Therefore, it is possible that this combination may have a future use for sustained release applications, where treatment is desired over an extended period, however further work is needed to ensure the complete filling of the channels with the hydrogel, to prevent closure and appropriately locate the pDNA, to enhance the subsequent release of the pDNA cargo and to modify the cargo to ensure cellular uptake following release.

A recent advance in design has been the advent of motorised microneedle devices, which may offer greater control of needle penetration depth and frequency, and several such devices are available commercially from Bomtech Electronics (Seoul, Korea) [105, 148, 149]. These devices are capable of puncturing the skin automatically when the power is on, at a consistent speed and depth, enabling a large number of punctures (>5000 punctures/s) over a relatively short space of time, using disposable needle cartridges [149]. Park *et al.* (2013) reported on the safe use of two of these commercially available devices (Digital Hand® and Digital Pro®) in a mouse model [148]. The motorised Digital Hand and Digital Pro devices were fixed to a needle of length of 1500 μm and applied to an area of skin on skh-hairless mice 56 times. For comparison purposes mice were also untreated or received five rolls with a handheld disk type microneedle therapy system (DTS®, DTS Lab, Korea), a rolling device with a needle length of 1500 μm. At 0, 1 and 7 days following application, skin was inspected for adverse reactions, transepidermal water loss (TEWL) was determined and skin samples were taken for microscopy to inspect for abnormalities.

Immediately following microneedle application, all groups, with the exception of the control mice, showed similar signs of slight trauma in the form of erythema and needle marks, which resolved spontaneously the following day. Biopsy samples revealed no major abnormalities between the epidermis or dermis of skin treated with microneedles, and there was no evidence of immune infiltration to the treated sites. Directly following treatment with each of the devices there was an increase in TEWL compared with untreated samples, which is to be expected due to a breach in the skin barrier and is desirable for gene delivery. A significantly larger increase in TEWL was observed in skin samples treated with the Digital Pro, suggesting that this device causes an increased disruption of the *stratum corneum*; however, 1 and 7 days following treatment, TEWL levels were comparable to the control.

Results from this study therefore indicated that motorised microneedles could successfully disrupt the *stratum corneum*, in a safe, controlled and reversible manner, which removes the variability in penetration associated with handheld devices [148]. However, although the skin of all mice recovered fully following treatment, the length of the needle used in this study may potentially activate nerve receptors, causing pain, which increases drastically as microneedle length is increased beyond 700 μm [135]. Subsequently, motorised microneedle devices have been utilised to deliver both pDNA [149] and siRNA to the skin [105]. Yan *et al.* (2014) utilised the My-M® microneedle device (Bomtech Electronics, Korea), which has 12 needles with an adjustable exposable needle length of 250–2000 μm, and adjustable punch-speed settings of 20–90 turns/s, to deliver pDNA coding GFP and firefly luciferase (pEGFP-Luc) to the skin of mice [149]. The group explored two methods of pDNA-microneedle treatment: firstly, puncturing the skin with the My-M device (40 punctures/s for 30 s with needle length 500 μm), and subsequently applying 100 μl of 100 μg/ml pEGFP-Luc solution (10 μg) to the treated skin surface; and secondly, applying the solution to an area of skin, followed by puncturing the skin with the motorised device using the same conditions [149]. The group found that although both methods were able to significantly increase luciferase expression within treated skin samples compared with untreated samples, the method of pre-applying pDNA solution and subsequently puncturing the skin induced significantly (87 times) higher levels of gene expression than puncturing skin prior to pDNA application ($p<0.001$, Student's t-test). The group further demonstrated that gene expression levels could be improved further by increasing the length of the exposed needle (to 750 μm) and application time (up to 60 s), although with longer application times and needle lengths increased redness and slight bleeding were observed, pointing to the need for a trade-off between gene expression and patient acceptability [149].

Hickerson *et al.* (2013) investigated the similar Triple-M® microneedle device (Bomtech Electronics, Korea) to deliver self-delivery (sd)-siRNA (siRNA technology from Dharmacon modified to enhance cellular uptake in the absence of vectors), targeting reporter gene expression in a transgenic murine model [105]. Similarly to the My-M device, the Tri-M device consists of 12 oscillating needles (capable of speeds up to 100 oscillations/s) with an adjustable length of from 0 to 2000 μm; however this device was adjusted to incorporate a cartridge above the needles to store and release the siRNA during microneedle application. This cartridge allowed the flow of siRNA from the reservoir through channels between the needles while the device was operational, and could be loaded with up to 300 μl of solution, removing the need for the cumbersome two-step application process. The group loaded the device with 50 μl of 100 μg/ml Cy3-labelled sd-siRNA, set the needle length to 100 μm and applied the solution to the flanks of hr/hr-hairless mice or excised human skin for 10 s at an oscillation speed of 100 oscillations/s. One hour after application, the skin was harvested for fluorescence microscopy, and imaging revealed similar levels of fluorescence distribution within the epidermis and dermis of mouse skin treated with the Tri-M device or hypodermic needle, indicating that the motorised device could successfully introduce siRNA to the viable skin layers [105]. Within treated human skin, a greater amount of Cy3-labelled siRNA was detected in the epidermis of Tri-M treated skin, especially around the points of needle penetration, compared with skin treated with a hypodermic needle, indicating that the solid microneedles may be better suited to treatment of epidermal skin disorders [105]. Finally, the group demonstrated that this solid motorised device was capable of silencing gene expression in the GFP-expressing hairless tg-CBL/gMGFP transgenic murine model. The authors used the device, as previously, to deliver 100 μl of 5 μg/μl CBL3 sd-siRNA to the flanks of mice six times over a 10-day period (although the exact quantity of sd-siRNA delivered was unknown). This treatment regimen was capable of silencing reporter gene by approximately 79% compared with skin treated with a non-specific siRNA control at 24 h [105].

These studies provide encouraging support for the use of motorised solid microneedles for gene therapy. An additional advantage of these motorised devices is that the protruding needle length can be adjusted to allow for targeting different layers of the skin or compensating for differences in skin thickness [105]. However, these devices would be less patient friendly and would require administration from a healthcare professional. There is also evidence that motorised devices result in slight tissue damage at the site of penetration, resulting in inflammation and immune infiltrate [105]. However, such "damage" may be beneficial for vaccination purposes, as the device may act as a natural adjuvant, increasing vaccine immunogenicity in a similar way to the tattoo gun [150]. Currently, the main limitation to solid microneedle progression remains the fact that there is no way to even estimate the quantity of nucleic acid being introduced into the epidermis or dermis [126]. Furthermore, the wastage of cargo from solution remaining on the skin and sticking to the device is likely to be high, and studies by Hickerson *et al.* (2013) estimated that despite the Triple-M device being capable of storing up to 300 μl of cargo solution, only a maximum of 40 μl could be delivered to the skin from the device, and even this may have been an overestimate [105, 141, 151]. As such, there has been a shift towards the use of second generation microneedle devices to improve reliability and consistency.

5.5.2 Coated Microneedles

The term coated microneedles describes solid microneedles that have been coated with a cargo-loaded solution which is left to dry in place, creating a solid, thin coating [152]. This coating solution commonly contains polymers or sugars to increase viscosity and stabilise the cargo, and may be applied to the needle surface via dip-coating [153], layer-by-layer application [154] or spray-coating [152, 155]. Following insertion into the skin the coating formulation dissolves in contact with the interstitial fluid, enabling cargo deposition within the epidermis and dermis. Pre-coating cargo onto the microneedle surface removes the need for a two-step application process, making the device more suitable for self-application. Importantly, storing cargo in the dry state in the coating formulation may improve stability, decreasing the need for cold chain storage and hence the cost [21, 156].

Gill and Prausnitz (2007) developed a simple dip-coating process which enabled the uniform coating of stainless-steel microneedles with a range of sensitive biopharmaceuticals, including pDNA [157]. This dip-coating process was later used in a DNA vaccination study to coat solid microneedles with 8 μg of pDNA encoding the hepatitis C virus protein, NS3/4A, prior to application into a shaved area of mouse abdominal or back skin for 1 min. A separate group of mice was immunised with the same plasmid via the gene gun to directly compare the efficacy of the delivery techniques [158]. Two weeks following immunisation, mice were sacrificed and the total spleen cell population, including cytolytic T cells, was harvested and stimulated for 5 days *in vitro* with an NS3-specific peptide. Target NS3-peptide pulsed RMA-S cells were subsequently incubated at several effector-to-target ratios with stimulated T cells and the specific cell lysis was determined [158]. This group found that a single immunisation via gene gun or microneedle array resulted in a significant increase in the target cell lysis compared with naïve mice ($p<0.05$). Although the specific cell lysis was greater following immunisation via gene gun, this was not significantly greater than that achieved following microneedle immunisation [158]. In an independent experiment the authors immunised BALB/c mice with 3.2 or 100 μg of N3/4A expressing pDNA via microneedle application or IM injection, respectively [158]. Two weeks following immunisation, mice were challenged with NS3/4A-expressing SP2 myeloma cells and monitored for tumour growth. Similarly to the cytolytic results achieved *in vitro*, immunisation of mice with the NS3/4A-expressing pDNA via microneedle application or IM injection significantly retarded tumour growth ($p<0.05$). Importantly, immunising via microneedle application with approximately 30-fold less pDNA achieved the same level of protection from tumour challenge as IM injection ($p>0.05$).

Kim *et al.* (2012) investigated utilising similar coated metal microneedles to immunise mice with a pDNA vaccine encoding a synthetic haemagglutinin (HA) gene from the influenza A/Viet Nam/1203/04 virus [159]. The group modified the coating formulation to include a sugar (carboxy methyl cellulose (CMC)) and surfactant (Lutrol F-68 NF), which had previously been utilised to coat viral influenza particles onto solid microneedles. Balb/c mice were immunised with 3 μg of pDNA via coated microneedle or IM injection (with pDNA recovered from dissolved microneedles) on weeks 0, 5 and 10. At 21 weeks following the final immunisation, mice were challenged intranasally with 20 times the lethal dose of wild-type VN/04 virus and monitored for general health for up to 14 days or humane end-point (25% weight loss), or sacrificed at 4 days post-challenge to determine viral titres in the major organs. The authors reported that all mice immunised via IM injection reached humane end-point by day 8, however 67% of mice immunised via microneedle survived to study completion. Additionally, DNA vaccination via microneedle was able to significantly reduce viral titres (300-fold) in lung compared with mice receiving mock vaccination. Despite these encouraging results the group were not satisfied with the incomplete protection that immunisation provided compared with previous studies using inactivated viruses. This prompted them to investigate the functionality of their pDNA cargo within the microneedle coating and the authors subsequently hypothesised that the presence of CMC in the solid microneedle coating diminished the integrity of the encapsulated DNA vaccine, highlighting the need to improve microneedle coating formulations to maintain the integrity of sensitive pDNA cargo during the drying process [159].

In a further study Kim *et al.* (2013) explored increasing the immunogenicity of the microneedle-based pDNA vaccine encoding haemagluttinin (HA) from the A/PR/8/34 H1N1 (A/PR8) virus by co-administering inactivated A/PR8 virus to augment the immune response [160]. In this experiment the authors fully determined the effect of differing coating formulations on the integrity of the sensitive biological cargo. By incorporating the disaccharide trehalose into the coating formulation the authors were able to retain 65% of the HA activity of the inactivated virus; however, by excluding

CMC from the coating formulation, the group were able to retain nearly 100% of the vaccine activity. They demonstrated that immunising BALB/c mice with 1 µg of inactivated virus and 3 µg pDNA coding HA generated significantly higher titres of A/PR8-specific antibodies than immunisation with placebo DNA or via simple IM injection. Furthermore, immunising mice with this heterologous virus/pDNA combination improved survival rates of mice challenged with a lethal dose of the H1N1 virus, compared with those treated with inactivated virus and placebo DNA. As such, the authors concluded that the combination of peptide antigen and DNA vaccine produced a synergistic effect and this may be a potent tool for vaccination purposes [160]. A limitation of this approach, however, is that the addition of inactivated virus to the formulation coating further complicates the fabrication process, as the stability of the viral proteins as well as that of the pDNA within the solution needs to be considered, necessitating the addition of one or more stabilising agents, which increases the cost and limits the loading of one or other of the active ingredients [160].

The Prausnitz group have further demonstrated that this coated microneedle device has the potential for dose-sparing in DNA vaccine studies, as delivery of 3 µg of pDNA through the microneedle device conferred better protection against influenza challenge than 10 µg of the same DNA vaccine delivered IM [121]. Importantly, the group have made a version of this microneedle system available commercially, allowing others to easily explore these devices as a means to increase the efficiency of gene delivery to the skin (Microneedle Systems, LLC, USA) [129]. After purchasing this microneedle system Hu *et al.* (2016) were able to test the potency of a DNA vaccine against cottontail rabbit papillomavirus (CRPV) delivered via microneedle or tattoo gun in a transgenic rabbit model [129]. The group delivered a 10 µg dose to each inner ear of the animals via the tattoo gun or coated microneedle device, three times at three-week intervals prior to challenge with the 5 µg CRPV. Immunisation with the tattoo gun and microneedles resulted in the development of significantly smaller tumours than animals treated with control pDNA ($p<0.05$, one-way ANOVA), with no significant difference being observed between microneedle treated rabbits and tattoo gun treated rabbits. Furthermore, three out of four rabbits receiving the DNA vaccine via the tattoo gun or microneedle were completely protected from CRPV infection. As such, the authors concluded that both strategies are promising for DNA vaccination, but that microneedles do not require the same equipment and were less invasive compared with the tattoo gun system, which resulted in trauma and swelling in the treated area [129].

The Kendall group based at The University of Queensland have developed an ultra-high density coated silicon microneedle array (>20 000 projections per cm^2 of <100 µm in length), termed the Nanopatch® [161], with the aim of directly contacting vaccine cargo to increased numbers of APCs within the skin, which has been investigated for DNA vaccination purposes [137, 162]. Chen *et al.* (2010) coated Nanopatch devices with 3364 projections within a central 4×4 mm section of an array, with 7 µl of coating solution composed of 10 mg/ml of methylcellulose and 72 µg/ml of pDNA coding herpes simplex virus 2 surface glycoprotein D2 (HSV-2-g2) labelled with the Mirus Label IT CX-Rhodamine nucleic acid labelling kit. The group applied these devices to the ventral side of mouse ears with a spring-based applicator, and left the patches in place for 5 min to allow coating dissolution, before harvesting the ears and fixing the ventral side in paraformaldehyde. The ears were subsequently stained with FITC-labelled anti-mouse MHC Class II to allow identification of APCs within the skin, and the co-localisation of the DNA vaccine and APCs was determined using fluorescent microscopy. Using this method the authors were able to identify that of 391 needle penetrations, 43±5% delivered the coated DNA vaccine into direct contact with an APC within the epidermis; extrapolating to the thousands of projections within the array this translates to approximately 3000 epidermal APCs being targeted per application [137].

As APCs are essential to the induction of both humoral and cellular immune responses it was hypothesised that utilising this device would increase DNA vaccine immunogenicity. Therefore, the group utilised the device to immunise mice against HSV-2 challenge. Female BALB/cJ mice were immunised with two devices (one on each ear) as described earlier, or with 1 μg of pHSV-2-g2 via IM injection, three times at three-week intervals. Three weeks after the final booster vaccination, following synchronisation of the mice oestrus cycles, mice were challenged intravaginally with a lethal dose of HSV-2 and monitored for survival. The group observed that six out of eight mice vaccinated via microneedles and seven out of eight mice vaccinated via IM injection survived until the end-point (21 days), indicating that the microneedles had at least similar efficacy to IM injection, which was an encouraging result considering the total delivery from the microneedle device was estimated to be about a tenth of the dose delivered via injection [137].

In a separate experiment, Kask *et al.* (2010) investigated the effect of increasing the quantity of pDNA (coding an immunogenic region of the gD2 HSV-2 viral fusion protein (pVAX1-gD2$_{1\text{-}340}$)) coated onto the Nanopatch [162]. The authors prepared patches coated with 0.05, 0.5 or 5 μg of pVAX1-gD2$_{1\text{-}340}$, and applied two patches (corresponding to 0.1, 1 or 10 μg total dose) to immunise female mice against the HSV-2 viral challenge. Separate groups of mice received equivalent doses of pVAX1-gD2$_{1\text{-}340}$ via IM injection to allow a comparison between the two vaccination routes. Mice were vaccinated and challenged with HSV-2 using the same schedule as reported by Chen *et al.* [137], but with differing doses of pDNA. Immunising mice with higher doses (10 μg) of pVAX1-gD2$_{1\text{-}340}$ via either route resulted in 90% survival rates from lethal challenge. Encouragingly, mice treated with medium (1 μg) or low (0.1 μg) doses of pVAX1-gD2$_{1\text{-}340}$ through the ID route demonstrated increased survival rates (78 and 22%, respectively) compared with mice vaccinated with the equivalent dose via the IM injection (50 and 6%, respectively) [162]. Again the authors demonstrated that these Nanopatch devices were capable of eliciting similar or more potent immune responses towards the encoded antigen than administration via the traditional IM route, despite the actual delivered dose being approximately ten times lower.

One of the potential problems with the Nanopatch DNA vaccine system developed by Chen *et al.* (2010) is that each array was estimated to be coated with 0.5 μg of pDNA, and of this only 8% of the cargo was considered to be delivered following application, which is considerably less than the quantity of pDNA able to be delivered by injection [137]. This is a common problem for coated microneedles where the quantities of pDNA coated on to the microneedle devices are in the region of 0.5–5 μg, and the quantity delivered to the skin is low [121, 160, 162]. Pearton *et al.* (2012) succeeded in coating up to 105 μg of pDNA onto the outside of a row of five steel metal microneedles, 750 μm in length, utilising a similar method to Gill and Prauznitz [157, 163]. To enable such a high loading, microneedles were dipped into a concentrated pDNA solution (6 μg/μl) 100 times, allowing 30 s between each immersion for complete drying. However, using such a high number of immersions resulted in an uneven distribution of cargo deposition, with the majority of the pDNA deposited near the microneedle baseplate, whereas following 10 or 40 immersions into the coating reservoir the cargo was located around the central portion of microneedle shaft [163]. Although arrays were still able to penetrate the *stratum corneum* following 100 coatings, the maximum penetration depth of the needles was reported to be ~300 μm of the total 750 μm length and therefore the majority of the coated pDNA may not be delivered following application. This coating process is cumbersome and therefore not easily translatable to the clinic, and resulted in quite a high variability in the coating efficiency at a higher number of immersions (105.7±22.9 μg of pDNA following 100 immersions in the coating solution). Additionally, storage of coated microneedle formulations led to a loss of the supercoiled pDNA structure from over 80 to ~50% in seven days, although the tertiary structure was mostly conserved by the subsequent addition of disaccharide stabilisers (trehalose, maltose and sucrose) [163]. Transfection studies in excised human skin finally revealed that microneedles coated with ~20 μg of pDNA induced significantly fewer points of gene expression compared with applying a solution to the skin and subsequently using a metal array to puncture the *stratum corneum*, possibly due to the dried formulation requiring an excessive time for dissolution [163]. As such, loading capacity onto solid microneedles seems to be limited by a trade-off with increasing the coating thickness, which may in turn significantly impact the cargo dissolution, needle sharpness, penetration and cargo delivery efficiency [153]. Hence vaccination strategies are effectively limited to highly potent antigens, and the quantity of loaded antigen may not be sufficient to elicit a potent CTL response. Additionally, although storing cargo in the solid state may increase stability, there is also a need to ensure that the sensitive genetic cargo retains functionality during the formulation and drying process, which is not a consideration for the same cargo when it is in solution.

5.5.3 Hollow Microneedles

Conceptually, hollow microneedles resemble hypodermic needles more than other solid counterparts, where single or multiple hollow needles in the micron range are used to administer liquid vaccine into the skin, with the distinct advantage for gene therapy being that cargo is brought into contact with a far greater number of cells than with a single injection point in order to exert its therapeutic effect [164, 165]. The main advantage of hollow microneedles over other microneedle devices is that as liquid is injected into the skin, in a similar manner to hypodermic needles, these devices may be compatible with currently available vaccine solutions and allow controlled dosage. Indeed, a hollow microneedle system (BD Soluvia Microinjection System™, BD) became the first commercial microneedle device to receive marketing authorisation from the European Commission for ID influenza vaccination, in 2009 [166].

Yan *et al.* (2014) compared the efficacy of injecting 1 μl of 100 μg/ml pEGFP-Luc at 100 points into the skin of rats (total volume 100 μl), with a 31G needle, modified to give an exposed needle length of 1500 μm, with the injection of 16 μl of the same solution at six points using an unmodified 31G needle (total volume 96 μl) [149]. Injecting the solution in small volumes at multiple points was designed to replicate gene delivery from a hollow microneedle array, and to determine whether introducing genetic cargo at multiple points in the skin results in higher gene expression. Despite the authors noting a loss of the solution through oozing following injection with the crude microneedle device, injecting the solution at 100 points within the skin resulted in significantly higher luciferase expression (approximately an order of magnitude) than injecting a similar volume at six points. This is likely because the genetic cargo has limited mobility once introduced into the skin, and it is often reported that nucleic acid expression is confined to the cells immediately within the vicinity of the penetration point [137, 142, 147, 163].

This result clearly illustrates the advantages of microneedles over the conventional hypodermic needle, where the introduction of genetic cargo at multiple points within the skin allows uptake and subsequent expression by a greater number of cells even when the same volume of solution is delivered to the same area [149]. Hollow microneedles may also provide a more controlled and reliable delivery of cargo to the layers of the dermis and/or epidermis compared with solid needles, as a pressure system can be attached to the device [167]. The existence of commercially available hollow microneedle devices, which can be adapted to attach on to conventional pen systems, such as the ADMINPEN 600 Liquid Injection System (AdminPen, USA), also provides researchers with a readily available device to deliver their genetic cargo to the skin via microneedles without having to formulate the cargo into a coating. However, there are some limitations to hollow microneedles. They are typically weaker than solid microneedles [168], which has previously necessitated the addition of solid microneedles to aid skin penetration [169]. It has previously been noted that hollow microneedles may experience bore blockage by the compressed dermal tissue [170]. Hollow microneedles have sometimes found to be prone to breakage within the skin, an inherent risk of all solid microneedles, which would compromise solution flow, but also result in potentially hazardous deposition of the needles into the patient's skin [171]. As with traditional hypodermic needles, with hollow microneedles the potential remains for needle re-use, and hence the transmission of infectious diseases. The need for a fluid reservoir and pressure applicator with the majority of systems also reduces the potential for the devices to be self-applied, and leaves the need for cold chain storage. One additional concern for DNA vaccines is that hollow microneedles release cargo largely into one area in the dermis at the end of the microneedle projections [169]; this largely bypasses the LCs of the epidermis and reduces the number of APCs that the pDNA may contact, as diffusion through the skin is likely to be limited. Hence arrays that release vaccine over a larger area may be preferable for DNA vaccination.

5.5.4 Dissolvable Microneedles

One concern with solid microneedles is the fate of these needles should breakage occur within the skin. To this end, microneedle platforms may be manufactured from dissolvable sugars (e.g. trehalose, maltose and galactose) or polymers (e.g. carboxymethylcellulose (CMC), poly(vinyl pyrrolidone) (PVP), or poly(vinyl alcohol) (PVA)) containing the cargo of interest, which dissolve in the dermis in the presence of the interstitial fluid. Importantly, unlike coated microneedles, the DNA cargo may be incorporated throughout the entire needle and not confined to a thin coating, increasing the potential payload while maintaining DNA stability within a solid matrix [171].

The first description of a dissolvable microneedle system to deliver nucleic acid cargo into the skin was reported by González-González *et al.* in 2010 [172]. The group developed a loadable, dissolvable protrusion array device (PAD) by bringing a 20% w/w PVA solution into contact with a commercially available "pin headers" template and slowly withdrawing to form a viscous polymer strand, which air-dried *in situ* to form needle-like hollow structures [172]. In order to load a nucleic acid cargo, these hollow projections were brought into contact with a DNA solution, which was drawn into the PAD channel via capillary action and imbibed into the polymer via hydration and swelling. Finally, these devices were hardened to facilitate skin penetration via drying at elevated temperatures (50°C) in a vacuum oven [172]. Using this technique, the group were able to load ~12 ng per needle of plasmid encoding the reporter gene luciferase (pGL3-CMV-Luc). In a proof-of-concept study PADs loaded with pGL3-CMV-Luc, composed of 3×4 needles (~144 ng/device), were applied to the ear and footpad skin of anaesthetised mice for 20 min. Despite delivering only ~10% of the genetic payload, luciferase gene expression was detected at 24 and 48 h following PAD application via the IVIS 200 imaging system in both skin types, demonstrating the ability of this PAD to successfully enable ID protein expression *in vivo* [172]. The group further demonstrated that the device could be used to silence gene expression in a transgenic (Tg CBL/hMGFP) mouse model. These mice express luciferase and GFP within the epidermis, and therefore provide an ideal model for proof-of-concept experiments targeting cutaneous monogenic mutations, such as pachyonychia congenita. PADs were loaded with Acell CBL3 siRNA, and Tg CBL/hMGFP mice were treated on the footpad with three PADs (with five needles) every other day for 12 days before reporter gene expression was analysed via qRT-PCR. On day 13, a reduction of 25–50% of CBL/hMGFP was observed in the treated footpads compared with those treated with control siRNA, indicating that these devices are appropriate for gene silencing *in vivo* [172].

Lara *et al.* (2012) also demonstrated that similar dissolvable PADs could be used to silence endogenous gene expression in a human epidermal model as well as in human skin xenografts [173]. The group loaded PADs composed of 5×5 needles with 2 μg of sd-siRNA (total 50 μg/array) targeting CD44, a gene expressed uniformly through the human epidermis. Immunodeficient mice bearing full thickness human skin equivalent xenografts were treated with three microneedle arrays daily for 10 days, after which xenografts were harvested for mRNA quantification via qRT-PCR. Xenografts treated with CD44 sd-siRNA loaded PADs displayed a reduction of CD44 mRNA levels of 45% compared with those treated with control siRNA, or 52% compared with non-treated xenografts. Encouragingly, this reduction in CD44 mRNA levels was similar to that seen in mice treated with CD44 sd-siRNA via ID injection (41% compared with the untreated control). A limitation of this study was the extensive treatment period required to bring about gene silencing, as well as the need for multiple microneedle treatments daily, which is possibly due to the low cargo delivery efficiency from the PAD (estimated ~10%) [173].

In a further study, the group directly compared the efficacy of the dissolvable PAD against a coated metal microneedle device [174]. Dissolvable PADs (3×5 needles/device) were trimmed to 750 μm and loaded with 20 ng of firefly luciferase encoding plasmid (pUb-luc2/eGFP) per needle. Similar metal microneedle arrays (1×5 needles/device) measuring 700 μm were coated with approximately 3 μg/device of DNA, via a dip-coating method with the same plasmid. Mice then had pUb-luc2/eGFP loaded arrays applied to their left paws using manual application, while devices loaded with mock DNA were inserted into the corresponding right paw; the mice were monitored for up to three months for reporter gene expression using the IVIS Imaging System [174]. Bioluminescent signals were detectable in mouse footpads for up to

72 days following application of both pUb-luc2/eGFP loaded devices [174]. However, bioluminescence levels following application of coated metal microneedles increased over the first week following application, and reached approximately an order of magnitude higher than that following application of PADs. This significantly higher reporter gene expression was attributed to differences in DNA loading between the arrays, where PAD arrays were loaded with 0.3 µg of DNA, although only five of the 15 needles on the array actually penetrated the skin during application, and hence only 0.1 µg of DNA was available for delivery, compared with 3 µg from the coated metal needles [174]. Additionally, while ~90% of DNA from the coated metal microneedles was likely to be deposited into the skin, only ~10% of DNA from the PADs has been reported to be deposited following device application, due to incomplete needle insertion using the PADs [172, 173].

As such, these studies demonstrate the potential of dissolvable microneedles for gene therapy but also highlight some of the areas that require optimisation, such as the formulation strength, loading capacity and payload loss. Furthermore, the individual loading of needles with cargo via a pipette is not viable for large-scale manufacture, and other methods that incorporate the cargo into the microneedle matrix prior to formulation may be more desirable [139]. Similar to coated formulations, the compatibility of the dried genetic cargo with the dissolvable matrix is a key consideration, as pDNA integrity has been found to be compromised within some dissolvable polymer matrices [175]. Lastly, the choice of the dissolvable matrix must be biocompatible, as the compound should be non-toxic, a non-irritant and eliminated safely following dissolution within the skin.

5.5.5 Microneedles in Combination with Other Delivery Technologies

Since the first use of microneedles to introduce DNA into the skin in 2002, several limitations of these platforms have been identified, including: highly variable delivery, the need for a two-step application process and a relatively low loading into microneedle patches. Several of these limitations have been addressed by the refinement of microneedle devices. Another major limitation of gene delivery from microneedle arrays has been that following deposition within the skin, as the genetic cargo remains vulnerable to degradation via nuclease attack and is inefficient at traversing the cellular membrane. Deng *et al.* (2016) found that following delivery of unmodified siRNA to the epidermis using a solid silicon microneedle device, there was no decrease in gene expression [126]. Similarly, it was shown that delivery of "naked" pDNA from a dissolvable microneedle array was inefficient at eliciting an immune response against the encoded E6/E7 cervical cancer antigens [176]. These realisations point to the need for additional aids to gene delivery following introduction to the skin to ensure enhanced intracellular uptake. To this end, researchers have explored combining microneedle technology with other gene delivery strategies.

5.5.5.1 In Combination with Physical Delivery Technologies

Of the current physical delivery technologies being developed for gene therapy, EP is the current gold standard. EP involves the local application of a series of short electrical pulses immediately following injection of payload to the target site. Nucleic acid entry is enabled following a transient destabilisation in the cell membranes of the host cells

[177]. EP has been well documented as a gene delivery technology, with reporter gene expression increasing up to 100–1000-fold *in vivo* following EP, compared with DNA injection alone [178, 179]. Traditionally, the electrodes used for EP are applied to the surface of the tissue following injection of the genetic cargo. However, owing to poor conduction and barrier properties of the *stratum corneum* the current reaching the cells beneath is not optimal [180]. As such, the pairing of EP with conductive metal or metal-coated microneedle arrays may prove a valuable tool for gene therapy within the skin. As the conductive needles penetrate the *stratum corneum*, this allows increased local voltages within the dermis and epidermis over the whole area of the array, enhancing pDNA uptake following introduction into the skin [180].

The first example of EP combined with microneedle technology for the purposes of gene therapy was in 2007, when Hooper *et al.* utilised the Easy Vax™ vaccine delivery system to immunise mice with the 4pox DNA vaccine (a DNA vaccine composed of four plasmids targeting two infectious forms of the orthopoxvirus) [181]. The device introduced the plasmid cargo into the skin, as a coating on electrically conductive microneedles, and applied electrical pulses to facilitate cellular uptake. Each of the 4pox plasmids were coated onto the microneedles of the device (composed of 8×10 conductive microneedles, <1000 μm in length) and dried at room temperature with a load of 30 μg of pDNA per array. Balb/c mice were then immunised with each plasmid (one per device, four in total) into the thigh, each at a separate site, followed by six electrical pulses (100 V, of 100 μs duration, 125 ms apart). Following three immunisations (weeks 0, 3 and 8), all mice generated detectable levels of neutralising antibodies against the smallpox virus, as determined by ELISA (enzyme-linked immunosorbent assay). Furthermore, mice immunised, and subsequently challenged five weeks following the final booster with ten times the lethal dose of IHD-J poxvirus strain, survived with minimal weight loss, whereas mice immunised with control DNA had all succumbed nine days post-challenge.

Daugimont *et al.* (2010) developed a conductive microneedle array composed of Radel R polymer coated with conductive gold. This array was composed of 16 hollow microneedles (4×4) surrounded by 20 solid microneedles to provide mechanical support [169]. Microneedle arrays were then attached to an EP device to create an array that was capable of simultaneously injecting pDNA into the dermis from a fluid reservoir, and applying an electrical pulse to enhance cell permeability [169]. After demonstrating that the device could be successfully utilised as a vaccination platform for protein/peptide vaccines, the group employed the device with high- and low-voltage EP to deliver 10 μg of luciferase expressing plasmid, pVAX-CMV-LUC, ID, for a proof-of-concept study. Then 48 h after pDNA injection and EP, using the microneedle array, the group harvested the mouse skin and utilised a luminescence detection assay to quantify the presence of luciferase within samples. Following microneedle delivery, the group were unable to quantify any reporter gene expression within the skin, whereas they were able to detect luciferase expression following delivery of 50 μg of pVAX-CMV-LUC via simple ID injection [169]. The authors speculated that this lack of gene expression may have been due to needle blockage, or the EP field not being ideally situated to enhance cell permeability in the areas surrounding pDNA deposition, and thus the design of the device would need to be reconsidered [169].

Lee *et al.* (2011) also described the use of a dual functioning hybrid electro-microneedle (HEM) to simultaneously introduce nucleic acids into the skin and subsequently apply transient electrical pulses to enhance transfection [85]. The cargo was incorporated into a dissolvable maltose solution, which was heated and coated on top of a 5×5 array of blunt electrodes (300 μm diameter, 1000 μm length) via drawing lithography. The 3D matrix was cured to give an ultra-sharp (5 μm tip), bell-shaped 3D dissolving microneedle with a length of 400 μm. HEMs were loaded with 20 μg of pDNA coding *Gaussia* luciferase (pCMV-Gluc), inserted into the skin of hairless mice and left in place for 20 min to allow cargo dissolution. Following dissolution, the electrode of the device was connected to an electroporator, and eight electrical pulses (70 V, 50 ms) were applied to facilitate transfection. Alternatively, to evaluate the effectiveness of the device and EP, mice received 20 μg of pCMV-Gluc followed by EP using a pair of electrode tweezers, or the microneedle was simply allowed to dissolve without EP. Four days post-treatment, significant increases in luciferase expression were detected within extracted skin that had been treated with *SC* injection + EP, and the HEM + EP, but not in skin that had simply been treated with the HEM, indicating that EP was essential to facilitate gene transfer into the cutaneous cells. Eight days following treatment, an increase in bioluminescence compared with the control was only detectable in skin treated with HEM + EP, but not in skin treated with injection + EP. This gene expression was maintained until 15 days post-treatment, indicating that the HEM + EP was more effective in inducing longer lasting gene expression than simple injection [85]. Subsequently, the group examined the efficacy of the HEM device in a subcutaneous melanoma skin model (B16F10). C57 BL/6 mice were challenged with 2×10^5 B16F10 melanoma cells, and when tumours reached 3–5 mm in diameter the group utilised the device to treat the tumours, with plasmid encoding for the pro-inflammatory cytokine IL-12 (p2CMVmIL-12), or with the control groups as given earlier. All mice receiving control plasmid or p2CMVmIL-12 without EP had reached the humane end-point 30 days after tumour challenge. In contrast, mice treated with intratumoural (IT) injection + EP, or with the HEM + EP had significantly delayed tumour growth, and increased survival times, with 25% of mice treated with HEM + EP surviving >45 days, while all mice treated with IT injection + EP reached the end-point 41 days post-challenge. Thus, EP following introduction of cargo using a microneedle device is more effective than using injection with EP, possibly because this brings the cargo into close proximity with a greater number of cells, and is clearly an effective way to overcome the limitations of simple dissolvable devices.

Recognising that the current devices combining EP and microneedle arrays may be limited by rigidity, which would not allow effective penetration and voltage application over a large tissue area, Wei *et al.* (2014) manufactured a novel flexible microneedle array electrode (MNAE) [182]. The MNAE was composed of silicon microneedles, 190 μm in height, coated with gold protruding from a parylene film, which give a strong device capable of penetrating the *stratum corneum*, conforming to the desired organ structure and delivering a uniform voltage over a large area of tissue [182]. The group demonstrated that the MNAE could be used to facilitate pDNA transfection following injection into the muscle of C57 BL/6 male mice. After 30 min from injection of 40 μg of hyaluronidase into the thigh muscle to increase tissue permeability, the group injected a further 40 μg of plasmid coding the red fluorescent protein (RFP) into the same area, and 10 min later inserted the MNAE into the mice and applied five electrical pulses (up to 35 V, 20 ms duration). It appeared that 48 h following treatment, RFP expression was detectable only in the thigh muscle treated with the MNAE, but not following delivery of pRFP only. It was observed that RFP expression was strong and even in almost the whole area where the MNAE had been applied, suggesting that the device had conformed well around the desired area of transfection [182]. The group also demonstrated that the flexible device could be used to facilitate siRNA uptake into tumours, with minimal tissue damage, indicating that the array could have multiple *in vivo* applications. A drawback of the device is that unlike those previously described, the cargo has to be injected into the desired area prior to EP, which complicates the administration process.

Overall, the combination of EP and microneedle technology is a promising emerging gene therapy strategy that shows limited tissue damage compared with traditional EP devices [182], enhanced skin permeability compared with using EP or microneedles alone [167] and improved cellular uptake of nucleic acids [85]. However, some studies have demonstrated that injection into the skin is a more reliable way to deliver cargo across the skin than the combination of MN and EP [167]. The combination of microneedles with iontophoresis [183] and gene gun technology have also been explored with varying levels of success [184, 185]. A limitation of all of these technologies is that they require specialist equipment and training, which is associated with increased costs and a reduction in the possibility of self-administration.

5.5.5.2 In Combination with Vector-based Delivery Technologies

Since the conception of gene therapy the most commonly employed gene delivery vectors have been viral, in fact the first successful gene therapy trial utilised retroviral vectors [122, 186]. The power of viral vectors comes from overcoming the barriers to gene delivery by employing specialised proteins to facilitate endocytosis, endosomal escape and nuclear trafficking [187]. Despite the popularity of viral vectors, a number of limitations exist with viral gene therapy. Viral particles are often damaged by the manufacturing process [188] and the preparation process is costly and time-consuming, due to uncertain reproducibility, the need for high purity and high safety requirements involved with working with biohazardous material [189]. Other issues with viral gene therapy include a limited carrying capacity [190], the potential for oncogene activation following integration into the host genome [191, 192] and toxicity concerns [193]. Additionally, viral vectors are inherently immunogenic, and thus repeated administrations are associated with inflammatory reactions and clearance, rendering long-term treatment impotent [194–196].

Despite these limitations, viral vectors remain popular due to eliciting superior transfection efficacies *in vivo* compared with non-viral vectors, and several studies have combined viral vectors with microneedles in preclinical models. Carey *et al* (2011) utilised solid silicon ImmuPatch devices to facilitate ID delivery of modified virus ankara (MVA) expressing *P. berghei* CSP (MVA-PbCSP) [145]. The group immunised BALB/c mice by applying 5 μl of solution onto both ears (total 1×10^6 pfu (plaque-forming unit) of MVA-PbCSP) and pressing ImmuPatch devices, of various densities and shapes, into the solution to create pores across the *stratum corneum*. Alternatively, mice were immunised with the same quantity of MVA-PbCSP via ID injection into the ear. Two weeks following immunisation mice were sacrificed and the cytokine secretion profile of the splenic CD8+ T-cell population specific for the dominant MHC I epitope from PbCSP was assessed using flow cytometry. The group found that applying devices into the MVA-PbCSP solution pre-applied to ears generated a similar distribution of antigen-specific IFN-γ^+TNFα^- (single positive) and

IFN-γ^+TNFα^+ (double positive) CD8+ T cells compared with ID injection. The authors noted that application of arrays with smaller total pore volumes (small size and/or low frequency) generated significantly fewer numbers of PbCSP-specific CD8+ T cells than ID injection, however, utilising devices with a high number of projections or of greater length induced a similar number of T cells to conventional injection [145].

The group then investigated whether the combination of ID injection and the ImmuPatch in a heterologous prime-boost model could increase antigen-specific CD8+ T cell responses. Mice were immunised with ID injection or ImmuPatch of various pore volumes as previously, and were boosted 14 days later with ID injection prior to analysing the antigen-specific T-cell population 14 days after boosting. Immunisation of mice with heterologous prime-boost protocols resulted in significantly higher numbers of PbCSP-specific CD8+ T cells than prime-boost with homologous ID injections, indicating that microneedles are a minimally invasive means to increase vaccine immunogenicity. Interestingly, in prime-boost protocols, utilising ImmuPatch devices with a smaller pore volume resulted in the generation of a significantly higher number of antigen-specific CD8+ T cells, indicating that immunising with a smaller pore volume results in the generation of a larger portion of memory T cells.

Next, this group determined whether the combination of the ImmuPatch and ID injection could confer greater protection of mice against *P. berghei* malarial challenge. BALB/c mice were immunised with ID injection, or ImmuPatch devices of small, medium or large pore volumes. Two weeks following priming mice were boosted with ID injection or ImmuPatch, and mice were subsequently challenged intravenous (IV) with 1000 infected red blood cells (RBCs) two weeks after boosting, and up to 20 days post-challenge the percentage parasitaemia in blood stains was determined. Immunising with the ImmuPatch and boosting with ID injection resulted in similar protection against sporozoite challenge, but priming and boosting with the ImmuPatch conferred the greatest protection against mice becoming parasitemic. Therefore, using the ImmuPatch alone, or in combination with ID injection, is just as efficacious as ID injection alone in protecting against malaria, and so microneedles could plausibly remove the need for vaccination with the hypodermic needle in this model [145].

In a similar study Carey *et al.* (2014) used the ImmunPatch alone or in combination with ID injection to deliver adenovirus human serotype 5 (HAdV5) based vaccines expressing the malarial antigen PyMSP1$_{42}$ to mice as described earlier [197]. In this study they found that immunising with the ImmuPatch and then boosting with ID injection conferred the greatest protection from RBCs infected with malarial parasite. Importantly, the group also demonstrated that the anti-vector immune response was lowered in mice immunised with the ImmuPatch compared with ID injection, while inducing similar responses to the encoded antigen, and this may allow for a higher number of administrations than ID injection [197].

These studies highlight that microneedle pore size can be manipulated to adjust the immune response, that the combination of microneedles of differing pore sizes may be used to replace conventional ID injection and that microneedle technology may provide other favourable outcomes for immunisation, such as reducing local tissue damage and decreasing the anti-vector adaptive immune response. However, this method of introducing live viruses to the skin using solid microneedles is not easy to translate into a clinical setting. As well as requiring a two-step application, the accessibility of these live viral vaccines may be hindered by the limited stability of the viral cargo, and arrays which store the vaccine in the dry state (such as coated and dissolvable arrays) may prove a more viable option for gene therapy, decreasing storage requirements and reducing cargo wastage. In 2012, Carey's group described the fabrication of such a solid array coated with live AdV or MVA for the purposes of vaccination [156]. One of the major challenges of this study was the selection of a coating solution that retained the integrity of the viral particles and was capable of coating the solid microneedle effectively. Of the nine different coating formulations screened, none were capable of retaining >50% survival of the MVA particles. Further to this, the coating and drying of the cargo leads to a further loss of viral particles. Despite this loss of cargo integrity, they were able to successfully spray-coat solid silicon microneedles with a 15% w/v trehalose solution encapsulating AdV and MVA, which could induce gene expression *ex vivo* when pressed into porcine skin. Furthermore, needles coated in AdV expressing the malarial antigen $PyMSP1_{42}$ were able to induce similar levels of serum anti-$PyMSP1_{42}$ antibodies to those for immunisation with ID injection, and needles coated with MVA expressing the malarial antigen PbCSP were able to generate similar levels of PbCSP-specific CD8+ T cells to immunisation with hypodermic needle [155].

Overall, this study was a good advancement for viral-based gene therapy, as storage in the dry state may provide a path to circumvent the poor stability of viral vaccines at ambient temperatures, which provides a major obstacle to clinical usage. The limitation of this study is in the dramatic loss of viral integrity during the coating and drying stages of formulation. Recently, Kines *et al.* (2015) also reported on the fabrication of a solid, coated microneedle array to deliver a DNA vaccine consisting of human papillomavirus pseudoviruses (HPV-PsV) encapsidating DNA expressing various antigens from the respiratory syncytial virus (RSV) (fusion protein (F), or M and M2 proteins) [188]. An advancement of this platform was that Kines *et al.* took advantage of the immunogenicity of the virus to induce robust immune responses against the virus capsid proteins, as well as against the encoded antigens, thus providing a means to immunise against two diseases simultaneously. However, owing to the low quantity of virus that could be incorporated into the microneedle coating, the group employed lyophilisation to dry HPV-PsVs, prior to direct reconstitution in the coating solution, to increase the HPV-PsV loading. These authors also noted a loss of functionality at the freeze-drying and coating stages of the formulation process, and thus it is likely that coating of viral solutions onto solid needles may always significantly compromise the integrity of live viruses. Although this may be limited by the careful screening of coating solutions or dissolvable matrices, this process is cumbersome and would need to be repeated and modified for every viral vector. To this end non-viral vectors may provide a superior alternative to viral gene therapy.

Non-viral vectors are generally cationic lipids, polymers or peptides that condense anionic DNA through electrostatic interactions into nanoparticles, simultaneously protecting DNA from degradation and increasing cellular uptake [187, 198]. Non-viral vectors are attractive in terms of low cost, ease of production, lack of immunogenicity and safety [125]. The main limitation to the use of these vectors is the lack of transfection efficiency compared with viral vectors, and thus development is focused on overcoming this issue [187]. A key advantage over viral vectors is that novel non-viral gene delivery strategies may allow selective tissue or cell targeting through incorporation of moieties that bind specific cell receptors. Of note is that siRNA therapeutics cannot be coded/delivered by viruses and so the enhancement of cellular uptake may only be achieved with non-viral vectors.

Numerous studies have investigated the delivery of nucleic acids complexed to non-viral vectors from solid, coated, hollow and dissolvable microneedles as a means of improving cargo penetration across the skin and/or increase transfection efficacy. For the most part, these studies demonstrate that non-viral vectors successfully improve the functionality of the encapsulated cargo. The findings of some key studies are discussed later, however there are too many to discuss extensively, and so we have summarised and referenced the remainder in Table 5.1.

The first example of solid microneedles being used to enhance delivery of pDNA, complexed with a non-viral vector, across the stratum corneum was reported in 2004 by Chabri *et al.* [199]. Solid silicon microneedles were applied for up to 30 s to the *stratum corneum* facing side of epidermal sheets obtained from excised human breast skin. The movement of lipid/pDNA complexes from the donor compartment of the Franz-cell apparatus, across treated or untreated epidermal sheets, was then determined using gel electrophoresis following a 24-h period. A clear loss of lipid/pDNA cargo could be seen in the donor compartment of microneedle-treated epidermal skin, compared with non-treated skin, indicating that microneedle treatment could facilitate movement of the lipid/pDNA cargo into the receptor compartment [199].

An improvement on this simple strategy has been the development of non-viral vectors designed to target uptake by APCs within the skin. For example, Hu *et al* (2014) modified 18 kDa polyethylenimine (PEI_{1800}) with a mannose ligand to enhance uptake via the mannose receptor (MR), which is expressed in high abundance on the surface of APCs [202]. To improve the transfection efficacies of the Man-PEI_{1800} further, and reduce cytotoxicity, the group further functionalised the vector with a cell-penetrating peptide (CPP) motif (Man-PEI_{1800}-CPP). Complexing this functionalised polymer to pDNA produced nanoparticles with ideal characteristics for transfection (nanoscale with cationic charge), which were able to protect pDNA from DNase degradation and were significantly less toxic than PEI_{2500}/pDNA complexes in DC 2.4 dendritic cells over a range of N:P ratios. Importantly, it was demonstrated that Man-PEI_{1800}-CPP/pDNA complexes produced significantly higher transfection efficacies in DC 2.4 cells than PEI_{2500}/pDNA complexes, due to receptor-mediated endocytosis via the MR (confirmed by competition assay). In a preclinical study the group utilised a solid microneedle array (Bomtech Electronic, Korea) to pre-treat shaven abdominal skin of BALB/c mice prior to application of polymer/pDNA complexes (N:P 10) coding GFP (pEGFP-N2). At pre-determined time points (12–72 h) the mice were sacrificed and the number of $C11c^+$ DCs expressing GFP was determined using flow cytometry. Treatment with Man-PEI_{1800}-CPP/pEGFP-N2 complexes resulted in significantly higher numbers of GFP^+C11c^+ DCs than treatment with PEI_{2500}/pEGFP-N2 or uncomplexed pEGFP-N2. In contrast, $C11c^-$ non-DCs showed limited GFP expression following treatment with any of the solutions, demonstrating that these functionalisations specifically target pDNA cargo to APCs within the skin, which may have a special relevance for vaccination [202].

In a preclinical melanoma model Hu *et al.* (2015) then demonstrated that this mannosylated vector could be used to immunise mice against tumour challenge [203]. Mice were treated with PEI_{2500}/pDNA or Man-PEI_{1800}-CPP/pDNA complexes expressing the melanoma tumour antigen Trp2, granulocyte-macrophage colony-stimulating factor

Table 5.1 Summary of studies combining non-viral vectors with microneedle technology.

Device	Vector/modification	Cargo	Model	Key findings	Reference
Solid silicon microneedles	DOTAP:protamine liposome	pEGFP-N1	Human skin	Microneedles enhance penetration of DOTAP:protamine:pDNA liposomes across excised human skin in Franz-cell diffusion model	[199]
Solid silicon microneedles	DOTAP:protamine liposome	pCMVβ	Human skin	Microneedles with a larger diameter enhance permeation of charged DOTAP:protamine:pDNA across excised human skin compared with smaller pores	[143]
Solid microneedle roller	PLGA; PLGA:DOTAP	pCMVβ pGPA	Balb/c mice C57 BL/6 mice SKH-1 mice	Cationic nanoparticles generate greater immune responses towards encoded antigens than anionic nanoparticles and are necessary to mediate cellular uptake in the skin; immunisation with cationic PLGA nanoparticles transcutaneously generates a similar cellular immune response than IM immunisation, but generates a greater humoral and mucosal response	[200]
Solid silicon microneedles	P123-PEI	pVAX-S	C57 BL/6 mice	Pre-treating abdominal skin of mice with solid microneedles and applying hepatitis DNA vaccine complexed into polyplexes generated greater quantities of antigen-specific antibodies and IFNγ secreting splenocytes than administering the same cargo IM	[201]
Motorised solid microneedle	Lipofectamine 2000; PEI	pEGFP-Luc	Rat	Motorised solid microneedles enhance pDNA penetration and transfection within the skin; complexing pDNA into lipoplexes and polyplexes may decrease skin permeation	[149]
Motorised solid microneedle	Man-PEI_{1800}-CPP; PEI_{2500}	pEGFP-N2	Balb/c mice	Functionalised PEI complexes allow enhanced uptake by cutaneous dendritic cells	[202]
Motorised solid microneedle	Man-PEI_{1800}-CPP; PEI_{2500}	pTrp2-GM-CSF-Fc-EGFP	Balb/c mice Melanoma B16	DNA vaccines complexed to mannosylated polyplexes provide greater protection from tumour challenge in both prophylactic and therapeutic models	[203]
Solid silicon microneedle	Cholesterol	*Gapdh* siRNA	C57 BL/6 mice	Application of *Gapdh* siRNA solution to the ear, followed by microneedle treatment, decreased targeted gene expression by up to 66%	[126]
Coated steel microneedle	Lipofectamine RNAiMAX	Lamin A/C siRNA; Accell CBL3 sd-siRNA	Transgenic hMGFP/CBL mice	Lipofectamine/siRNA complexes mediate significant reductions in gene expression *in vitro*, but are inactive following coating onto metal microneedles, highlighting the need to use other vectors/coating processes	[204]

(*Continued*)

Table 5.1 (Continued)

Device	Vector/modification	Cargo	Model	Key findings	Reference
Dry-coated Nanopatch	DOTAP:DOPE: Cholesterol:PEG (liposome)	CXCL1-siRNA	Balb/c mice	Liposome/siRNA complexes delivered via coated microneedle significantly reduced expression of a transiently expressed chemokine involved in neutrophil migration within the skin following damage/infection	[96]
Polycarbonate microneedle with pH-responsive coating	Man-DA3 (polyplex)	pTarget-Ig--Aβ-Fc	Balb/c mice	Administration of pTarget-Ig-Aβ-Fc polyplexes via microneedle generated higher levels of protein expression than via SC injection; immunisation of mice via microneedle generated significantly higher numbers of antigen-specific antibodies than immunisation via SC injection	[205]
Nanopass™ 33G Hollow microneedle	Lipofectamine 2000; PEI; Superfect	pOVA	Balb/c mice	Immunising mice with pOVA complexed into cationic nanoparticles with lipofectamine, PEI or Superfect generated significantly higher serum concentrations of OVA-specific antibodies than naked pOVA; immunising mice with cationic nanoparticles ID with hollow microneedles generated a significantly stronger humoral response than immunising SC	[167]
Triple-pronged array-type micro-needle head at-tached to micro-syringe (hollow)	Cholesterol	HPV16 E6-siRNA	Balb/c mice SiHa cervical cancer model	Intratumoural injection of cholesterol-modified HPV16 E6-siRNA mediated significant knockdown of the HPV oncogene in established SiHa xenografts; treatment of established SiHa tumours resulted in significant retardation of tumour growth	[206]
Dissolvable PVP microneedle	DPPC:cholesterol: DDA (liposome)	pVR-E2E	Balb/c mice	Immunisation of mice with VR-E2E liposomes using the dissolvable microneedle or IM injection generated similar numbers of antibodies against the encoded antigen; The addition of CpG ODN to VR-E2E liposomes significantly enhanced the anti-HBsAg IgG response	[207, 208]
Dissolvable PVP microneedle	RALA	pCMV-Luc	C57 BL/6 mice	Delivery of RALA/pCMV-Luc to the ears of mice via microneedle results in systemic expression of the encoded protein	[139]
Dissolvable PVP microneedle	RALA	pE6/E7	C57 BL/6 mice TC-1 cervical cancer model	Immunising mice with RALA/pE6/E7 nanoparticles via microneedles was significantly more effective in retarding tumour growth in both prophylactic and therapeutic models of cervical cancer than immunisation via IM injection or with "naked" pE6/E7	[176]

Dissolvable PVP and PVA microneedles	RALA	pEGFP-N1	NCTC-929 C57 BL/6 mice	Complexing pDNA into cationic nanoparticles prior to incorporation into dissolvable matrices increases transfection efficacies *in vitro*, and protects pDNA integrity during the fabrication process; the choice of dissolvable matrix significantly impacts cargo functionality and encapsulation and delivery efficiency following manufacture	[175]
Dissolvable PVA microneedles	PLGA-PLL/γPGA	EboDNA	Mice	Repeated immunisation of mice with cationic PLGA-PLL/γPGA/EboDNA nanoparticles generated significantly higher serum concentrations of anti-Ebo-GP antibodies than vaccination with naked EboDNA; mice immunised with dissolvable microneedles demonstrated higher antibody responses than those immunised via the IM route and showed the highest neutralising activity against Ebola GP-pseudovirions	[138]

(GM-CSF) and GFP (pTrp2-GM-CSF-Fc-EGFP) as in the previous study. Mice were immunised three times at weekly intervals with 50 μg of pTrp2-GM-CSF-Fc-EGFP, and one week following the final booster were challenged with 2×10^5 B16 melanoma cells. Results indicated that treatment with Man-PEI_{1800}-CPP/pDNA complexes significantly prolonged survival times of mice challenged with melanoma compared with PEI_{2500}/pDNA and PBS groups, highlighting the importance of targeting APCs within the skin for DNA vaccination purposes [203].

However, despite the success of these studies, others have reported mixed results using the combination of solid microneedles and non-viral vectors. Yan *et al.* reported that applying lipoplexes (Lipofectamine^{TM} 2000/pEGFP-Luc) or polyplexes (PEI/pEGFP-Luc) on to rat skin and subsequently creating punctures into the *stratum corneum* using the motorised My-M device, resulted in lower gene expression than utilising naked pEGFP-Luc [149]. A reason for this reduced gene expression may be that the cationic nanoparticles may interact with anionic molecules at the membrane surface or within the fluid-filled microchannel [143, 149, 209]. This may explain why other researchers using microneedles that actively localise cargo into the skin typically see improved results with vectored nucleic acids. With this in mind coated, hollow and dissolvable formulations may be more suitable for the delivery of this cargo to the skin.

Kim *et al.* (2014) developed a novel, two-tier delivery system incorporating polyplexes into a pH-responsive, rapidly dissolving microneedle coating [154]. This pH-responsive coating was designed to dissolve rapidly within the dermal interstitial fluid releasing APC-targeting polyplexes composed of mannosylated PEI (Man-DA3) and pDNA. Following dissolution, Man-DA3 polyplexes are expected to enhance pDNA uptake via mannose receptor (MR)-mediated endocytosis into dermal DCs [210]. Indeed, Man-DA3 complexes demonstrated enhanced transfection efficacy in the MR^+ RAW 264.7 macrophage cell line compared with non-mannosylated PEI polyplexes, and MR-mediated transfection was confirmed by a free mannose competition assay [205]. The authors demonstrated that delivery of Man-DA3 polyplexes encoding an Alzheimer's disease antigen, amyloid beta monomer (Aβ 1-42), via microneedle application resulted in greater Aβ expression than *SC* injection of the same polyplexes. In a separate study the group immunised Balb/c mice with 10 μg of pDNA encoding the Aβ 1-42 peptide via microneedle delivery or *SC* injection [154]. One week following immunisation microneedle treated mice exhibited significantly higher plasma concentrations of Aβ-specific antibodies compared with naïve or *SC* immunised mice [205]. This two-tier approach combining an easily used physical delivery technology to localise pDNA to a highly immunogenic site, and a vector to enhance cellular uptake within APC, is likely to be of key benefit for DNA vaccination purposes. An improvement over this system may be the use of dissolvable formulations to incorporate a larger quantity of cargo throughout the microneedle shaft and/or baseplate, as even with the use of vectors the limited loading capacity of coated devices is likely to show negative efficacy in larger animals.

Qiu *et al.* (2015) developed a dissolvable PVP microneedle system encapsulating cationic liposome/DNA lipoplexes encoding hepatits B virus (HBV) envelope proteins (VR-E2E) with or without adjuvant CpG ODN [207]. Balb/c mice were subsequently vaccinated with 10 μg of VR-E2E DNA ± liposome and/or CpG ODN adjuvant, via application of this novel microneedle device or IM injection on days 0 and 21, and the serum was assessed for anti-VR-E2E antibodies two weeks following the last immunisation, via ELISA. The group found that immunising with VR-E2E liposomes + CpG ODN via the dissolvable microneedle patch significantly increased serum anti-HBsAg IgG levels, although these were similar to those produced when immunising via IM injection [207]. Overall these results were encouraging in an antigenic model as only ~30% of the microneedle payload was deposited into the skin, and so if the loading and payload deposition could be optimised this is a promising platform for DNA vaccination [207]. Yang *et al.* (2017) also recently reported on the use of dissolvable microneedles to immunise mice with an Ebola DNA vaccine (EboDNA) [138]. These authors coated the EboDNA onto the surface of cationic polylactic-*co*-glycolic acid-poly-L-lysine/poly-γ-glutamic acid (PLGA-PLL/γPGA) nanoparticles. The group incorporated these PLGA-PLL/γPGA/pDNA nanoparticles into aqueous PVA gels and casted the solution into polydimethylsiloxane master moulds under vacuum to fill the indents, before casting a second PVA/PVP solution to form the supporting baseplate. Using this fabrication method, the group were able to fabricate strong, sharp microneedles with a height of 750 μm, which encapsulated ~45 μg of pDNA complexed with PLGA-PLL/γPGA. Immunisation of mice four times at four-week intervals with the dissolvable device encapsulating PLGA-PLL/γPGA/EboDNA generated significantly higher levels of antigen-specific IgG1 titres compared with immunising with PLGA-PLL/γPGA/EboDNA via IM injection. As such the authors successfully generated a safe and biocompatible device that could be applied with minimal training and equipment [138].

The use of lipid and polymer carriers for gene therapy presents several challenges. Owing to the high positive charge of the groups, lipoplexes and polyplexes frequently interact with negatively charged serum proteins [211, 212], compromising transfection efficacy *in vivo*, and are rapidly removed from circulation by the RES [213]. The addition of hydrophilic polymers, such as poly(ethylene glycol) (PEG) to lipoplexes and polyplexes provides steric hindrance to serum proteins [214, 215]. However, the addition of PEG often results in reduced transfection efficacies, the so-called "PEG dilemma" [216, 217], and the repeated administration of PEGylated nanoparticles has resulted in the formation of anti-PEG IgM, leading to PEG opsonisation and hence rapid clearance [218, 219]. Perhaps more importantly, high molecular weight lipids and polymers that are associated with greater transfection efficacies are known to be highly cytotoxic [220–222], and this may pose a significant barrier to translation to the clinic. Therefore, there is a need for the development of newer, non-cytotoxic vectors that do not require a trade-off between transfection efficiency and cytotoxicity. A third class of non-viral gene delivery vectors is that of cationic peptides, which hold the potential to mimic viral gene delivery. By identifying, replicating and improving the short peptide sequences responsible for virus uptake, endosomal escape and nuclear localisation, peptide chains can be rationally designed to carry out one or more of these actions [187, 198]. The versatility of these viral mimetic peptides, accompanied by good biocompatibility and low cytotoxicity profile make them clear forerunners as non-viral delivery agents.

Recently, McCaffrey and coworkers (2016) developed a two-tier DNA delivery system, complexing DNA with the novel non-toxic amphipathic delivery peptide, RALA [223], to produce nanoparticles that were encapsulated into a dissolvable PVP microneedle matrix [139]. The authors demonstrated that complexing pDNA with the RALA peptide produced nanoparticles, <100 nm in size, which were stable for up to 28 days in solution and able to significantly increase gene expression *in vitro* compared with naked pDNA alone. These RALA/pDNA nanoparticles were encapsulated into a microneedle system using a two-step manufacturing process, whereby a PVP-RALA/pDNA solution was used to fill the conical indents of a silicon master mould via centrifugation, followed by the addition of a second non-loaded PVP solution to form a baseplate for support [139]. Using this system McCaffrey's group (2016) were able to deliver ~16 μg of DNA encoding firefly luciferase (pCMV-Luc) to the ears of C57BL/6 mice, and although no local gene expression was detected using whole body bioluminescence imaging, gene expression was detectable systemically in the liver and kidney up to 48 h post microneedle application [139].

The functionality of this dissolvable PVP microneedle system was further assessed by Ali *et al.* (2017) in a cervical cancer model [176]. The authors utilised the device encapsulating RALA/pDNA nanoparticles coding the cervical cancer antigens E6 and E7 to immunise female C57 BL/6 mice in both prophylactic and therapeutic cervical cancer models. Mice immunised with RALA/pE6-E7 nanoparticles via microneedle survived significantly longer than mice immunised with pE6-E7 only, or with RALA/pE6-E7 administered via IM injection, in both the prophylactic and therapeutic studies. Immunisation of mice three times with RALA/pE6-E7 via microneedle was able to prevent the establishment of TC-1 tumours in four out of the nine mice challenged and 55% of these mice survived to 100 days post-challenge (when the study was concluded), compared with 10% of the mice immunised with pE6/E7 only via microneedle, or 25% of mice immunised with RALA/pE6/E7 IM. In the therapeutic study three immunisations with RALA/pE6-E7 via microneedle were able to cause regression of established tumours in two mice (out of nine), and 40% of mice survived until the conclusion of the study (100 days post-challenge) compared with 10% of mice immunised with pE6/E7 only via microneedle or RALA/pE6/E7 nanoparticles administered via simple IM injection. Ali *et al.* also demonstrated that the RALA/pE6-E7 cargo remained functional in the dissolvable PVP matrix for up to 28 days when stored at 4°C, room temperature or 37°C. This manufacturing process represents a simple, cost-effective way to produce a clinically relevant prototype DNA delivery system that can be stored at elevated temperatures prior to use, and self-applied [176].

Given the increased transfection efficacies gained by complexing pDNA to non-viral vectors, and that this is relatively inexpensive, future studies in microneedle gene therapy are likely to continue to focus on developing novel vectors.

5.6 Conclusions

- Taken together these studies demonstrate that the most promising microneedle technologies are those which incorporate physical or vector-based gene delivery systems into the microneedle device.
- Solid and hollow microneedles are amenable to combination with electroporation, and these devices may prove popular in clinical settings where specialist equipment and highly trained personnel are available.
- Coated and dissolvable microneedles may be particularly efficacious in combination with specialised non-viral vectors, which can be integrated easily within the device. These platforms are likely to be valuable where minimal expertise, equipment or refrigeration is available, for example, mass vaccination purposes.

- An area which remains to be clarified is the requirements that these devices will have to comply with to gain regulatory approval, particularly coated and dissolvable formulations where the dissolvable matrix is deposited into the skin, as well as devices incorporating viral cargo, which may be the subject of additional safety requirements.
- More advice would be helpful from regulatory authorities as to whether devices and/or genetic cargo will require complete sterility. Should complete sterility be a requirement this will significantly increase the cost of microneedle manufacture as the nucleic acid and the nanoparticle cargo are vulnerable to degradation by conventional sterilisation processes, and therefore the need for manufacture under aseptic conditions is likely.
- Another key area that requires consideration is the development of manufacturing processes that are suitable for the scale-up of these sensitive biopharmaceuticals.
- Future directions that should take priority include: increasing the quantity of nucleic acid which can loaded into coated or dissolvable devices; ensuring that the formulation process is thoroughly optimised to protect the genetic cargo; incorporating adjuvants into the device or nucleic acid to improve immunogenicity; and establishing efficacy in larger animal models.

References

1 R.C. Mulligan, The basic science of gene therapy (1993). *Science* 260 (5110): 926–932.

2 E.W.F.W. Alton, *et al.* (2015). Repeated nebulisation of non-viral CFTR gene therapy in patients with cystic fibrosis: a randomised, double-blind, placebo-controlled, phase 2b trial. *Lancet Respir. Med.* 3 (9): 684–691.

3 H. Yin, *et al.* (2014). Genome editing with Cas9 in adult mice corrects a disease mutation and phenotype. *Nat. Biotechnol.* 32 (6): 551–553.

4 S.A. Leachman, *et al.* (2010). First-in-human mutation-targeted siRNA phase Ib trial of an inherited skin disorder. *Mol. Ther.* 18 (2): 442–446.

5 C. Xu and J. Wang (2015). Delivery systems for siRNA drug development in cancer therapy. *Asian J. Pharm. Sci.* 10 (1): 1–12.

6 M. Collins and A. Thrasher (2015). Gene therapy: progress and predictions. *Proc. R. Soc. London B Biol. Sci.* 282 (1821): doi:10.1098/rspb.2014.3003.

7 L.E. Gracey Maniar, J.M. Maniar, Z.-Y. Chen, *et al.* (2013). Minicircle DNA vectors achieve sustained expression reflected by active chromatin and transcriptional level. *Mol. Ther.* 21 (1): 131–138.

8 A. Gluch, M. Vidakovic and J. Bode (2008) Scaffold/matrix attachment regions (S/MARs): relevance for disease and therapy. In *Handbook of Experimental Pharmacology* 186, 67–103. Springer.

9 A. Stoecklinger *et al.* (2011). Langerin+ dermal dendritic cells are critical for CD8+ T cell activation and IgH γ-1 class switching in response to gene gun vaccines. *J. Immunol.* 186 (3): 1377–1383.

10 P. Stoitzner, *et al.* (1999). Migration of langerhans cells and dermal dendritic cells in skin organ cultures: augmentation by TNF-alpha and IL-1beta. *J. Leukoc. Biol.* 66 (3): 462–470.

11 J. Valladeau and S. Saeland (2005). Cutaneous dendritic cells. *Semin. Immunol.* 17 (4): 273–283.
12 G. Cole, J. McCaffrey, A.A. Ali and H.O. McCarthy (2015). DNA vaccination for prostate cancer: key concepts and considerations. *Cancer Nanotechnol.* 6 (1): 2.
13 J.A. Wolff, *et al.*, Direct gene transfer into mouse muscle in vivo (1990). *Science* 247 (4949) Pt 1: 1465–1468.
14 L. Redding and D.B. Weiner (2009). DNA vaccines in veterinary use. *Expert Rev. Vaccines* 8 (9): 1251–1276.
15 J.A.C. Schalk, F.R. Mooi, G.A.M. Berbers, *et al.* (2006). Preclinical and clinical safety studies on DNA vaccines. *Hum. Vaccin.* 2 (2): 45–53.
16 B. Langer, M. Renner, J. Scherer, *et al.* (2013). Safety assessment of biolistic DNA vaccination. *Methods Mol. Biol.* 940: 371–388.
17 M.A. Kutzler and D.B. Weiner (2008). DNA vaccines: ready for prime time? *Nat. Rev. Genet.* 9 (10): 776–788.
18 A. Suhrbier (1997). Multi-epitope DNA vaccines. *Immunol. Cell Biol.* 75 (4): 402–408.
19 J. Gummow, *et al.* (2015). A multiantigenic DNA vaccine that induces broad hepatitis C virus-specific T-cell responses in mice. *J. Virol.* 89 (15): 7991–8002.
20 R.A. Madan, P.M. Arlen, M. Mohebtash, *et al.* (2009). Prostvac-VF: a vector-based vaccine targeting PSA in prostate cancer. *Expert Opin. Investig. Drugs* 18 (7): 1001–1011.
21 J. Bonnet, *et al.* (2010). Chain and conformation stability of solid-state DNA: implications for room temperature storage. *Nucleic Acids Res.* 38 (5): 1531–1546.
22 D. Clermont, S. Santoni, S. Saker, *et al.* (2014). Assessment of DNA encapsulation, a new room-temperature DNA storage method. *Biopreserv. Biobank.* 12 (3): 176–183.
23 B. Ferraro, M.P. Morrow, N.A. Hutnick, *et al.* (2011). Clinical applications of DNA vaccines: current progress. *Clin. Infect. Dis.* 53 (3): 296–302.
24 J.S. Blum, P.A. Wearsch and P. Cresswell (2013). Pathways of antigen processing. *Annu. Rev. Immunol.* 31: 443–473.
25 F. Kotsias, E. Hoffmann, S. Amigorena and A. Savina (2013). Reactive oxygen species production in the phagosome: impact on antigen presentation in dendritic cells. *Antioxid. Redox Signal.* 18 (6): 714–729.
26 S. Burgdorf and C. Kurts (2008). Endocytosis mechanisms and the cell biology of antigen presentation. *Curr. Opin. Immunol.* 20 (1): 89–95.
27 J. Neefjes, M.L.M. Jongsma, P. Paul and O. Bakke (2011). Towards a systems understanding of MHC class I and MHC class II antigen presentation. *Nat. Rev. Immunol.* 11 (12): 823–836.
28 P. Cresswell, A.L. Ackerman, A. Giodini, *et al.* (2005). Mechanisms of MHC class I-restricted antigen processing and cross-presentation. *Immunol. Rev.* 207: 145–157.
29 M. Chemali, K. Radtke, M. Desjardins and L. English (2011). Alternative pathways for MHC class I presentation: a new function for autophagy. *Cell. Mol. Life Sci.* 68 (9): 1533–1541.
30 A.R. Mantegazza, *et al.* (2008). NADPH oxidase controls phagosomal pH and antigen cross-presentation in human dendritic cells. *Blood* 112 (12): 4712–4722.
31 A. Savina, *et al.* (2006). NOX2 controls phagosomal pH to regulate antigen processing during crosspresentation by dendritic cells. *Cell* 126 (1): 205–218.

32 J.M. Rybicka, D.R. Balce, S. Chaudhuri, *et al.* (2012). Phagosomal proteolysis in dendritic cells is modulated by NADPH oxidase in a pH-independent manner. *EMBO J.* 31 (4): 932–944.
33 L. Delamarre, M. Pack, H. Chang, *et al.* (2005). Differential lysosomal proteolysis in antigen-presenting cells determines antigen fate. *Science* 307 (5715): 1630–1634.
34 M. Kovacsovics-Bankowski and K.L. Rock (1995). A phagosome-to-cytosol pathway for exogenous antigens presented on MHC class I molecules. *Science* 267 (5195): 243–246.
35 L. Shen, L.J. Sigal, M. Boes and K.L. Rock (2004). Important role of cathepsin S in generating peptides for TAP-independent MHC class I crosspresentation in vivo. *Immunity* 21 (2): 155–165.
36 J.D. Pfeifer, M.J. Wick, R.L. Roberts, *et al.* (1993). Phagocytic processing of bacterial antigens for class I MHC presentation to T cells. *Nature* 361 (6410): 359–362.
37 S.W. Howland and K.D. Wittrup (2008). Antigen release kinetics in the phagosome are critical to cross-presentation efficiency. *J. Immunol.* 180 (3): 1576–1583.
38 P. Guermonprez, L. Saveanu, M. Kleijmeer, *et al.* (2003). ER-phagosome fusion defines an MHC class I cross-presentation compartment in dendritic cells. *Nature* 425 (6956): 397–402.
39 S. Burgdorf, A. Kautz, V. Böhnert, *et al.* (2007). Distinct pathways of antigen uptake and intracellular routing in CD4 and CD8 T cell activation. *Science* 316 (5824): 612–616.
40 M. Lakadamyali, M.J. Rust and X. Zhuang (2006). Ligands for clathrin-mediated endocytosis are differentially sorted into distinct populations of early endosomes. *Cell* 124 (5): 997–1009.
41 A. Engering, *et al.* (2002). The dendritic cell-specific adhesion receptor DC-SIGN internalizes antigen for presentation to T cells. *J. Immunol.* 168 (5): 2118–2126.
42 S. Burgdorf, V. Lukacs-Kornek and C. Kurts (2006). The mannose receptor mediates uptake of soluble but not of cell-associated antigen for cross-presentation. *J. Immunol.* 176 (11): 6770–6776.
43 K.J. Farrand, N. Dickgreber, P. Stoitzner, *et al.* (2009). Langerin+ CD8alpha+ dendritic cells are critical for cross-priming and IL-12 production in response to systemic antigens. *J. Immunol.* 183 (12): 7732–7742.
44 N. Monu and E.S. Trombetta (2007). Cross-talk between the endocytic pathway and the endoplasmic reticulum in cross-presentation by MHC class I molecules. *Curr. Opin. Immunol.* 19 (1): 66–72.
45 J. Sabatté, *et al.* (2007). Interplay of pathogens, cytokines and other stress signals in the regulation of dendritic cell function. *Cytokine Growth Factor Rev.* 18 (1–2): 5–17.
46 O. Berthier-Vergnes, F. Bermond, V. Flacher, *et al.* (2005). TNF-alpha enhances phenotypic and functional maturation of human epidermal Langerhans cells and induces IL-12 p40 and IP-10/CXCL-10 production. *FEBS Lett.* 579 (17): 3660–3668.
47 W.B. Bowne, *et al.* (1999). Injection of DNA encoding granulocyte-macrophage colony-stimulating factor recruits dendritic cells for immune adjuvant effects. *Cytokines. Cell. Mol. Ther.* 5 (4): 217–225.
48 S. Miwa, *et al.* (2012). TNF-α and tumor lysate promote the maturation of dendritic cells for immunotherapy for advanced malignant bone and soft tissue tumors. *PLoS One* 7 (12): e52926.

49 S. Corinti, C. Albanesi, A. la Sala, *et al.* (2001). Regulatory activity of autocrine IL-10 on dendritic cell functions. *J. Immunol.* 166 (7): 4312–4318.
50 Y. Kodaira, S.K. Nair, L.E. Wrenshall, *et al.* (2000). Phenotypic and functional maturation of dendritic cells mediated by heparan sulfate. *J. Immunol.* 165 (3): 1599–1604.
51 O. Joffre, M.A. Nolte, R. Spörri and C. Reis e Sousa (2009). Inflammatory signals in dendritic cell activation and the induction of adaptive immunity. *Immunol. Rev.* 227 (1): 234–247.
52 R. Förster, A.C. Davalos-Misslitz and A. Rot (2008). CCR7 and its ligands: balancing immunity and tolerance. *Nat. Rev. Immunol.* 8 (5): 362–371.
53 F. Sallusto, C. Nicolò, R. De Maria, *et al.* (1996). Ceramide inhibits antigen uptake and presentation by dendritic cells. *J. Exp. Med.* 184 (6): 2411–2416.
54 A. Chow, D. Toomre, W. Garrett and I. Mellman (2002). Dendritic cell maturation triggers retrograde MHC class II transport from lysosomes to the plasma membrane. *Nature* 418 (6901): 988–994.
55 R.M. Steinman (2001). Dendritic cells and the control of immunity: enhancing the efficiency of antigen presentation. *Mt. Sinai J. Med.* 68 (3): 160–166.
56 R.S. Alli and A. Khar (2004). Interleukin-12 secreted by mature dendritic cells mediates activation of NK cell function. *FEBS Lett.* 559 (1–3): 71–76.
57 J. Charles, A. Janeway, P. Travers, *et al.* (2001). T-cell Receptor Gene Rearrangement. Garland Science.
58 B. Malissen, C. Grégoire, M. Malissen and R. Roncagalli (2014). Integrative biology of T cell activation. *Nat. Immunol.* 15 (9): 790–797.
59 P.A. van der Merwe and O. Dushek (2011). Mechanisms for T cell receptor triggering. *Nat. Rev. Immunol.* 11 (1): 47–55.
60 D.D. Billadeau, J.C. Nolz and T.S. Gomez (2007). Regulation of T-cell activation by the cytoskeleton. *Nat. Rev. Immunol.* 7 (2): 131–143.
61 T.S. Gomez and D.D. Billadeau (2008). T cell activation and the cytoskeleton: you can't have one without the other. *Adv. Immunol.* 97: 1–64.
62 A. Le Floc'h and M. Huse (2015). Molecular mechanisms and functional implications of polarized actin remodeling at the T cell immunological synapse. *Cell. Mol. Life Sci.* 72 (3): 537–556.
63 L. Chen and D.B. Flies (2013). Molecular mechanisms of T cell co-stimulation and co-inhibition. *Nat. Rev. Immunol.* 13 (4): 227–242.
64 D.-M. Zhu, M. L. Dustin, C. W. Cairo, *et al.* (2006). Mechanisms of cellular avidity regulation in CD2-CD58-mediated T cell adhesion. *ACS Chem. Biol.* 1 (10): 649–658.
65 D. Li, J.J. Molldrem and Q. Ma (2009). LFA-1 regulates CD8+ T cell activation via T cell receptor-mediated and LFA-1-mediated Erk1/2 signal pathways. *J. Biol. Chem.* 284 (31): 21001–21010.
66 D.V Dolfi, *et al.* (2011). Dendritic cells and CD28 costimulation are required to sustain virus-specific CD8+ T cell responses during the effector phase in vivo. *J. Immunol.* 186 (8): 4599–4608
67 N.L. Harris and F. Ronchese (1999). The role of B7 costimulation in T-cell immunity. *Immunol. Cell Biol.* 77 (4): 304–311.
68 A.G. Wingren, *et al.* (1995). T cell activation pathways: B7, LFA-3, and ICAM-1 shape unique T cell profiles. *Crit. Rev. Immunol.* 15 (3–4): 235–253.

69 R.M. Steinman, D. Hawiger and M. C. Nussenzweig (2003). Tolerogenic dendritic cells. *Annu. Rev. Immunol.* 21: 685–711.

70 R.A. Maldonado and U.H. von Andrian (2010). How tolerogenic dendritic cells induce regulatory T cells. *Adv. Immunol.* 108: 111–165.

71 F. Broere, S.G. Apasov, M.V Sitkovsky and W. Van Eden (2011). Principles of Immunopharmacology. *Princ. Immunopharmacol.* 15–28.

72 N. Zhang and M.J. Bevan (2011). CD8(+) T cells: foot soldiers of the immune system. *Immunity* 35 (2): 161–168.

73 M. Catalfamo and P. A. Henkart (2003). Perforin and the granule exocytosis cytotoxicity pathway. *Curr. Opin. Immunol.* 15 (5): 522–527.

74 R. Kennedy and E. Celis (2008). Multiple roles for CD4+ T cells in anti-tumor immune responses. *Immunol. Rev.* 222: 129–144.

75 Y.-P. Lai, C.-J. Jeng and S.-C. Chen (2011). The roles of CD4+ T Cells in tumor immunity. *ISRN Immunol.* 2011: 1–6.

76 S. Lee and K. Margolin (2011). Cytokines in cancer immunotherapy. *Cancers (Basel)* 3 (4): 3856–3893.

77 S.A. Quezada, *et al.* (2010). Tumor-reactive CD4(+) T cells develop cytotoxic activity and eradicate large established melanoma after transfer into lymphopenic hosts. *J. Exp. Med.* 207 (3): 637–650.

78 D.M. Brown (2010). Cytolytic CD4 cells: Direct mediators in infectious disease and malignancy. *Cell. Immunol.* 262 (2): 89–95.

79 K. Palucka and J. Banchereau (2012). Cancer immunotherapy via dendritic cells. *Nat. Rev. Cancer* 12 (4): 265–277.

80 K.L. Knutson and M.L. Disis (2005). Tumor antigen-specific T helper cells in cancer immunity and immunotherapy. *Cancer Immunol. Immunother.* 54 (8): 721–728.

81 L. Senovilla, *et al.* (2013). Trial watch: DNA vaccines for cancer therapy. *Oncoimmunology* 2 (4): e23803.

82 R.T. Kenney, S.A. Frech, L.R. Muenz, *et al.* (2004). Dose sparing with intradermal injection of influenza vaccine. *N. Engl. J. Med.* 351 (22): 2295–2301.

83 J.B. Alarcon, A.W. Hartley, N.G. Harvey and J.A. Mikszta (2007). Preclinical evaluation of microneedle technology for intradermal delivery of influenza vaccines. *Clin. Vaccine Immunol.* 14 (4): 375–381.

84 Y.M. Jo, *et al.* (2009). Dose sparing strategy with intradermal influenza vaccination in patients with solid cancer. *J. Med. Virol.* 81 (4): 722–727.

85 K. Lee, J.D. Kim, C.Y. Lee, *et al.* (2011). A high-capacity, hybrid electro-microneedle for in-situ cutaneous gene transfer. *Biomaterials* 32 (30): 7705–7710.

86 S.A. Eming, T. Krieg and J.M. Davidson (2007). Gene therapy and wound healing. *Clin. Dermatol.* 25 (1): 79–92.

87 B. Wightman, I. Ha and G. Ruvkun (1993). Posttranscriptional regulation of the heterochronic gene lin-14 by lin-4 mediates temporal pattern formation in C. elegans. *Cell* 75 (5): 855–862.

88 R.C. Lee, R.L. Feinbaum and V. Ambros (1993). The C. elegans heterochronic gene lin-4 encodes small RNAs with antisense complementarity to lin-14. *Cell* 75 (5): 843–854.

89 C.C. Mello and D. Conte (2004). Revealing the world of RNA interference. *Nature* 431 (7006): 338–342.

90 R.W. Carthew and E.J. Sontheimer (2009). Origins and mechanisms of miRNAs and siRNAs. *Cell* 136 (4): 642–655.
91 P. Hydbring and G. Badalian-Very (2013). Clinical applications of microRNAs. *F1000Research* 2: 136.
92 T. Yamamoto, *et al.* (2013). Six-transmembrane epithelial antigen of the prostate-1 plays a role for in vivo tumor growth via intercellular communication. *Exp. Cell Res.* 319 (17): 2617–2626.
93 J.K.W. Lam, M.Y.T. Chow, Y. Zhang and S.W.S. Leung (2015). siRNA versus miRNA as therapeutics for gene silencing. *Mol. Ther. Nucleic Acids* 4: e252.
94 J. Tabernero, *et al.* (2013). First-in-humans trial of an RNA interference therapeutic targeting VEGF and KSP in cancer patients with liver involvement. *Cancer Discov.* 3 (4): 406–417.
95 M.E. Davis, *et al.* (2010). Evidence of RNAi in humans from systemically administered siRNA via targeted nanoparticles. *Nature* 464 (7291): 1067–1070.
96 O. Haigh, *et al.* (2014). CXCL1 gene silencing in skin using liposome-encapsulated siRNA delivered by microprojection array. *J. Control. Release* 194: 148–156.
97 M. Jakobsen, *et al.* (2009). Amelioration of psoriasis by anti-TNF-α RNAi in the xenograft transplantation model. *Mol. Ther.* 17 (10): 1743–1753.
98 L.V. Depieri, *et al.* (2016). RNAi mediated IL-6 in vitro knockdown in psoriasis skin model with topical siRNA delivery system based on liquid crystalline phase. *Eur. J. Pharm. Biopharm.* 105: 50–58.
99 J. Allouche, *et al.* (2015). In vitro modeling of hyperpigmentation associated to neurofibromatosis type 1 using melanocytes derived from human embryonic stem cells. *Proc. Natl. Acad. Sci. U. S. A.* 112 (29): 9034–9039.
100 M. Aldawsari, M.B. Chougule and R.J. Babu (2015). Progress in topical siRNA delivery approaches for skin disorders. *Curr. Pharm. Des.* 21 (31): 4594–4605.
101 D. Trochet, B. Prudhon, S. Vassilopoulos and M. Bitoun (2015). Therapy for dominant inherited diseases by allele-specific RNA interference: successes and pitfalls. *Curr. Gene Ther.* 15 (5): 503–510.
102 W.H. Irwin McLean and C.B. Moore (2011). Keratin disorders: From gene to therapy. *Hum. Mol. Genet.* 20 (R2): 189–197.
103 S.A. Leachman, *et al.* (2010). First-in-human mutation-targeted siRNA phase Ib trial of an inherited skin disorder. *Mol. Ther.* 18 (2): 442–426.
104 W.H. I. McLean, C.D. Hansen, M.J. Eliason and F.J.D. Smith (2011). The phenotypic and molecular genetic features of pachyonychia congenita. *J. Invest. Dermatol.* 131 (5): 1015–1017.
105 R.P. Hickerson, *et al.* (2013). Gene silencing in skin after deposition of self-delivery siRNA with a motorized microneedle array device. *Mol. Ther. Nucleic Acids* 2 (10): e129.
106 K. Kawabata, Y. Takakura and M. Hashida (1995). The fate of plasmid DNA after intravenous injection in mice: involvement of scavenger receptors in its hepatic uptake. *Pharm. Res.* 12 (6): 825–830.
107 K.B. Meyer, M.M. Thompson, M.Y. Levy, *et al.* (1995). Intratracheal gene delivery to the mouse airway: characterization of plasmid DNA expression and pharmacokinetics. *Gene Ther.* 2 (7): 450–460.

108 T. Nomura, S. Nakajima, K. Kawabata, *et al.* (1997). Intratumoral pharmacokinetics and in vivo gene expression of naked plasmid DNA and its cationic liposome complexes after direct gene transfer. *Cancer Res.* 57 (13): 2681–2686.
109 M.Y. Levy, L.G. Barron, K.B. Meyer and F.C. Szoka (1996). Characterization of plasmid DNA transfer into mouse skeletal muscle: evaluation of uptake mechanism, expression and secretion of gene products into blood. *Gene Ther.* 3 (3): 201–211.
110 M. Yoshida, R.I. Mahato, K. Kawabata, *et al.* (1996). Disposition characteristics of plasmid DNA in the single-pass rat liver perfusion system. *Pharm. Res.* 13 (4): 599–603.
111 N. Kobayashi, T. Kuramoto, K. Yamaoka, *et al.* (2001). Hepatic uptake and gene expression mechanisms following intravenous administration of plasmid DNA by conventional and hydrodynamics-based procedures. *J. Pharmacol. Exp. Ther.* 297 (3): 853–860.
112 J. Hisazumi, N. Kobayashi, M. Nishikawa and Y. Takakura (2004). Significant role of liver sinusoidal endothelial cells in hepatic uptake and degradation of naked plasmid DNA after intravenous injection. *Pharm. Res.* 21 (7): 1223–1228.
113 J. Snoeys, G. Mertens, J. Lievens, *et al.* (2005). 413. Functional differences of the reticuloendothelial system in C57BL/6 and Balb/c mice mediate differential effects of Kuppfer cell blockade, rag-deficiency and splenectomy on transgene expression after adenoviral transfer. *Mol. Ther.* 11 (S1): 160–160.
114 I. Mellman (1996). Endocytosis and molecular sorting. *Annu. Rev. Cell Dev. Biol.* 12: 575–625.
115 A. Sasaki and M. Kinjo (2010). Monitoring intracellular degradation of exogenous DNA using diffusion properties. *J. Control. Release* 143 (1): 104–111.
116 D. Lechardeur and G.L. Lukacs (2002). Intracellular barriers to non-viral gene transfer. *Curr. Gene Ther.* 2 (2): 183–194.
117 F. Eriksson, T. Tötterman, A.-K. Maltais, *et al.* (2013). DNA vaccine coding for the rhesus prostate specific antigen delivered by intradermal electroporation in patients with relapsed prostate cancer. *Vaccine* 31 (37): 3843–3848.
118 R.D. Cassaday, *et al.* (2007). A phase I study of immunization using particle-mediated epidermal delivery of genes for gp100 and GM-CSF into uninvolved skin of melanoma patients. *Clin. Cancer Res.* 13 (2) Pt 1: 540–549.
119 T. Nguyen-Hoai, *et al.* (2012). HER2/neu DNA vaccination by intradermal gene delivery in a mouse tumor model: Gene gun is superior to jet injector in inducing CTL responses and protective immunity. *Oncoimmunology* 1 (9): 1537–1545.
120 B.E. Verstrepen, *et al.* (2008). Improved HIV-1 specific T-cell responses by short-interval DNA tattooing as compared to intramuscular immunization in non-human primates. *Vaccine* 26 (26): 3346–3351.
121 J.-M. Song, Y.-C. Kim, E. O, R. W. Compans, M. R. Prausnitz, and S.-M. Kang (2012). DNA vaccination in the skin using microneedles improves protection against influenza. *Mol. Ther.* 20 (7): 1472–1480.
122 M. Giacca and S. Zacchigna (2012). Virus-mediated gene delivery for human gene therapy. *J. Control. Release* 161 (2): 377–388.
123 P. Saccardo, A. Villaverde and N. González-Montalbán (2009). Peptide-mediated DNA condensation for non-viral gene therapy. *Biotechnol. Adv.* 27 (4): 432–438.

124 M.D. Brown, A.G. Schätzlein and I.F. Uchegbu (2001). Gene delivery with synthetic (non viral) carriers. *Int. J. Pharm.* 229 (1–2): 1–21.
125 N. Nayerossadat, T. Maedeh and P. A. Ali (2012). Viral and nonviral delivery systems for gene delivery. *Adv. Biomed. Res.* 1: 27.
126 Y. Deng, *et al.* (2016). Transdermal delivery of siRNA through microneedle array. *Sci. Rep.* 6: 21422.
127 A. I. Daud, *et al.* (2008). Phase I trial of interleukin-12 plasmid electroporation in patients with metastatic melanoma. *J. Clin. Oncol.* 26 (36): 5896–5903.
128 X. Zhou, L. Zheng, L. Liu, *et al.* (2003). T helper 2 immunity to hepatitis B surface antigen primed by gene-gun-mediated DNA vaccination can be shifted towards T helper 1 immunity by codelivery of CpG motif-containing oligodeoxynucleotides. *Scand. J. Immunol.* 58 (3): 350–357.
129 J. Hu, C.E. Bounds, L.R. Budgeon and N.M. Cladel (2016). DNA vaccines delivered by microneedle and tattoo gun induce protective immune responses to Hla-A2. 1 Restricted CRPV E1 and HPV16e7 epitopes in HLA-A2. 1 Transgenic rabbits. *SM Vaccines Vaccin. J.* 2 (1): 1–8.
130 B.A. Ginsberg, *et al.* (2010). Immunologic response to xenogeneic gp100 DNA in melanoma patients: comparison of particle-mediated epidermal delivery with intramuscular injection. *Clin. Cancer Res.* 16 (15): 4057–4065.
131 J. Villemejane and L.M. Mir (2009). Physical methods of nucleic acid transfer: general concepts and applications. *Br. J. Pharmacol.* 157 (2): 207–219.
132 I.F. Ciernik, B.H. Krayenbühl and D.P. Carbone (1996). Puncture-mediated gene transfer to the skin. *Hum. Gene Ther.* 7 (8): 893–899.
133 J.H. van den Berg, *et al.* (2009). Optimization of intradermal vaccination by DNA tattooing in human skin. *Hum. Gene Ther.* 20 (3): 181–189.
134 J. McCaffrey, R.F. Donnelly and H.O. McCarthy (2015). Microneedles: an innovative platform for gene delivery. *Drug Delivery Transl. Res.* 5 (4): 424–437.
135 H.S. Gill, D.D. Denson, B.A. Burris and M.R. Prausnitz (2008). Effect of microneedle design on pain in human volunteers. *Clin. J. Pain* 24 (7): 585–594.
136 R.F. Donnelly, *et al.* (2014). Hydrogel-Forming Microneedle Arrays Can Be Effectively Inserted in Skin by Self-Application: A Pilot Study Centred on Pharmacist Intervention and a Patient Information Leaflet. *Pharm. Res.* 31 (8): 1989–1999.
137 X. Chen, *et al.* (2010). Improved DNA vaccination by skin-targeted delivery using dry-coated densely-packed microprojection arrays. *J. Control. Release* 148 (3): 327–333.
138 H.-W. Yang, L. Ye, X.D. Guo, *et al.* (2017). Ebola vaccination using a DNA vaccine coated on PLGA-PLL/γPGA nanoparticles administered using a microneedle patch. *Adv. Healthc. Mater.* 6 (1): 1600750.
139 J. McCaffrey, *et al.* (2016). Transcending epithelial and intracellular biological barriers; A prototype DNA Delivery device. *J. Control. Release.* 226: 238–247.
140 S. Henry, D.V McAllister, M.G. Allen and M.R. Prausnitz (1998). Microfabricated microneedles: a novel approach to transdermal drug delivery. *J. Pharm. Sci.* 87 (8): 922–925.
141 J.A. Mikszta, J.B. Alarcon, J.M. Brittingham, *et al.* (2002). Improved genetic immunization via micromechanical disruption of skin-barrier function and targeted epidermal delivery. *Nat. Med.* 8 (4): 415–419.

142 J. Birchall, *et al.* (2005). Cutaneous DNA delivery and gene expression in *ex vivo* human skin explants via wet-etch microfabricated microneedles. *J. Drug Target.* 13 (7): 415–421.

143 S. Coulman, *et al.* (2006). Minimally invasive cutaneous delivery of macromolecules and plasmid DNA via microneedles. *Curr. Drug Delivery* 3 (1): 65–75.

144 S. Doddaballapur (2009). Microneedling with dermaroller. *J. Cutan. Aesthet. Surg.* 2 (2): 110–111.

145 J.B. Carey, *et al.* (2011). Microneedle array design determines the induction of protective memory CD8+ T cell responses induced by a recombinant live malaria vaccine in mice. *PLoS One* 6 (7): e22442.

146 N. Wilke, A. Mulcahy, S.-R. Ye and A. Morrissey (2005). Process optimization and characterization of silicon microneedles fabricated by wet etch technology. *Microelectronics J.* 36 (7): 650–656.

147 M. Pearton, *et al.* (2008). Gene delivery to the epidermal cells of human skin explants using microfabricated microneedles and hydrogel formulations. *Pharm. Res.* 25 (2): 407–416.

148 K.Y. Park, *et al.* (2013). Safety evaluation of stamp type digital microneedle devices in hairless mice. *Ann. Dermatol.* 25 (1): 46–53.

149 G. Yan, N. Arelly, N. Farhan, *et al.* (2014). Enhancing DNA delivery into the skin with a motorized microneedle device. *Eur. J. Pharm. Sci.* 52: 215–222.

150 A.D. Bins, *et al.* (2005). A rapid and potent DNA vaccination strategy defined by in vivo monitoring of antigen expression. *Nat. Med.* 11 (8): 899–904.

151 M. Pearton, *et al.* (2008). Gene delivery to the epidermal cells of human skin explants using microfabricated microneedles and hydrogel formulations. *Pharm. Res.* 25 (2): 407–416.

152 R. Haj-Ahmad, *et al.* (2015). Microneedle coating techniques for transdermal drug delivery. *Pharmaceutics* 7 (4): 486–502.

153 Y.-C. Kim, F.-S. Quan, R.W. Compans, *et al.* (2010). Formulation and coating of microneedles with inactivated influenza virus to improve vaccine stability and immunogenicity. *J. Control. Release* 142 (2): 187–195.

154 N. W. Kim, *et al.* (2014). Polyplex-releasing microneedles for enhanced cutaneous delivery of DNA vaccine. *J. Control. Release* 179: 11–17.

155 A. Vrdoljak, *et al.* (2012). Coated microneedle arrays for transcutaneous delivery of live virus vaccines. *J. Control. Release* 159 (1): 34–42.

156 S.B. Lee, C.A. Crouse and M.C. Kline (2010). Optimizing storage and handling of DNA extracts. *Forensic Sci. Rev.* 22 (2): 131–144.

157 H.S. Gill and M.R. Prausnitz (2007). Coated microneedles for transdermal delivery. *J. Control. Release* 117 (2): 227–237.

158 H.S. Gill, J. Söderholm, M.R. Prausnitz and M. Sällberg (2010). Cutaneous vaccination using microneedles coated with hepatitis C DNA vaccine. *Gene Ther.* 17 (6): 811–814.

159 Y.-C. Kim, *et al.* (2012). Increased immunogenicity of avian influenza DNA vaccine delivered to the skin using a microneedle patch. *Eur. J. Pharm. Biopharm.* 81 (2): 239–247.

160 Y.-C. Kim, D.-G. Yoo, R. W. Compans, *et al.* (2013). Cross-protection by co-immunization with influenza hemagglutinin DNA and inactivated virus vaccine using coated microneedles. *J. Control. Release* 172 (2): 579–588.

161 X. Chen, *et al.* (2009). Dry-coated microprojection array patches for targeted delivery of immunotherapeutics to the skin. *J. Control. Release* 139 (3): 212–220.

162 A.S. Kask, *et al.* (2010). DNA vaccine delivery by densely-packed and short microprojection arrays to skin protects against vaginal HSV-2 challenge. *Vaccine* 28 (47): 7483–7491.

163 M. Pearton, *et al.* (2012). Microneedle delivery of plasmid DNA to living human skin: Formulation coating, skin insertion and gene expression. *J. Control. Release* 160 (3): 561–569.

164 P.M. Wang, M. Cornwell, J. Hill and M.R. Prausnitz (2006). Precise microinjection into skin using hollow microneedles. *J. Invest. Dermatol.* 126 (5): 1080–1087.

165 M.S. Lhernould (2012). Optimizing hollow microneedles arrays aimed at transdermal drug delivery. *Microsyst. Technol.* 19 (1): 1–8.

166 BioPharm International (2009) First intradermal influenza vaccine approved in the European Union. [online] Biopharminternational.com. Retrieved from: http://www.biopharminternational.com/first-intradermal-influenza-vaccine-approved-european-union (accessed 23 January 2018).

167 B. Pamornpathomkul, A. Wongkajornsilp, W. Laiwattanapaisal, *et al.* (2017). A combined approach of hollow microneedles and nanocarriers for skin immunization with plasmid DNA encoding ovalbumin. *Int. J. Nanomedicine* 12: 885–898.

168 S.P. Davis, B.J. Landis, Z.H. Adams, *et al.* (2004). Insertion of microneedles into skin: measurement and prediction of insertion force and needle fracture force. *J. Biomech.* 37 (8): 1155–1163.

169 L. Daugimont, *et al.* (2010). Hollow microneedle arrays for intradermal drug delivery and DNA electroporation. *J. Membr. Biol.* 236 (1): 117–125.

170 W. Martanto, *et al.* (2006). Microinfusion using hollow microneedles. *Pharm. Res.* 23 (1): 104–113.

171 M.G. McGrath, *et al.* (2014). Production of dissolvable microneedles using an atomised spray process: effect of microneedle composition on skin penetration. *Eur. J. Pharm. Biopharm.* 86 (2): 200–211.

172 E. Gonzalez-Gonzalez, *et al.* (2010). Silencing of reporter gene expression in skin using siRNAs and expression of plasmid DNA delivered by a soluble protrusion array device (PAD). *Mol. Ther.* 18 (9): 1667–1674.

173 M.F. Lara, *et al.* (2012). Inhibition of CD44 gene expression in human skin models, using self-delivery short interfering RNA administered by dissolvable microneedle arrays. *Hum. Gene Ther.* 23 (8): 816–823.

174 E. González-González, *et al.* (2011). Visualization of plasmid delivery to keratinocytes in mouse and human epidermis. *Sci. Rep.* 1: 158.

175 G. Cole, *et al.* (2017). Dissolving microneedles for DNA vaccination : Improving functionality via polymer characterization and RALA complexation. *Hum. Vaccin. Immunother.* 13 (1): 50–62.

176 A.A. Ali, *et al.* (2017). DNA vaccination for cervical cancer; a novel technology platform of RALA mediated gene delivery via polymeric microneedles. *Nanomedicine Nanotechnology, Biol. Med.* 13 (3): 921–932.

177 N.Y. Sardesai and D.B. Weiner (2011). Electroporation delivery of DNA vaccines: prospects for success. *Curr. Opin. Immunol.* 23 (3): 421–429.

178 S. van Drunen Littel-van den Hurk and D. Hannaman (2010). Electroporation for DNA immunization: clinical application. *Expert Rev. Vaccines* 9 (5): 503–517.

179 S. van Drunen Littel-van den Hurk, *et al.* (2008). Electroporation-based DNA transfer enhances gene expression and immune responses to DNA vaccines in cattle. *Vaccine* 26 (43): 5503–5509.

180 C. O'Mahony, *et al.* (2016). Design, modelling and preliminary characterisation of microneedle-based electrodes for tissue electroporation *in vivo*. *J. Phys. Conf. Ser.* 757 (1): 12040.

181 J.W. Hooper, J.W. Golden, A.M. Ferro and A.D. King (2007). Smallpox DNA vaccine delivered by novel skin electroporation device protects mice against intranasal poxvirus challenge. *Vaccine* 25 (10): 1814–1823.

182 Z. Wei, *et al.* (2014). A flexible microneedle array as low-voltage electroporation electrodes for in vivo DNA and siRNA delivery. *Lab Chip* 14 (20): 4093–4102.

183 W. Lin, *et al.* (2001). Transdermal delivery of antisense oligonucleotides with microprojection patch (Macroflux) technology. *Pharm. Res.* 18 (12): 1789–1793.

184 D. Zhang, D.B. Das and C.D. Rielly (2014). Potential of microneedle-assisted micro-particle delivery by gene guns: a review. *Drug Delivery* 21 (8): 571–587.

185 D. Zhang, D.B. Das and C.D. Rielly (2014). Microneedle assisted micro-particle delivery from gene guns: experiments using skin-mimicking agarose gel. *J. Pharm. Sci.* 103 (2): 613–627.

186 R.M. Blaese, *et al.* (1995). T Lymphocyte-directed gene therapy for ADA- SCID: initial trial results after 4 years. *Science* 270 (5235): 475–480.

187 H.O. McCarthy, Y. Wang, S.S. Mangipudi and A. Hatefi (2010). Advances with the use of bio-inspired vectors towards creation of artificial viruses. *Expert Opin. Drug Delivery* 7 (4): 497–512.

188 R.C. Kines, *et al.* (2015). Vaccination with human papillomavirus pseudovirus-encapsidated plasmids targeted to skin using microneedles. *PLoS One* 10 (3): e0120797.

189 N. Wu and M.M. Ataai (2000). Production of viral vectors for gene therapy applications. *Curr. Opin. Biotechnol.* 11 (2): 205–208.

190 S. Daya and K.I. Berns (2008). Gene therapy using adeno-associated virus vectors. *Clin. Microbiol. Rev.* 21 (4): 583–593.

191 S. Hacein-Bey-Abina, *et al.* (2003). LMO2-associated clonal T cell proliferation in two patients after gene therapy for SCID-X1. *Science* 302 (5644): 415–419.

192 S. Hacein-Bey-Abina, *et al.* (2008). Insertional oncogenesis in 4 patients after retrovirus-mediated gene therapy of SCID-X1. *J. Clin. Invest.* 118 (9): 3132–3142.

193 E. Marshall (1999). Clinical trials: Gene therapy death prompts review of adenovirus vector. *Science* 286 (5448): 2244–2245.

194 S. Nayak and R.W. Herzog (2010). Progress and prospects: immune responses to viral vectors. *Gene Ther.* 17 (3): 295–304.

195 F. Mingozzi, *et al.* (2007). CD8(+) T-cell responses to adeno-associated virus capsid in humans. *Nat. Med.* 13 (4): 419–422.

196 T.M. Kündig, C.P. Kalberer, H. Hengartner and R. M. Zinkernagel (1993). Vaccination with two different vaccinia recombinant viruses: long-term inhibition of secondary vaccination. *Vaccine* 11 (11): 1154–1158.

197 J.B. Carey, A. Vrdoljak, C. O'Mahony, *et al.* (2014). Microneedle-mediated immunization of an adenovirus-based malaria vaccine enhances antigen-specific antibody immunity and reduces anti-vector responses compared to the intradermal route. *Sci. Rep.* 4: 6154.

198 S.P. Loughran, C.M. McCrudden and H.O. McCarthy (2015). Designer peptide delivery systems for gene therapy. *Eur. J. Nanomedicine* 7 (2): 85–96.
199 F. Chabri, *et al.* (2004). Microfabricated silicon microneedles for nonviral cutaneous gene delivery. *Br. J. Dermatol.* 150 (5): 869–877.
200 A. Kumar, P. Wonganan, M.A. Sandoval, *et al.* (2012). Microneedle-mediated transcutaneous immunization with plasmid DNA coated on cationic PLGA nanoparticles. *J. Control. Release* 163 (2): 230–239.
201 D. Yin, *et al.* (2013). Hepatitis B DNA vaccine-polycation nano-complexes enhancing immune response by percutaneous administration with microneedle. *Biol. Pharm. Bull.* 36 (8): 1283–1291.
202 Y. Hu, *et al.* (2014). A mannosylated cell-penetrating peptide-graft-polyethylenimine as a gene delivery vector. *Biomaterials* 35 (13): 4236–4246.
203 Y. Hu, *et al.* (2015). Microneedle-assisted dendritic cell-targeted nanoparticles for transcutaneous DNA immunization. *Polym. Chem.* 6 (3): 373–379.
204 R.H.E. Chong, *et al.* (2013). Gene silencing following siRNA delivery to skin via coated steel microneedles: In vitro and in vivo proof-of-concept. *J. Control. Release* 166 (3): 211–219.
205 N.W. Kim, *et al.*, (2014). Polyplex-releasing microneedles for enhanced cutaneous delivery of DNA vaccine. *J. Control. Release* 179: 11–17.
206 T. Tang, *et al.* (2016). Local administration of siRNA through microneedle: optimization, bio-distribution, tumor suppression and toxicity. *Sci. Rep.* 6: 30430.
207 Y. Qiu, *et al.* (2015). DNA-based vaccination against hepatitis B virus using dissolving microneedle arrays adjuvanted by cationic liposomes and CpG ODN. *Drug Delivery* 23 (7) 2391–2398.
208 Y. Qiu, L. Guo, P. Mao and Y. Gao (2015). Dissolving microneedle arrays for intradermal immunization of hepatitis B virus DNA vaccine. *Procedia Vaccinol.* 9: 24–30.
209 A. Kumar, P. Wonganan, M.A. Sandoval, *et al.* (2012). Microneedle-mediated transcutaneous immunization with plasmid DNA coated on cationic PLGA nanoparticles. *J. Control. Release* 163 (2): 230–239.
210 A.J. Engering, *et al.* (1997). The mannose receptor functions as a high capacity and broad specificity antigen receptor in human dendritic cells. *Eur. J. Immunol.* 27 (9): 2417–2425.
211 C.W. Pouton and L.W. Seymour (2001). Key issues in non-viral gene delivery. *Adv. Drug Delivery Rev.* 46 (1–3): 187–203.
212 O. Zelphati, L.S. Uyechi, L.G. Barron and F.C. Szoka (1998). Effect of serum components on the physico-chemical properties of cationic lipid/oligonucleotide complexes and on their interactions with cells. *Biochim. Biophys. Acta* 1390 (2): 119–133.
213 S.C. Semple, A. Chonn and P. R. Cullis (1998). Interactions of liposomes and lipid-based carrier systems with blood proteins: Relation to clearance behaviour in vivo. *Adv. Drug Delivery Rev.* 32 (1–2): 3–17.
214 A.L. Klibanov, K. Maruyama, V.P. Torchilin and L. Huang (1990). Amphipathic polyethyleneglycols effectively prolong the circulation time of liposomes. *FEBS Lett.* 268 (1): 235–237.
215 A.L. Klibanov, K. Maruyama, A.M. Beckerleg, *et al.* (1991). Activity of amphipathic poly(ethylene glycol) 5000 to prolong the circulation time of liposomes depends

on the liposome size and is unfavorable for immunoliposome binding to target. *Biochim. Biophys. Acta* 1062 (2): 142–148.

216 S. Mishra, P. Webster and M. E. Davis (2004). PEGylation significantly affects cellular uptake and intracellular trafficking of non-viral gene delivery particles. *Eur. J. Cell Biol.* 83 (3): 97–111.

217 T. Gjetting, *et al.* (2010). In vitro and in vivo effects of polyethylene glycol (PEG)-modified lipid in DOTAP/cholesterol-mediated gene transfection. *Int. J. Nanomedicine* 5: 371–383.

218 T. Ishida, *et al.* (2006). Injection of PEGylated liposomes in rats elicits PEG-specific IgM, which is responsible for rapid elimination of a second dose of PEGylated liposomes. *J. Control. Release* 112 (1): 15–25.

219 X. Wang, T. Ishida and H. Kiwada (2007). Anti-PEG IgM elicited by injection of liposomes is involved in the enhanced blood clearance of a subsequent dose of PEGylated liposomes. *J. Control. Release* 119 (2): 236–244.

220 H. Hosseinkhani, F. Abedini, K.-L. Ou and A.J. Domb (2015). Polymers in gene therapy technology. *Polym. Adv. Technol.* 26 (2): 198–211.

221 H. Lv, S. Zhang, B. Wang, *et al.* (2006). Toxicity of cationic lipids and cationic polymers in gene delivery. *J. Control. Release* 114 (1): 100–109.

222 W.T. Godbey, K.K. Wu and A.G. Mikos (1999). Size matters: molecular weight affects the efficiency of poly(ethylenimine) as a gene delivery vehicle. *J. Biomed. Mater. Res.* 45 (3): 268–275.

223 M.T.C. McCrudden, *et al.* (2014). Design and physicochemical characterisation of novel dissolving polymeric microneedle arrays for transdermal delivery of high dose, low molecular weight drugs. *J. Control. Release* 180: 71–80.

6

Delivery of Nanomedicines Using Microneedles

Eneko Larrañeta and Lalit Vora

School of Pharmacy, Queen's University Belfast, Belfast BT9 7BL, UK

6.1 Introduction

Nanotechnology as a scientific discipline involves the design, synthesis and application of materials with at least one dimension in the size range of 1–100 nm. The National Nanotechnology Initiative, USA defines a nanoparticle (NP) as a particle with all dimensions between 1 and 100 nm [1]. However, in the scientific literature it is common to find the term "nanoparticle" referring to particles larger than 100 nm.

Major applications of NPs include encapsulation of poorly soluble drugs, protection of therapeutic molecules, modification of blood circulation and tissue distribution and targeting of cancer with therapeutics and diagnostic agents [1, 2]. NPs have been used to deliver a wide variety of biologically active compounds, such as conventional drugs, vaccines, proteins and nucleotides [2, 3]. These types of carriers can be made from a wide variety of materials, including proteins, lipids, sugars, polymers, metals and other inorganic compounds [1]. In addition to NPs, microparticles (MPs), which are similar agents but in the micron size range, have also been used extensively.

NPs and MPs present numerous advantages over conventional drug delivery systems. They allow sustained drug release over a prolonged period of time while providing protection for encapsulated materials against proteolytic or chemical degradation [1]. Additionally, their surfaces can be modified to include certain ligands, allowing targeting of the delivery to certain parts of the body [4].

Formulations containing NPs and MPs can be administered via various routes including oral, intravenous, pulmonary, nasal and ocular [5]. NPs and MPs have also been investigated for use in the development of formulations that can successfully overcome skin barriers. This transdermal route of administration presents several advantages over other methods.

Transdermal delivery systems are non-invasive and can be easily self-administered, as with oral delivery systems. However, transdermal routes do not present the limitations associated with the oral route, such as the instability of the drug in the gastrointestinal tract or the first-pass effect [6, 7]. The main barrier to the administration of drugs through the skin is the *stratum corneum* (*SC*), thus the number of drugs capable of being administered through this route is limited due to the presence of this barrier [8, 9]. In order to permeate through the *SC* successfully, drugs should have moderately low

Microneedles for Drug and Vaccine Delivery and Patient Monitoring, First Edition.
Edited by Ryan F. Donnelly, Thakur Raghu Raj Singh, Eneko Larrañeta, and Maelíosa T.C. McCrudden.

molecular weights, octanol–water partition coefficients that heavily favour lipids and exhibit reasonable potency (low dose) [10–13]. Despite all of these limitations, several medicinal preparations have been commercialised since the first transdermal product was approved in the United States in 1979 [14]. However, the transdermal administration of macromolecules, peptides, hydrophilic drugs or nanomedicines has remained moderately unexploited [14].

A good approach to increasing the permeability of these types of molecules through the skin is the use of microneedles (MNs). The development of combinatorial approaches to maximise the advantages of both NP and MN delivery systems have increased over the past 15 years. This chapter aims to present the main MN/NP combinatory approaches developed to date.

6.2 Skin Structure and Barrier Properties Which Impact on Nanoparticle and Microparticle Penetration

Skin is the largest and most complex organ of the human body and its principal physiological function is to defend the body from the external environment [15]. In addition to this, the skin carries out a wide variety of alternative functions [15]: it protects and contains the internal body organs and fluids; it acts as a sensory organ for heat, cold, touch and pressure through connection to the central nervous system; it provides temperature and, to a certain degree, humidity control for the body.

Skin is composed of three layers: epidermis, dermis and hypodermis [16]. The epidermis is classified histologically as a stratified squamous epithelial layer in which keratinocytes are interspersed in five different strata: *stratum basale* (attached to dermis), *stratum spinosum, stratus granulosum, stratum lucidium* and *stratum corneum* (*SC*), which is in contact with the external environment.

The *SC* (horny layer) is the outermost layer of the epidermis, which is responsible for its defensive functions. This layer has been described as the main barrier to the penetration of most drugs and nanomedicines (NMs) [17, 18]. The typical thickness of the *SC* is around 10 μm. However, factors such as the location of the skin and its hydration have a direct impact on the thickness of the *SC*. In order to illustrate this variability, it is important to note that the thickness of the *SC* on the palms of the hands and soles of the feet can reach values of 400–600 μm [18] and that hydration can lead to a fourfold increase in the thickness [19]. In the *SC*, keratinocytes are completely differentiated into enucleated and keratin-flaggrin filled cells, called corneocytes. The macrostructure of the *SC* in humans generally refers to the cross-sectional organisation of the corneocytes, described by the bricks-and-mortar model [19], where "bricks" refer to alternate corneocytes (15–30%) and "mortar" refers to a lipidic matrix surrounding the corneocytes [11]. The *SC* microstructure in humans is a supramolecular organisation of intercorneocyte lipids that are assembled in parallel head–head and tail–tail repeating bilayers. These repeating bilayers consist of both hydrophilic and lipophilic regions. The barrier action of the *SC* is a result of collective cooperation and interactions between the *SC* macro- and microstructures [20]. In addition, these bilayers have an alternative function, as they prevent the desiccation of the underlying tissues by hindering water loss [20]. There are also aqueous pores (hydrophilic regions) in the *SC*, approximately 36 nm in diameter, which define the trans-epidermal intercellular polar (hydrophilic) route of skin absorption [14]. The main mechanism of transport of substances through this layer is passive diffusion. The three reported routes of passive diffusion are: transcellular, intracellular and appendageal [14]. The transcellular route consists in diffusion through the lipidic matrix that is located in the intercellular spaces of the keratinocytes [14]; permeation through the keratinocytes constitutes the intracellular route [14]; and finally, the appendageal route entails permeation through sebaceous glands, sweat glands and hair follicles [14]. The permeation of nanoparticles and microparticles through the skin is influenced by their size. In general, larger agents (10–210 μm) penetrate the skin through the trans-follicular route, while smaller agents (5–36 nm) use the intracellular route [17]. However, the permeation of NPs through the skin is influenced by many factors, so generalisations cannot be made.

The *SC* is not the only barrier to NP penetration, as the viable epidermis plays an important role here too. The epidermis, a stratified epithelium that lies directly above the dermo–epidermal junctions, is hydrophilic in nature and consequently limits the permeation of lipophilic agents. Moreover, this layer contains proteolytic enzymes that can potentially degrade foreign substances [14]. Finally, the viable epidermis contains tight junctions between the skin layers that limit the permeation of NPs [14].

Other skin appendages, such as sweat glands and pilosebaceous units, which open on the skin surface can allow the penetration of nanoparticles. Sweat glands are coiled tubular glands extending from the *SC* to the dermis, which are 2–5 mm in length and are involved in thermoregulation and excretion of body waste [14]. Pilosebaceous units consist of hair infundibulum, that is, a hair follicle with one hair serving as a route to expel the sebum of associated sebaceous glands. Even though sweat glands and hair follicles provide large openings on the skin surface, they are of low density (sweat glands 0.01%, hair follicles 0.1%) [14], which may be a limiting factor for the utilisation of trans-follicular routes for the delivery of nanoparticles.

6.3 Conventional Nanocarriers for Topical and Transdermal Delivery

In this section the main types of nanocarriers used for passive permeation through the *SC* will be described. The term "conventional" will be used to refer to this type of system as opposed to delivery strategies that breach the *SC*. In addition, the term "nanoparticle" will be used to refer to particles with sizes of less than 1000 nm.

The delivery of NPs using the transdermal route has mainly been used to elicit local effects [1]. However, there is an ongoing debate around the NP permeation mechanism, as to date it remains unclear [1, 17]. The skin sites with potential for targeting with NPs are furrows, hair follicles and the *SC* (Figure 6.1) [1]. Drug delivery to deeper layers of the skin (epidermis and dermis) using conventional strategies based on NPs has met with limited success [1, 17, 21, 22].

6.3.1 Lipidic Vesicles

The main types of lipidic vesicles are: liposomes, Niosomes®, Ethosomes®, Transfersomes®, Proliposomes®, Pharmacosomes® and Vesosomes®. The key common features of these vehicles are that they present a certain degree of deformability and particle sizes larger than 100 nm [17].

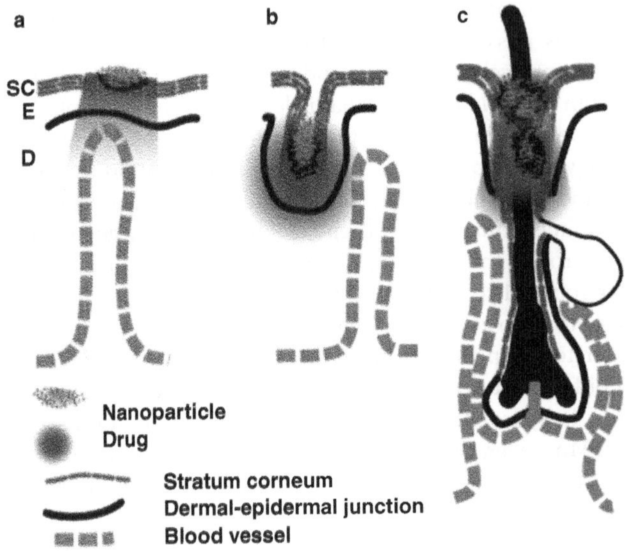

Figure 6.1 Sites in skin for nanoparticle delivery. Topical nanoparticle drug delivery takes place at three major sites: (a) *SC* surface, (b) furrows and (c) openings of hair follicles. The nanoparticles are shown in dotted gray and the drug in gray shade. Other sites for delivery are the viable epidermis (E) and dermis (D). Reproduced with permission from: [1] Prow T.W., Grice J.E., Lin L.L., *et al*. (2011). Nanoparticles and microparticles for skin drug delivery. *Adv. Drug Deliv. Rev*. 63: 470–491.

The most popular type of nanocarriers are liposomes [22]. They are vesicles enclosing an aqueous environment surrounded by single or multiple lipid bilayers composed of phospholipids and/or cholesterol [23]. Niosomes are similar but they include a non-ionic surfactant in their structure, showing better stability than liposomes [17]. Ethosomes are another variation of the traditional liposomes, they present enhanced skin penetration due to their elastic lipidic vesicle [23]. Transferosomes are vesicles made from phospholipids supplemented with surfactants to produce ultra-deformable particles [23]. Proliposomes are particles made from a mixture of phospholipids, drugs and a water soluble porous powder [24], the last acting as a support to increase the surface of dry lipids [24]. Pharmacosomes are a colloidal dispersion of phospholipids covalently linked to drugs [24]. The last type of lipidic vesicles are Vesosomes, which are vesicles inside another vesicle designed for transcutaneous immunisation purposes [24].

Over the last 35 years, these types of nanocarriers have been studied extensively for use in transdermal delivery. One of the main applications has been the topical delivery of anaesthetics [25, 26] or anti-inflammatory [27–29] compounds. Liposomes have been used as delivery systems to the skin by topical application for antibiotics, antifungal agents and for a broad range of peptides [24]. The nanocarriers also present promising properties for the treatment of skin cancer, as they can be used to deliver DNA repair enzymes or prodrugs, such as 5-aminolevulinic acid, for photodynamic therapy [24]. Liposomes are good candidates for topical delivery but they are not able to cross the *SC* to act as carriers for transdermal drug delivery [24]. On the other hand, Niosomes, Ethosomes and Transfersomes allow deeper penetration into the skin than traditional liposomes, thus showing more potential for transdermal delivery.

6.3.2 Lipid Nanoparticles

Liposomes present some stability problems and so lipid nanoparticles were developed to overcome these problems [30]. The main types of lipid nanoparticles are solid lipid nanoparticle (SLNs) and nanostructured lipid carrier (NLCs). SLNs are made from a lipid coated with a surfactant as a stabiliser [31]. In this type of carrier, the lipid should be solid at body temperatures [31]. On the other hand, NLCs are composed of a solid lipidic matrix, formed by a mixture of solid and liquid lipids, encapsulating a liquid lipidic nanocompartment [31]. SLNs and NLCs can be positively or negatively charged, they are rigid and their sizes range from 50 to 1000 nm [17]. The main applications of these carriers are cosmetic and dermatological, mainly for the treatment of acne and eczema [17, 30, 31]. The drug permeation enhancement exhibited by these nanocarriers has been suggested to be due to the formation of occlusive films on the surface of the skin rather than to the permeation of the particles through the *SC* [17].

6.3.3 Polymeric Nanoparticles and Microparticles

Micro/nanospheres and micro/nanocapsules can be included within this category. The main difference between both types of particles is that the former are particle matrices while the latter are vesicles (normally oily) covered with polymers.

These types of vehicles can display a wide range of sizes and charges, either positive or negative [17]. They have been used widely in different locations of the body as drug delivery systems [17]. Owing to their rigidity these types of particles do not penetrate the *SC* but their ability to penetrate into hair follicles has been reported [17, 32–34]. A wide range of examples of polymeric NPs as topical delivery systems have been reported [35]. Biopolymers such as chitosan, alginate, albumin or gelatine have been used to prepare NPs as topical delivery systems. Chitosan has been frequently used to formulate NPs for topical delivery of molecules, for example, topical delivery of quercetin, retinol, acyclovir and of macromolecules, such as plasmid DNA and antisense oligonucleotides [35–37].

Alternatively, synthetic polymers have been used for the preparation of topical NP delivery systems. Two main types of synthetic polymers have been used for this purpose: biodegradable and non-biodegradable. The main type of biodegradable polymer used to formulate nanoparticulate systems is poly(D,L-lactic-*co*-glycolic acid) (PLGA). PLGA NPs have been used for the delivery of several therapeutic agents such as ketoprofen, spantide II and indomethacin, among others [38, 39]. Non-degradable polymers have not been used in the same way as the biodegradable ones. However, several examples can be found in the literature [35]. One example is the use of polyacrylate for cutaneous delivery of active molecules [35].

6.3.4 Microemulsions

Microemulsions are formulations made from an emulsion of an aqueous and an oil phase in the presence of a surfactant or cosurfactant [40]. The formed droplets are highly flexible and their sizes are normally less than 150 nm [17]. This type of formulation has been used to enhance the permeation of a wide range of compounds, including drugs and macromolecules [41]. However, the permeation of intact droplets into the skin has not been reported [17]. The release of their cargoes arises when particles collide with skin structures [17]. As microemulsions shows good retention in the skin layers they have been used extensively for cutaneous delivery rather than for percutaneous drug administration [32]. This type of vehicle has been used for therapeutic and cosmetic applications to the skin [32, 42, 43]. Examples of cosmetic applications are the skin delivery of ascorbic acid and lycopene [42, 43]. On the other hand, a wide range of therapeutics such as cyclosporine, alpha-tocopherol, temozolomide hexyl ester, lidocaine ascorbyl palmitate, desmopressin acetate and paclitaxel have been delivered to the skin using microemulsions [32].

It is important to note that despite its limited capacity for transdermal delivery, microemulsions have been used successfully for transdermal delivery of molecules such as testosterone, lidocaine, estradiol, nicotinic acid and sodium diclofenac [32].

6.3.5 Metallic and Mineral Nanoparticles

This group includes a wide range of types of NPs, including magnetic NPs, titanium and zinc oxide NPs, carbon nanotubes, fullerenes and quantum dots (QDs).

Magnetic NPs are mainly particles of iron derivatives with a size distribution ranging between 2 and 100 nm [17]. This type of particle is able to permeate into the deeper layers of the SC, hair follicles, the *SC–stratum granulosum* interface and exceptionally in some cases into the viable epidermis [44]. Applications of magnetic NPs include: diagnosis of skin diseases through imaging, magnetic drug delivery/targeting, magnetic transfection of cells and hyperthermia treatment of tumours [14, 17].

QDs are nanomaterials with a core-shell structure formed mainly by elements from the Periodic Table from Groups IV–VI or Group IV alone [45]. They have unique spectral and optoelectronic properties that make them suitable for a wide range of biomedical applications [45]. The surface of the QDs is often coated with different compounds depending on their intended applications. The skin permeation of QDs have been investigated by several research groups [46–50], with the permeation of QDs through murine and porcine skin being reported [47, 49]. However, alternative studies showed no conclusive QD permeation [48, 50]. All these studies were not developed with human skin and, consequently, before extracting any final conclusions the results should be evaluated carefully.

Titanium oxide and zinc oxide nanomaterials have the ability to scatter UV radiation. Hence, they are widely used in the cosmetics industry for the production of sunscreens [14]. A common procedure is to coat them with silicon dioxide, silicon oils or aluminium oxide to increase their size and dispersion stabilities [17]. The size of zinc oxide NPs ranges between 30 and 200 nm, while the size of titanium oxide particles can be up to several microns [14]. Nohynek *et al.* reported that both types of particles, zinc and titanium oxides, only penetrated into the outermost layers of the *SC* [51]. However, Kimura *et al.* showed that zinc and titanium oxide NPs did not migrate beyond the skin surface into deeper layers [52].

Fullerenes and carbon nanotubes are nanomaterials based on carbon with the shape of hollow spheres or single/multi-walled hollow tubes, respectively [53]. Rouse *et al.* showed that fullerene coupled with a peptide with a size of 3.5 nm was able to diffuse into the skin by passive penetration after applying mechanical stress to the skin [54]. Alternatively, carbon nanotubes are able to enhance transdermal permeation of different compounds such as siRNA [55] or indomethacin [56]. However, and despite enhancing the permeation of some drugs, carbon nanotubes are not able to permeate through the skin [56].

6.4 Microneedle-mediated Transdermal Delivery of Nanoparticles and Microparticles

As described in previous chapters, MN devices are composed of arrays of micron-size needles able to bypass the *SC* without causing pain or bleeding due to the short length of the needles [57, 58]. The pores created can be used to deliver drugs from the skin surface to the rich dermal microcirculation or to deeper layers of the skin [57, 58]. They can be produced in a wide range of shapes and materials, as described in earlier in this book. Therefore, MNs are an ideal candidate to enhance the permeation of NPs and MPs inside the skin.

6.4.1 Microneedle-assisted Nanoparticle/Microparticle Permeation

Multiple examples of MN and NP/MP combined systems for intradermal and transdermal drug delivery can be found in the literature [59]. This section of the chapter will be centred around the studies focused mainly on optimisation of the permeation of NPs and MPs when used in combination with MNs. Therapeutic applications of this combinatory technology will be described in subsequent sections of the chapter.

In 2003 McAllister *et al.* published the first research work describing the permeation of NPs through the skin using MNs [60]. In this study, the permeations of compounds with different molecular radii were evaluated after using solid MN arrays. Two types of polystyrene latex nanospheres (25 and 50 nm radius) were employed as a model to study NP permeation. The permeation was shown to be related to the molecular radius of the compound. Further, the idea of the transport taking place via diffusion through water-filled holes in the skin was supported by mathematical modelling. Similarly, Coulman *et al.* studied the diffusion of fluorescent polystyrene nanospheres with sizes ranging from 100 to 150 nm in diameter across skin pre-treated with solid silicon MN arrays [61]. The results obtained in this study are consistent with those obtained by McAllister *et al.* [60]. Moreover, this work describes the influence of additional factors on the permeation of the particles, such as the charge in the surface of the particles. The permeation of NPs across human skin was tested, with mixed results. The complexity of the tissue due to the multi-layered human epidermal membranes could be responsible for this. NPs were found on the interior of microchannels created after MN insertion. Additionally, they were found adsorbed to the corneocyte surfaces present in the disrupted epidermal membrane (Figure 6.2).

Zhang *et al.* used PLGA NPs loaded with coumarin 6 combined with solid silicon MN arrays [62]. The results were consistent with the findings reported previously by other authors. The main findings of this study showed that larger amounts of NPs were found in the epidermis rather than in the dermis and that the permeation rate for the NPs was dependent on the NP concentration. The permeation rate reached a limit when higher concentrations of NPs were used (about 1.6 mg/ml).

The use of solid MN arrays with NPs was alternatively evaluated by Gomaa *et al.* [63]. Again, the findings of this work are in line with those described in previous paragraphs. In this case the permeation of PLGA NPs of different sizes containing encapsulated dyes (Rhodamine B and fluorescein isothiocyanate) across excised porcine skin treated with polymeric MN arrays was studied. A set of parameters, such as the particle size, NP composition, surface charge of the particles or the chemical nature of the encapsulated dye,

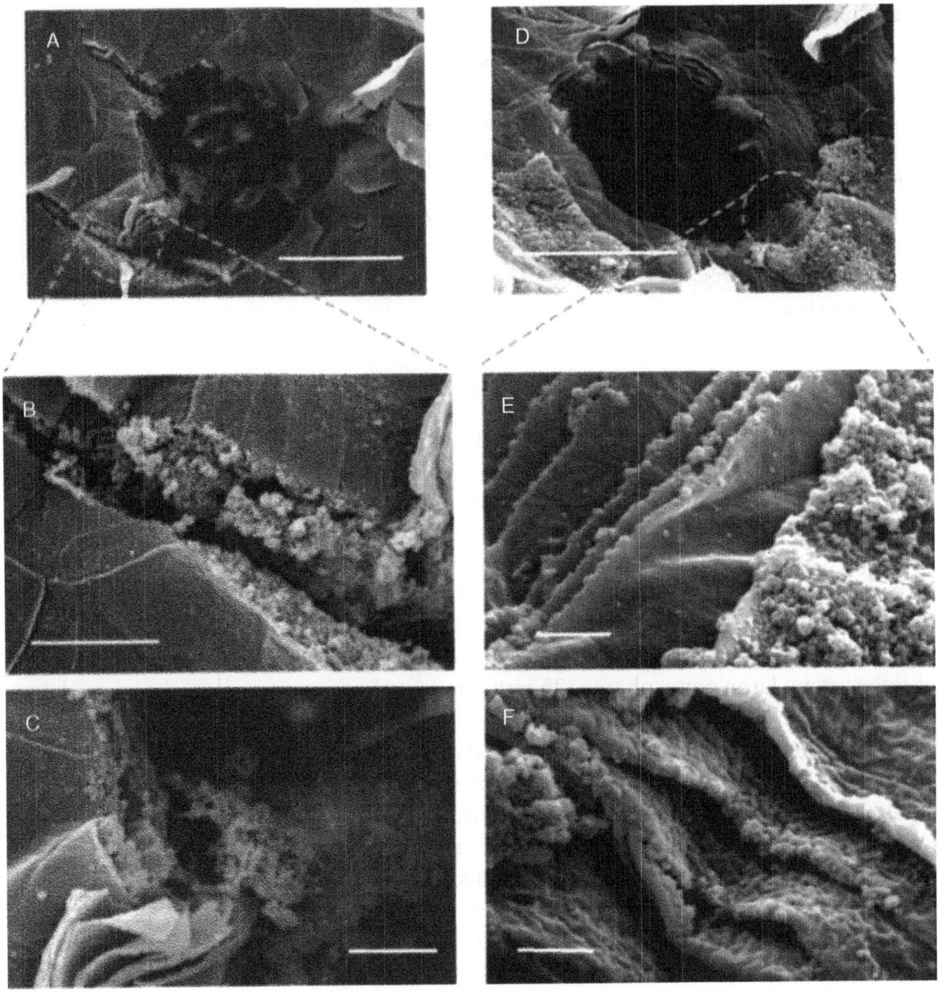

Figure 6.2 Scanning electron micrographs of those human epidermal membranes treated with a frustum tipped wet-etch microneedle device following topical application of a fluorescent nanosphere formulation in diffusion experiments. (A and D) scale bar = 50 μm; (B and C) scale bar = 10 μm; (E and F) scale bar = 2 μm. Reproduced with permission from: [61] Coulman S.A., Anstey A., Gateley C., *et al.* (2009). Microneedle mediated delivery of nanoparticles into human skin. *Int. J. Pharm.* 366: 190–200.

had a strong influence on the permeation of the particles. Not all these types of studies were conducted on biological tissue. Zhang *et al.*, in two separate studies, evaluated the MN-assisted permeation of MPs into an agarose gel skin model by using metallic MNs [64, 65].

In addition to solid MNs, there have been several studies evaluating the use of hollow MNs for the delivery of NPs and MPs across the skin or alternative biological membranes [59]. This strategy was followed by Wang *et al.* who studied the intradermal injection of polymeric MPs using a single hollow glass MN [66]. In this study, the MPs presented sizes of around 2.5–2.8 μm. The experiment was carried out *ex vivo* and *in vivo* in hairless rats skin. The MPs were successfully injected but in order to obtain the optimal

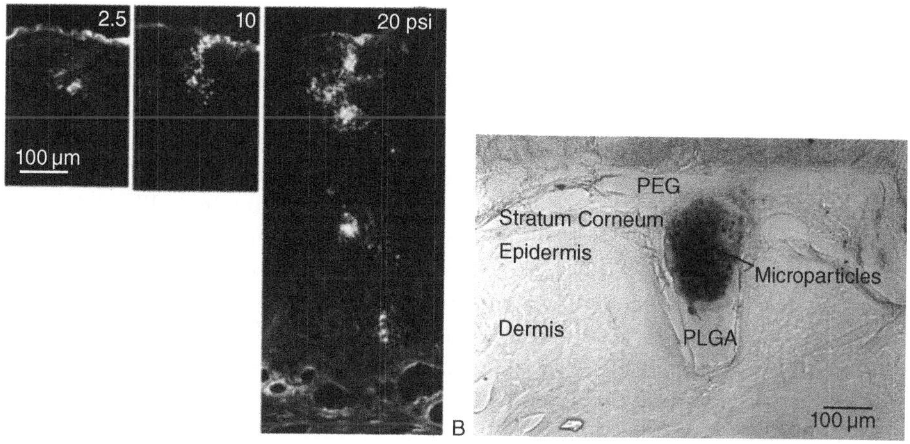

Figure 6.3 (A) Microinjection of polymeric particles at different pressures. Fluorescence micrographs of histological sections after microinjection of 2.5-µm fluorescent microspheres into hairless rat skin *in vivo* under pressures ranging from 2.5 to 20 psi via the same needle and loading for the same time periods. The volume and depth of injection increased with increasing pressure. (B) Multi-layered microneedles inserted into pig cadaver skin, cryo-sectioned vertically and viewed by brightfield microscopy. MNs were inserted into the skin allowing the deposition of microparticles (black dots) into the epidermis. *Source*: [66] Wang P.M., Cornwell M., Hill J. and Prausnitz M.R. (2006). Precise microinjection into skin using hollow microneedles. *J Invest Dermatol.* 126: 1080–1087; [67] Park J.H., Choi S.O., Kamath R., *et al.* (2007). Polymer particle-based micromolding to fabricate novel microstructures. *Biomed. Microdevices* 9: 223–234.

results, different aspects had to be adjusted, predominantly the injection pressure and the retraction of the MN after insertion (Figure 6.3A).

Häfeli *et al.* developed an injection device based on hollow silicon MNs [68]. The devices were tested by injecting several suspensions containing blue polystyrene microspheres (0.93 µm in diameter) and fluorescent polystyrene microspheres (0.7 µm in diameter). Confocal microscopy was used to evaluate the penetration depths of the particles after injection, which was around 70 µm. Nevertheless, the maximum concentration of the particles was found at around 20 µm underneath the skin surface.

Ocular delivery is another area of application for MN arrays. Research is providing interesting challenges that can be addressed with the use of MNs, such as the delivery of therapeutics to the back of the eye. Jiang *et al.* demonstrated the ability of hollow glass MNs to inject NPs and MPs into the sclera [69]. In this study, poly(lactic acid) (PLA) NPs loaded with Nile Red were injected using an insertion–retraction method. On the other hand, when latex fluorescent MPs were injected, the addition of enzymes to the tissue to disrupt the scleral structure was required. In a similar way Patel *et al.* studied the injection of NPs (size ranging from 20 nm to 1 µm) into the suprachoroidal space of the eye [70]. By adjusting the infusion pressure, MN length and intraocular pressure, the injection of the particles was optimised. Additionally, decreasing NP size eased the injection process. Kim *et al.* showed that hollow MNs can be used to deliver particle-stabilised emulsion droplets to the back of the eye in a rabbit animal model [71]. The injected formulation consisted of fluorescein-tagged polystyrene NPs containing high-density perfluorodecalin in the core. At least 50% of the injected NPs were found close to the macula or the ciliary body after infusion. The final target of the NPs depends on the eye cornea orientation.

In addition to solid and hollow MNs, alternative types of MNs have been used for the delivery of NPs and MPs. Park *et al.* developed an alternative method for the manufacture of MNs containing stacked layers in the needle tips and MN arrays with complex geometries [67]. One of the designs proposed needle tips made from PLGA and a needle shaft made from poly(ethylene glycol) (PEG) containing PLGA MPs with vitamin B (riboflavin-5′-phosphate sodium salt dehydrate) encapsulated inside. These MN arrays were used to create slow drug release reservoirs, as after insertion the PEG layer dissolved leaving the needles containing the MPs inside the tissue (Figure 6.3B). Alternatively, a different MN delivery strategy of coated MNs was used by DeMuth *et al.* to deliver Plasmid DNA or PLGA NPs into the viable epidermis [72]. In this study the authors showed the potential of this type of MN for vaccine delivery and the delivery of nanoencapsulated drugs [72].

Research work considering MN-NP combined systems is not just focused on evaluating the ability of MN to enhance the permeation of NP through the skin or other biological barriers. Recently, Kennedy *et al.* showed the biodistribution of NPs loaded with Rhodamine B after administration using dissolving MN arrays in a murine animal model [73]. This was an exploratory study which showed that dissolving MNs can be used to deliver NPs intradermally and that subsequently they are able to enter into the lymphatic system.

In addition, MN systems with shorter needle tips, termed nanoneedles, have been developed for the delivery of nanomedicines [74]. This technology allows cell membranes to be bypassed, enhancing drug and biological delivery. Moreover, it allows single-cell stimulation and intracellular sensing [74]. Using this technology, Chiappini *et al.* developed nanoneedle systems for the cytosolic delivery of CdTe QDs using a porous silico nanoneedle [74]. These systems were successfully used to deliver the QDs to cell cytosol. The same system was also used for the *in vivo* superficial intracellular delivery in a murine model. The conclusions of the study were that nanoneedles allowed finely tuned and localised delivery.

6.4.2 Drug Delivery

An approach combining MN arrays and NP or MP formulations has been explored for a different type of therapeutics for targeted transdermal/intradermal delivery with the help of solid, hollow and polymeric dissolving MNs.

The nature of the administration of insulin, the most commonly explored molecule for targeted transdermal delivery, has been considered using a combination of MNs and NPs/MPs by various research groups [75–77]. Ito *et al.* designed insulin absorbing porous silicon dioxide and calcium silicate microparticles and then incorporated these into chondroitin sulfate dissolving MNs for the delivery of insulin to mice [75]. MNs containing free insulin were also used for the control. On application of both types of MN arrays to mice for 8 h, a greater hypoglycemic effect was recorded, in contrast to that obtained following treatment with the MN-insulin arrays. In addition, MN-MP-insulin systems required longer times to reach the minimum glucose level.

Recently, Yu *et al.* developed a "smart insulin patch." This patch contained glucose-responsive vesicles loaded with insulin within the tips of an MN array [77].

These glucose responsive vesicles are formed by a hypoxia-sensitive hyaluronic acid conjugate containing 2-nitroimidazole (a hydrophobic component). Under hypoxic conditions (high glucose levels), the vesicles are bio-reduced and dissociated, releasing their insulin cargo (Figure 6.4A and B). This novel MN-based system was evaluated *in vivo* using a chemically induced Type 1 diabetes murine model. The results showed that these MN arrays were able to control glucose levels over the period of several hours.

Another research group, Chen *et al.*, studied the use of iontophoresis to assist the transdermal delivery of insulin encapsulated in nanovesicles through holes created in the skin by solid MN arrays [76]. Insulin was encapsulated in nanovesicles prepared using soybean lecithin and propylene glycol. These nanovesicles of various sizes (ranging between 91 and 176 nm) and zeta potentials (ranging between –51 and 28 mV) and with 89.05% insulin entrapment efficiencies were obtained by different pressure homogenisation cycles or ultrasound methodologies. Higher insulin permeation into the skin was obtained with iontophoresis plus MN pre-treatment than for passive permeation of the vesicles or free insulin. Considering the different types of nanovesicles that were evaluated, a positive zeta potential (zeta potential of +27.8 mV and 107.4 nm in diameter) provided the highest insulin permeation. This formulation was then tested in a mouse model, in combination with MN assisted iontophoresis, and the reduction in blood glucose levels were shown to be comparable to those obtained after subcutaneous injection of insulin.

Traditional surgical methods are not always feasible for the treatment of certain body parts with cancers, such as cancer of the oral cavity. One such novel treatment approach involved the solid MN assisted delivery of doxorubicin-loaded PLGA NPs in the treatment of oral cancer [78]. In this work, stainless-steel MN (700 μm in length and 200 μm in width) arrays were coated with a doxorubicin-NP (average diameter of 137 nm) formulation for intratumoural doxorubicin delivery in a minimally invasive and painless fashion. It was applied into model porcine buccal tissue and tested, using confocal microscopy, to evaluate the distribution of the NPs. It was shown that the particles could diffuse to a depth of more than 1 mm into the tissue at the insertion point. In addition to this, Paleco *et al.* developed a system to deliver the antioxidant quercetin encapsulated in lipidic MPs across the skin [79]. This lipidic MP cream formulation was applied to the porcine skin after pre-treatment with silicon MN arrays (36 MNs over 1 cm^2; MNs of height 200 μm), presenting a significant increase in the permeation of quercetin when compared with free quercetin. In addition to quercetin, the same research group designed poly(D,L-lactic acid) NPs (sizes 115–150 nm) containing ketoprofen, which were applied onto excised porcine skin after pre-treatment with silicon MN arrays; the system was able to release the drug over 24 hours [80].

The use of dyes and pigments as model drugs is common practice in drug delivery research. These types of analytically potent molecules permit rapid detection, and function as a proof of concept before beginning explorations with therapeutic drugs. Donnelly *et al.* studied the use of dissolving MN arrays as a platform technology to deliver PLGA NPs containing a model lipophilic molecule, Nile Red [81]. This pre-formed hydrophobic photosensitiser, Nile Red, was encapsulated into a PLGA NP formulation by combining emulsion and salting out approaches, yielding particles of approximately 150 nm in diameter. The NPs containing Nile Red were uniformly suspended into poly(methylvinylether/maleic anhydride) (Gantrez® AN-139) gel and then cast into a mould to form MN arrays.

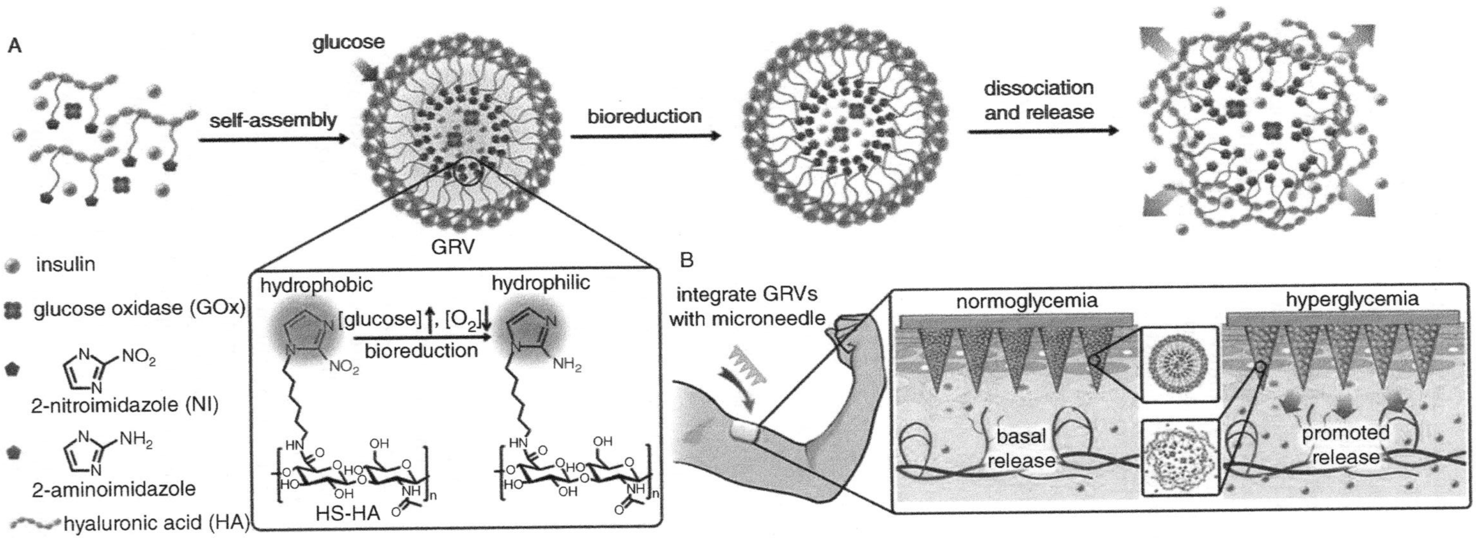

Figure 6.4 (A) Schematic representation of the formation and triggered insulin delivery of the glucose-responsive vesicles. (B) Schematic of the release from integrated MN/glucose responsive vesicles patch mechanism of action. Reproduced with permission from: [77] Yu J., Zhang Y., Ye Y., *et al*. (2015). Microneedle-array patches loaded with hypoxia-sensitive vesicles provide fast glucose-responsive insulin delivery. *Proc. Natl. Acad. Sci. U. S. A.* 112 (27): 8260–8265.

The resulting arrays were inserted into excised neonatal porcine skin for 6 h to check the Nile Red NP permeation by using a standard Franz-cell diffusion apparatus. MN free patches that did not contain needles were also employed as controls. The permeation results demonstrated that NP permeation was only possible when the NP formulation was applied using dissolving MN arrays. Alternatively, Gomaa *et al.* showed the ability of this platform to deliver a different type of dye, Rhodamine, using the same type of NPs and MN arrays [82]. In a similar way, Zhang *et al.* studied the permeation of coumarin-6 and R-phycoerythrin (a fluorescent probe) loaded PLGA NPs (about 160 nm in diameter) in hairless mouse skin pre-treated with solid MNs. In this instance, NPs diffused through the created microconduits but were not detected in the receptor compartment. Furthermore, Ke *et al.* designed a dissolving MN system containing PLGA MPs loaded with two dyes (Alexa 488 and Cyanine 5) as model drugs and with $NaHCO_3$, which allows pH-triggered release [83]. Once the $NaHCO_3$ comes into contact with the acidic pH, this stimulates the production of CO_2 bubbles to generate pores in the PLGA structure, which subsequently enhance the release of the dye molecules. These PLGA MP embedded poly(vinylpyrrolidone) (PVP) MN arrays were tested in an *in vivo* rat model and it was determined that this delivery platform was capable of co-delivering two compounds.

In further studies along a similar line, Lee *et al.* used Nile Red encapsulated in a nanostructured lipid carrier (~270 nm in diameter) loaded inside dissolving MN arrays [84]. The lipidic nanocarriers were formulated by high-pressure homogenization and were mixed with a water soluble biopolymer, hyaluronic acid, to form the dissolving MN arrays, following a drawing lithography technique. This MN system was tested in a Franz-cell diffusion experiment using minipig dorsal skin to confirm the ability of the system to enhance the permeation of Nile Red into the upper layer of the skin. Similarly, Vora *et al.* designed a bilayer dissolving MN array loaded with PLGA nano- and microparticles (NMP). These particles were located in the needle tips and they were loaded with a model drug, vitamin D_3 (VD_3). This system showed potential for enhanced transdermal drug delivery and targeted intradermal administration. The prepared NMP were mixed directly with 20% w/v poly(vinyl pyrrolidone) (PVP) gel, and the mixture then filled laser engineered micromoulds using 30 min high-speed centrifugation to concentrate the NMP into the MN tips. The particle sizes of the PLGA NMP ranged from 300 nm to 3.5 μm, and they retained their particle sizes after moulding of the MN arrays. Innovatively, these workers used this wide particle size distribution of PLGA NMP to produce a compact structure in the bilayer MN, as shown in Figure 6.5. *Ex vivo* intradermal neonatal porcine skin penetration of VD_3 NMP from bilayer MN was quantitatively analysed after cryostatic skin sectioning, with a good amount of the VD_3 loading being delivered intradermally. Thus, this two-step novel formulation design provides a simple and easy method for localising particulate delivery systems into dissolving MNs. It was demonstrated that biodegradable nano-microparticle embedded bilayer MN arrays significantly assist skin deposition for the effective delivery of the nanoparticles and microparticles to the viable skin layers to achieve the intradermal targeted delivery [85].

In addition to these previously described categories of NPs and MPs, another interesting nanocarrier is QDs. The use of QDs in combination with MNs has been developed as a promising drug release system. Justin *et al.* developed a biodegradable MN formulation that contains lidocaine hydrochloride loaded graphene QDs (50–55 nm in diameter) [86]. The photoluminescent, electrically conductive graphene QDs were used

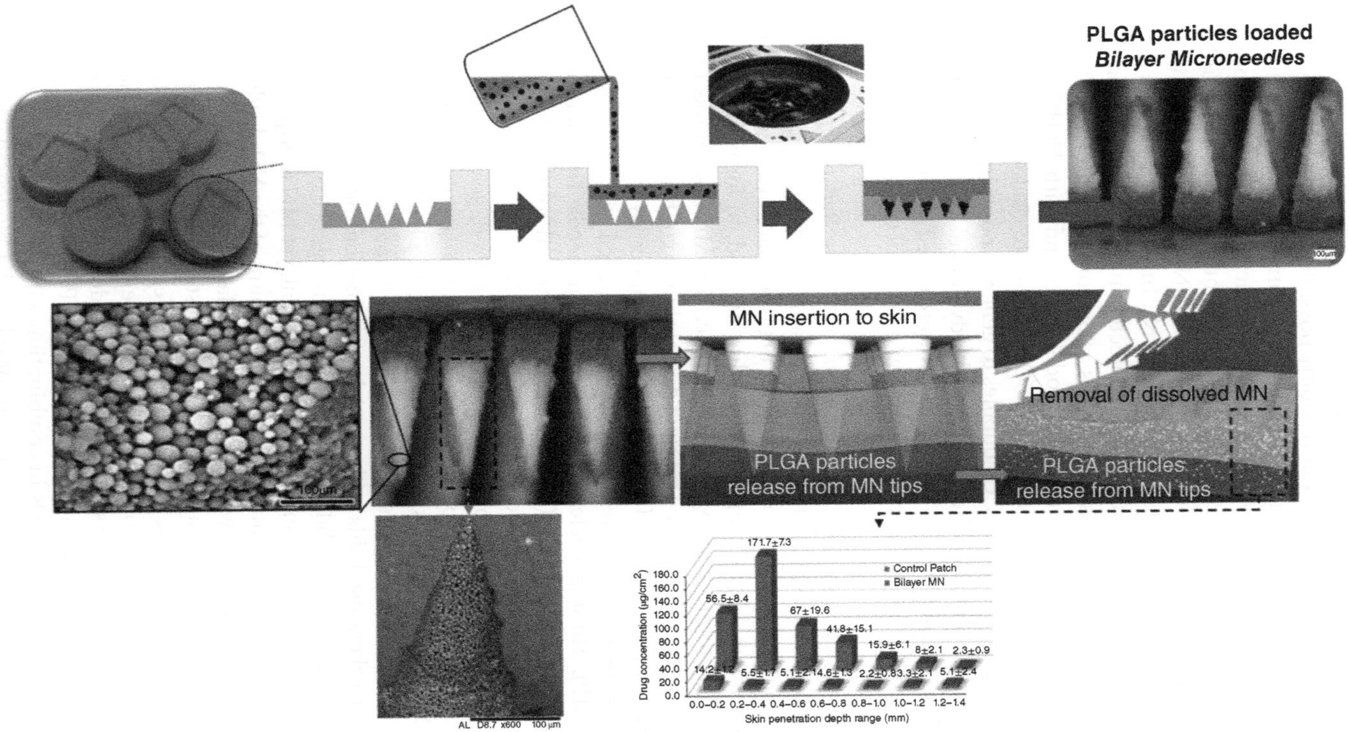

Figure 6.5 Schematic representation of the fabrication of PLGA nano-microparticle-loaded bilayer microneedle arrays, digital and scanning electron microscopic images and the concentration of drug in different depths of excised neonatal porcine skin following the application of control patches and bilayer microneedles containing drug-loaded PLGA NMP. Reproduced with permission from: [85] Vora L., Donnelly R., González-Vázquez P., *et al*. (2017). Novel bilayer dissolving microneedle arrays with concentrated PLGA nano-microparticles for targeted intradermal delivery: Proof of concept. *J. Control. Release* 265: 93–101.

to prepare a composite in combination with chitosan, which was used to formulate MN arrays. These MN arrays showed good mechanical properties and they were successfully inserted manually inside chicken skin. Furthermore, the arrays were capable of releasing between 50 and 70% of the total amount of lidocaine loaded in an *in vitro* setup. The ability to deliver different molecules from this type of MN arrays was tested using BSA as a model molecule. The system was used in combination with iontophoresis, showing promising results. This multifunctional nanocomposite enabled MN arrays to provide new ways to deliver small and large molecules intradermally.

6.4.3 Vaccine Delivery

As indicated earlier, MN arrays have been used in combination with NP/MP formulations to deliver different classes of therapeutics across the skin, and these have also included protein antigen and vaccine bio-therapeutics.

The epidermis and dermis contain vast quantities of antigen presenting cells (APCs). Their activation can lead to a strong humoral, cellular and mucosal immune response. Therefore, the skin is a very attractive site for vaccination, and an intradermal application of antigen may be much more effective than an intramuscular or subcutaneous injection [87].

Vaccine formulations are the most potent form of drug delivery design and successful vaccination is accomplished through activation of the adaptive immune system and induction of long-lasting memory responses. Dendritic cells (DCs), one of the most effective antigen-presenting cell types, can enable robust antigen-specific adaptive immunity. DCs process antigenic fragments, presenting these to T cells as peptide sequences bound to MHC (Class I and II), resulting in the activation of CD8+ and CD4+ cytotoxic T cells. These triggered T cells can act to modulate infection within the biological system [88].

As the upper layers of the skin contain antigen-presenting cells, such as DCs and Langerhans cells, MN arrays assist the pain-free delivery of vaccines/antigens to these layers of the skin. MN arrays with sharp tips help to penetrate through the *SC* layer and allow the deposition of vaccine-loaded NP formulations into the immune cell rich skin layer (Figure 6.6).

In the recent past, the amalgamation of vaccines into particle-based adjuvant/systems has been developed as a novel immunisation approach for the successful delivery of vaccine therapeutics [89]. The use of particle-based vaccination systems can aid in the stabilisation of protecting labile antigens from proteolysis and improves uptake and processing by APCs *in vivo*, in addition to providing controlled and sustained release of the same at the administration site to avoid booster doses.

Various group of researchers have shown that NPs have inherent immunogenic properties. It has been confirmed that antigen encapsulated NPs can trigger the T cell immune responses, with the NPs employed in the studies themselves exhibiting immunogenic properties, comparable to those of traditional adjuvants, such as aluminium hydroxide and Freund's complete adjuvant [90, 91]. Using NP formulations, the antigenic material can be adsorbed or conjugated onto the surface of the particles, or directly encapsulated within the polymeric matrix [92].

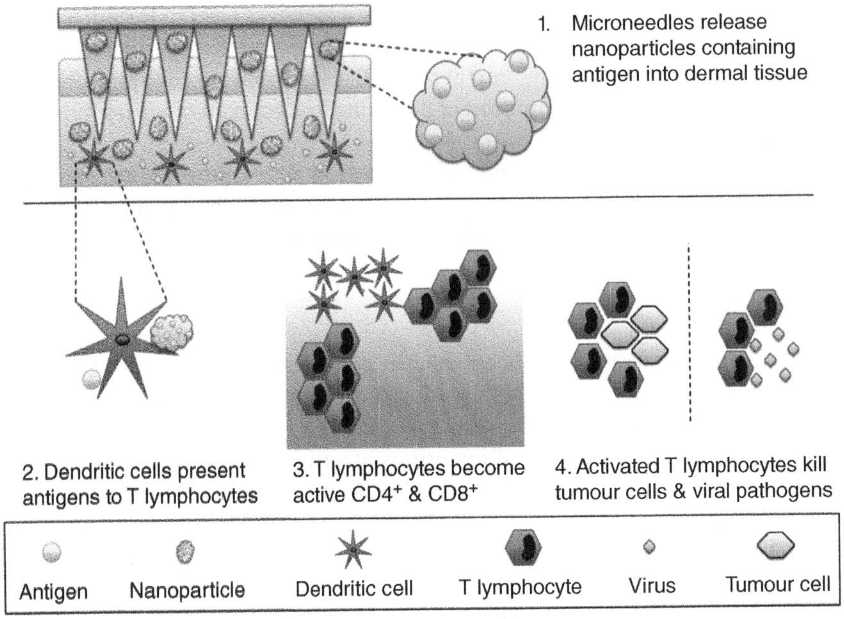

Figure 6.6 Schematic representation of microneedle arrays penetrating the skin layers releasing nanoparticles, containing vaccine antigens. Antigens are processed by dendritic cells and antigen fragments are presented to T lymphocytes. T lymphocytes become active to CD4+ and CD8+, which then help to destroy tumour cells and viral pathogens.

In the case of vaccine delivery, different types of MNs have been designed to date, but there are mainly two type of MN: dissolvable and non-dissolvable. Typically, the non-dissolvable MNs include hollow microneedles made out of metal, which are only for injection of the enclosed contents, and solid ones, which are made out of solid inorganic materials, such as silicon, metal, glass or ceramics; the active vaccine components (with or without NP) on the microneedle surfaces, therefore, function mainly as a poking device. Dissolvable MN arrays can be produced using biocompatible materials to achieve dissolution or biodegradation in body fluids after insertion. The antigen or antigen/adjuvant combination are normally incorporated during the formulation procedure. However, more sophisticated approaches such as the delivery of vaccines encapsulated in micro/nanoparticles require extra steps. In this case the micro/nanoparticulate formulation should be prepared prior to MN manufacturing. Subsequently, they have to be included within the MN matrix [93].

Considering the first use of NP formulations for vaccine delivery specifically in conjugation with MN arrays, in 2010 Bal *et al.* demonstrated the effective delivery and immunogenicity of *N*-trimethyl chitosan (TMC) NP adjuvanted diphtheria toxoid (DT) [57].

The DT was initially encapsulated into positively charge TMC NPs, which were then coated onto solid metallic MNs, to enable their delivery to the antigen presenting skin layers. As TMC has a permanent positive charge with good water solubility, this allowed the researchers to obtain good encapsulation of anionic antigen, based on ionic interactions, and then subsequently the DT loaded NPs. This research work served as an interesting early example of the vaccine, NP and MN combinatorial approach, suggesting that

NPs could act as a depot cum adjuvant for vaccine antigens, which could be transported across the skin layers using novel MN arrays as a delivery device [94].

Ovalbumin (OVA) is a widely explored model antigen used to test any kind of novel vaccine delivery system. In another study, in 2011, by Bal *et al.*, OVA conjugated TMC NPs were applied to the skin surface in a comparison of liquid formulations of OVA with roller microneedling devices, serving as a means of enhancing intradermal delivery. In this mouse study, significantly higher anti-OVA IgG titres were recorded with OVA conjugated NP treated mice, in comparison with those which had been treated with free OVA solution [94]. The authors successfully demonstrated that incorporation of protein antigens into the NP can serve as an adjuvant to increase their immunogenicity.

Ning *et al.* have developed an effective mucosal vaccine delivery method via the vaginal route with two types of multifunctional liposomes, the 200 nm sized mannosylated lipid A-liposomes (MLLs) and the 50 nm sized stealth lipid A-liposomes (SLLs). Both of these were loaded with a model antigen and fabricated together into microneedles, forming the pro-SL/MLL-constituted microneedle array (proSMMA), which upon rehydration dissolved rapidly, recovering the initial MLLs and SLLs. They studied the antigen response in mice by vaginal mucosa patch application with proSMMAs rather than conventional intradermal administration. The authors demonstrated that the MLLs from the administered proSMMAs were mostly captured by vaginal mucosal dendritic cells, whereas the recovered SLLs trafficked directly to draining lymph nodes, where they were taken up by macrophages. Innovatively, this group concurrently encapsulated sodium bicarbonate into liposomes. The authors described a system that contained a model antigen and NH_4CO_3 encapsulated into liposomes. The administered lyposomes were mostly taken up by vaginal mucosal dendritic cells, where they were trafficked directly to draining lymph nodes wherein they were picked up by macrophages. As a consequence of lysosome escape and ROS stimulation, the liposomes were cross presented on APCs, resulting in a mixed Th1/Th2 type response. This response was also enhanced by liposomal lipid A activation of TLR4. This investigation confirmed that vaginal vaccination of the engineered HSV_2 antigen-loaded proSMMAs successfully protected mice from the virus challenge at a lethal dose of HSV_2, which is typically transmitted via sexual contact. Thus, the proSMMAs are in fact a vaccine adjuvant-dual delivery system, which can elicit a robust humoral and cellular response against the invading pathogens, especially, the sexually transmitted ones [95].

Zhen *et al.* developed MNs containing mannose-PEG-cholesterol liposomes for the delivery of a model antigen protein, bovine serum albumin (BSA) [96]. The lyposomes were prepared following an emulsification–lyophilisation process. Subsequently, these carriers were loaded into PVP/sucrose microneedles. The MN arrays were stable at room temperature and could be inserted successfully into porcine skin. The system was designed to release the liposomes after insertion as the needle tips dissolve rapidly releasing their cargo. This system was tested in a murine animal model. For this purpose, the arrays were applied into the oral mucosa of the animals. After administration, high levels of BSA-specific IgG in the sera and IgA in the salivary, intestinal and vaginal secretions of mice were observed. Additionally, enhanced levels of IgG2a and IFN were obtained. These results suggested that the system induced a mixed Th1/Th2 immunoresponse.

In 2013, Demuth *et al.* employed dissolving MN platform and NP formulations in a research strategy to study the rapid implantation of controlled-release polymer depots into the cutaneous tissue. Readily water soluble polyacrylic acid (PAA) supporting matrix-based MN arrays were prepared with either PLGA NPs or solid PLGA. Upon insertion into the skin, the PAA backbone rapidly disintegrated, releasing the OVA cargo within 5 min. The PLGA NP sustained release of the subunit vaccine for a number of weeks following patch removal [97].

To this end, a microneedle materials platform has been demonstrated for rapid implantation of controlled-release polymer depots into the cutaneous tissue. Arrays of microneedles composed of drug-loaded poly(lactide-*co*-glycolide) (PLGA) microparticles or solid PLGA tips were prepared with a supporting and rapidly water-soluble poly(acrylic acid) (PAA) matrix. Upon application of microneedle patches to the skin of mice, the microneedles perforate the *stratum corneum* and epidermis. Penetration of the outer skin layers is followed by rapid dissolution of the PAA binder on contact with the interstitial fluid of the epidermis, implanting the microparticles or solid polymer microneedles in the tissue, which are retained following patch removal. These polymer depots remain in the skin for weeks following application and sustain the release of encapsulated cargoes for systemic delivery. Sustained release of vaccine component for weeks, localised within the skin layers, presents an opportunity for APC targeting and stimulation of long-lasting immunity. The outcome of this work was subsequently supported by studies performed by Zaric *et al.* [88]. PGLA NPs were mixed in aqueous blends of 20% w/w PMVE/MA to fabricate dissolving MN arrays, accomplishing the controlled release of OVA antigen into the skin layers. This NP embedded MN delivery strategy augmented antigen delivery and supported the generation of CD4+ Th1 immune responses and CD8+ cytotoxic T-cell responses. This study was significant as it not only demonstrated successful prophylactic vaccination against both tumour development and viral challenge, but did so with a single dosing regimen [88]. In terms of specific formulation considerations, in this same study nanoencapsulation was shown to improve antigen stability in the MN devices with potential ramifications for the reduction of cold chain storage costs. In the latter study, the significant role of skin-resident murine Langerhans cells, a sub-set of DCs, in skin immunisation strategies via this same NP/MN delivery mechanism were elucidated [98].

It is worth recognising, however, that the formulation of NP containing protein antigen will not essentially result in an immunological response, such as those shown previously. The physicochemical properties of the antigen and NPs influence the immunological response obtained after administration. It has also been noted that the optimal vaccine delivery differs for each administration site, with the diffusion of antigen into the skin potentially the most critical rate-limiting step [94, 99].

Although protein-based antigens have been the mainstay of vaccination practises for many years, DNA (deoxyribonucleic acid) vaccination is a realistic novel technology that utilises genetically engineered DNA to produce an immunologic response [100]. Thus, an amalgamation of DNA vaccine and NP formulations for intradermal targeted delivery is a logical next step to advance this field. For example, Kumar *et al.* formulated plasmid DNA coated cationic charged PLGA NPs. This DNA vaccine was able to permeate through the *SC* after treating the skin with a solid MN roller device. This vaccine delivery system was successfully tested in a murine animal model providing immune response against anthrax [101].

An interesting finding of this study was the fact that it emphasises how positively charged plasmid DNA coated NPs produced a stronger immune response in mice than negatively charged plasmid DNA coated NPs or plasmid DNA alone [101]. In comparison with studies using NP decorated with plasmid DNA, DNA can also be encapsulated into the NPs. One such recent study by McCaffrey *et al.* illustrated the inherent potential of a nano-encapsulated plasmid DNA vaccine to generate detectable levels of reporter gene expression *in vivo,* when delivered from novel dissolving MN arrays [102]. This technology was particularly innovative. It uses a self-assembling amphipathic peptide (RALA) to form nanocarrier containing DNA [103, 104], which improves the uptake and processing by cells to facilitate the endosomal escape of the DNA cargo and promotes nuclear localisation of the DNA for transcription [102–104]. This work perfectly exemplifies how the combination of two formulation platforms can yield the development of a true state-of- the-art technology platform for the delivery of, in this instance, nucleic acids.

Hung-Wei Yang *et al.* presented a novel method of Ebola vaccination using a DNA vaccine coated on poly(lactic-*co*-glycolic acid)–poly-lysine/poly-γ-glutamic acid (PLGA-PLL/γPGA) NPs administered using a microneedle patch. To enhance cellular uptake and expression of EboDNA, the vaccine was coated onto the surface of cationic PLGA-PLL NPs. The authors investigated whether PLGA-PLL-DNA NPs maintain integrity after encapsulation and release from the MNs. MNs encapsulating PLGA-PLL-GFP-pDNA NPs were dissolved and then assayed for transfection efficiency using HeLa cells. The transfection efficiency showed that naked GFP-pDNA reconstituted from MNs was significantly lower than naked GFP-pDNA, while PLGA-PLL-GFP-pDNA NPs reconstituted from MNs were not significantly different from PLGA-PLL-GFP-pDNA NPs. These cationic NP and MN excipients were not cytotoxic to HeLa cells. Mice were vaccinated using MN patches containing PLGA-PLL/PGA-EboDNA NP, which dissolved within 5 min of insertion without any side-effects. Four groups of mice were each immunised four times at four-week intervals, with EboDNA vaccine with either naked EboDNA or PLGA-PLL/PGA-EboDNA NPs administered either by IM injection or MN patch. Two weeks after the final dose, PLGA PLL/γPGA-EboDNA NPs administered either by IM or by MN patch showed similar total antigen-specific IgG titre responses compared with the positive control, but vaccination with the naked EboDNA containing MN patch yielded significantly lower total antigen-specific IgG titres [105].

Moving on from the aforementioned studies to more complex NP approaches, Kim *et al.* developed polyplex DNA vaccines layered onto MN arrays through a pH-responsive polyelectrolyte multilayer assembly. In detail, solid polycarbonate MN arrays were designed to deliver a range of NP cargoes with confirmed robust humoral immune responses, in comparison to the delivery of the same cargoes via *SC* injection [106]. Following this, Hu *et al.* conducted studies to exemplify MN-assisted dendritic cell targeting of NP for transcutaneous DNA immunisation in immunocompromised BALB/c mice [107]. The MN-assisted *in vivo* skin administration of mannosylated grafted cell-penetrating peptide low molecular weight copolymer NPs was investigated; the DC targeting efficiency was analysed and the stimulation of protective and therapeutic anti-tumour immunity was observed. In this detailed study, the authors reported that the process efficiently promoted Trp2-specific cellular immune responses, resulting in effective protection against B16 melanoma challenge. The MN-assisted NP formulation strongly induced CD8+ cytotoxic T cells and CD4+ T cells, which secreted interferon-gamma and interleukin 12 cytokines, against melanoma cells. This study, along with others, once again points out that combinatorial NP/MN strategies have the potential to deliver immunotherapeutic agents.

It has been well recognised in the last two decades that MN assisted technologies can deliver therapeutic agents to the skin. Therefore, combining this platform technology with many of the positive aspects of novel NP formulation, unambiguously in terms of the potential for stabilising the antigen, depot release and the augmentation of immune responses, serves to make this combinatorial approach an appealing and exhilarating area for future research. Initial formulation strategies featuring antigen laden NPs have been published, and since then the arena of research has extended to the integration of NP-based therapeutics into MN devices. Further clinical work on individualised formulations for antigen and DNA vaccine components is undoubtedly required before a vaccine MN product, incorporating NPs, will be realised. However, a well-founded base of literature now exists for future research to build upon and develop.

6.4.4 Other Uses

MN arrays used in combination with NPs and MPs have been designed for diagnostic as well as theranostic purposes, as a second option for drug and vaccine delivery. Their other main field of use is in the delivery of metallic NPs for the diagnostic/theranostic application.

Optical coherence tomography (OCT) is a promising diagnostic tool for the detection of cancer at an early stage. The main limitation of this OCT technique, however, is the low contrast levels in biological tissue, especially between normal and neoplastic tissue. To overcome this limitation, Kim *et al.* studied the delivery of gold NPs (71 nm in diameter) across the *SC* and the epithelial barriers as an OCT contrast agent after treating the skin with an MN roller device [108]. The NP diffusion through the microchannels created by the MN roller was subsequently enhanced by the use of ultrasound. This novel local diagnosis system was effectively used in a model for oral carcinogenesis, increasing the contrast level of the OCT technique by nearly 150%. Furthermore, the same research group designed a dissolving MN array containing gold NPs for a similar application [109]. The dissolving sodium carboxymethyl cellulose (CMC) and sucrose were used to prepare the gold NP loaded MN arrays. After application of the arrays to hamster skin, the MN arrays dissolved and started releasing the gold NPs. In order to enhance the permeation of the NPs, ultrasound was employed. The use of the MN/NP system in the treated tissue noticeably enhanced the optical contrast of the OCT images.

In another interesting application for theranostic purposes, MN technology can be used when combined with QDs. Gittard *et al.* used MN arrays made from an acrylate-based polymer to inject QDs into the deep epidermis and dermis of porcine skin [110]. Multiphoton microscopy was then employed to track the trajectory of the QDs and to visualise the needles inserted into the tissue. Owing to the ability to inject QDs without pain, this approach is suitable for further theranostic applications. The same research group also designed organically modified ceramic MNs for the targeted intradermal delivery of PEG-amine QD NP solutions into porcine skin [111]. The ceramic MN arrays were used to create micropore channels in the living tissue and subsequently a solution containing QDs was applied to the treated skin. After the application of this formulation, nanometric QD particles were found in the deep layers (epidermal and dermal) of the porcine skin. The use of MN arrays has hence been shown to be a possible alternative for QD delivery through the skin. However, more *in vivo* and toxicological studies should be carried out to demonstrate the clinical effectiveness of QDs in theranostic applications.

From a formulation point of view, NPs and MPs have been used to increase the mechanical properties of MNs. For example, Raja *et al.* designed a biopolymer, silk protein MN array loaded with silk MPs to improve the mechanical strength of MN arrays [112]. In addition, the MPs were encapsulated with BSA and sulforhodamine and subsequently the release of these model molecules was assessed in a 3D collagen gel and in human cadaver skin, respectively. Following a similar concept, Yan *et al.* developed a nanocomposite to prepare mechanically strong MN arrays [113]. The presence of 5% (w/w) of layered double hydroxide NPs densely packed inside CMC MNs resulted in dramatically improved mechanical properties without any concession to the dissolution rate of the MNs in the skin and release of the payload within one minute. These nanocomposite-strengthened MNs were tested for *in vivo* vaccine delivery and the results showed that considerably stronger antibody response could be induced when compared with subcutaneous injection. These early reports suggest that nanomaterials could be useful to produce densely packed polymeric MNs. The presence of nanomaterials within the MN structure provides improved mechanical properties and dissolution rates.

In addition to improvement to the mechanical properties of MN, NPs/MPs can be embedded inside MN arrays to trigger drug release under certain biological conditions. Kim *et al.* designed hydrogel MPs containing biodegradable PLGA MNs that enhance drug delivery from the needle tips in biological environments [114]. Hydrogel MPs (10–40 μm) were synthesised by an emulsification method (radical cross-linking) using *N*-isopropylacrylamide and embedded into PLGA MN arrays. The authors demonstrated that microneedle dissolution into the skin was driven by hydrogel MP swelling in response to contact with interstitial fluid after the MN arrays were inserted into the skin (Figure 6.7A).

Similarly, it should be possible to design triggerable drug delivery systems to facilitate drug dosing-on-demand to accommodate patient regimens as well as the delivery of multiple dosages at one time. Some photo/radio-sensitive NP variants possess interesting radiation responsive properties that have been used to design light-triggerable drug release from MN arrays. Chen *et al.* designed silica-coated lanthanum hexaboride (LaB6@SiO2) nanostructures (serving as near-infrared radiation absorbers), which were subsequently incorporated into polycaprolactone microneedles. This NP variant acted as a local heat source when the array was irradiated with external near-infrared radiation (NIR) [115, 116]. After irradiation, the temperature that was reached within the arrays was 50 °C, thus leading to the melting of the polycaprolactone to trigger the drug release from the array needles (Figure 6.7B). Rhodamine 6G was incorporated into MNs as a model molecule to ascertain this remote trigger behaviour. Furthermore, this NIR-light activated device was tested for *in vivo* controlled release of doxorubicin in a rat model [116].

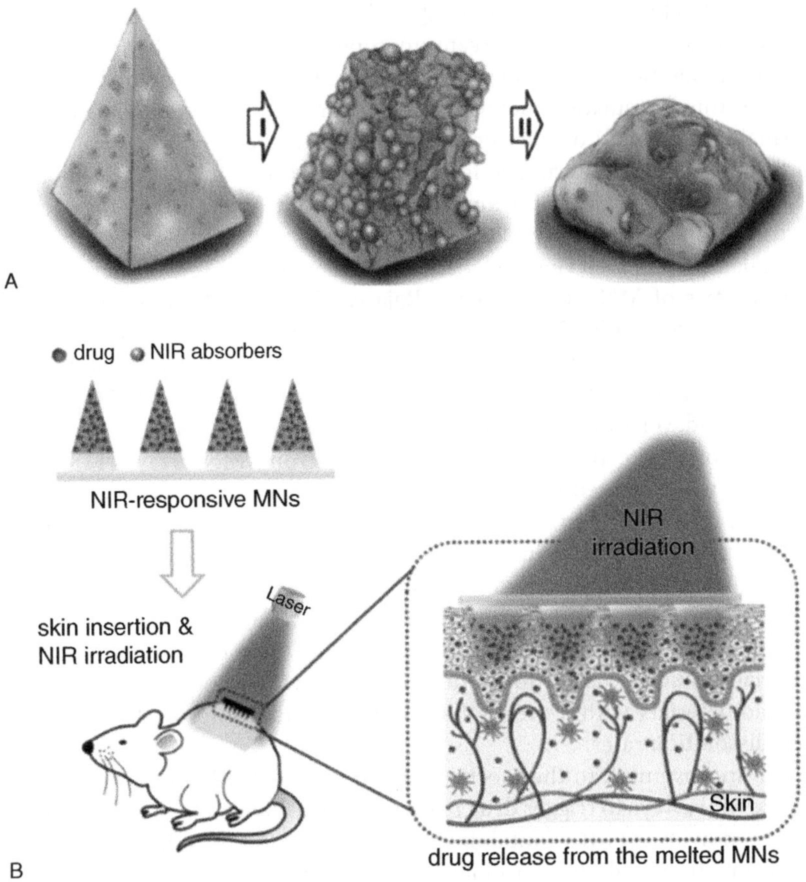

Figure 6.7 (A) Schematic representation of the mechanism of responsive release of MPs from MN arrays containing hydrogel MPs. (B) Schematic diagram of triggered intradermal drug delivery using near-infrared light-responsive MN. *Source*: adapted from [114] Kim M., Jung B. and Park J.H. (2012). Hydrogel swelling as a trigger to release biodegradable polymer microneedles in skin. *Biomaterials* 33 (2): 668–678; [115] Chen M.C., Ling M.H., Wang K.W., *et al*. (2015). Near-infrared light-responsive composite microneedles for on-demand transdermal drug delivery. *Biomacromolecules* 16 (5): 1598–1607.

6.5 Conclusions

Considering the complexity of biological systems and diseases, a synergistic drug delivery platform will always be needed to achieve the desired outcome for the overall wellbeing of patients. The combination of nanomedicine and MN technology strategies for patient-friendly transdermal delivery of complex molecules has been made possible due to the exponential rate of development of new technologies in the respective areas over the course of the last three decades. This chapter has focused on early synergistic developments of nanomedicine with MN technology, in an effort to provide a comprehensible and conversant understanding of this niche scientific area. To date, most of the studies have been on the feasibility of MN technology to deliver NPs/MPs into the skin with model NPs or the release of model molecules, such as dyes, from

NPs. Considering the specific requirements for protein/DNA based vaccine delivery, the NP/MN platform is an emerging area of research. As with multiple facets of science and technological developments, it is abundantly clear that nanomedicine and MN technologies hold remarkable potential, but it is essential that further research is carried out in order to comprehensively understand, and then exploit, their inherent abilities for effective drug delivery.

References

1 Prow T.W., Grice J.E., Lin L.L., *et al.* (2011). Nanoparticles and microparticles for skin drug delivery. *Adv. Drug Deliv. Rev.* 63: 470–491.

2 Parveen S., Misra R. and Sahoo S.K. (2012). Nanoparticles: a boon to drug delivery, therapeutics, diagnostics and imaging. *Nanomedicine* 8: 147–166.

3 Moghimi S.M., Hunter A.C. and Murray J.C. (2005). Nanomedicine: current status and future prospects. *FASEB J.* 19: 311–330.

4 Patravale V., Dandekar P. and Jain R. (2012). *Nanoparticulate Drug Delivery: Perspectives on the Transition from Laboratory to Market.* Elsevier Science.

5 Kumar C.S.S.R. (2007). *Nanomaterials for Medical Diagnosis and Therapy.* Wiley.

6 Prausnitz M.R. and Langer R. (2008). Transdermal drug delivery. *Nat. Biotechnol.* 26: 1261–1268.

7 Donnelly R.F., Singh T.R.R., Morrow D.I.J. and Woolfson A.D. (2012). *Microneedle-mediated Transdermal and Intradermal Drug Delivery.* Wiley.

8 Smith H.W, Clowes G.H.A. and Marshall E.K. (1919). On dichloroethylsulphide (mustard gas). IV, The mechanism of absorption by the skin. *J. Pharm. Exp. Ther.* 13: 1–30.

9 Scheuplein R.J. (1967). Mechanism of percutaneous absorption. II. Transient diffusion and the relative importance of various routes of skin penetration. *J. Invest. Dermatol.* 48: 79–88.

10 Hadgraft J. (2002). *Transdermal Drug Delivery Systems: Revised and Expanded.* CRC Press.

11 Bronaugh R.L. and Maibach H.I. (1999). *Percutaneous Absorption: Drugs–Cosmetics–Mechanisms–Methodology: Drugs–Cosmetics–Mechanisms–Methodology*, 3rd edn. CRC Press.

12 Williams A. (2003). *Transdermal and Topical Drug Delivery from Theory to Clinical Practice.* Pharmaceutical Press.

13 Prausnitz M.R., Mitragotri S. and Langer R. (2004). Current status and future potential of transdermal drug delivery. *Nat. Rev. Drug Discov.* 3: 115–124.

14 Donnelly R.F. and Singh T.R.R. (2015). *Novel Delivery Systems for Transdermal and Intradermal Drug Delivery.* Wiley.

15 Chuong C.M., Nickoloff B.J., Elias P.M., *et al.* (2002). What is the 'true' function of skin? *Exp. Dermatol.* 11: 159–163.

16 Krieg T., Bickers D.R. and Miyachi Y. (2010). *Therapy of Skin Diseases: A Worldwide Perspective on Therapeutic Approaches and Their Molecular Basis.* Berlin, Heidelberg: Springer.

17 Baroli B. (2010). Penetration of nanoparticles and nanomaterials in the skin: Fiction or reality? *J. Pharm. Sci.* 99: 21–50.

18 Wiechers J.W. (1989). The barrier function of the skin in relation to percutaneous absorption of drugs. *Pharm. Weekblad Sci. Ed.* 11: 185–198.
19 Michaels A.S., Chandrasekaran S.K. and Shaw J.E. (1975). Drug permeation through human skin: Theory andinvitro experimental measurement. *AIChE J.* 21: 985–996.
20 Bouwstra J.A., Gooris G.S., van der Spek J.A. and Bras W. (1991). Structural investigations of human stratum corneum by small-angle X-ray scattering. *J. Invest. Dermatol.* 97: 1005–1012.
21 Schneider M., Stracke F., Hansen S. and Schaefer U.F. (2009). Nanoparticles and their interactions with the dermal barrier. *Dermatoendocrinology* 1: 197–206.
22 Cevc G. and Vierl U. (2010). Nanotechnology and the transdermal route: A state of the art review and critical appraisal. *J. Control. Release* 141: 277–299.
23 Elsayed M.M.A., Abdallah O.Y., Naggar V.F. and Khalafallah N.M. (2007). Lipid vesicles for skin delivery of drugs: Reviewing three decades of research. *Int. J. Pharm.* 332: 1.
24 Pierre M.B. and Dos Santos Miranda Costa I. (2011). Liposomal systems as drug delivery vehicles for dermal and transdermal applications. *Arch. Dermatol. Res.* 303: 607–621.
25 Taddio A., Soin H.K., Schuh S., *et al.* (2005). Liposomal lidocaine to improve procedural success rates and reduce procedural pain among children: a randomized controlled trial. *Can. Med. Assoc. J.* 172: 1691–1695.
26 Mura P., Maestrelli F., González-Rodríguez M.L., Michelacci I., *et al.* (2007). Development, characterization and in vivo evaluation of benzocaine-loaded liposomes. *Eur. J. Pharm. Biopharm.* 67: 86.
27 Simões S.I., Delgado T.C., Lopes R.M., *et al.* (2005). Developments in the rat adjuvant arthritis model and its use in therapeutic evaluation of novel non-invasive treatment by SOD in transfersomes. *J Control. Release* 103: 419.
28 Maestrelli F., González-Rodríguez M.L., Rabasco A.M. and Mura P. (2006). Effect of preparation technique on the properties of liposomes encapsulating ketoprofen-cyclodextrin complexes aimed for transdermal delivery. *Int. J. Pharm.* 312: 53.
29 Paolino D., Lucania G., Mardente D., *et al.* (2005). Ethosomes for skin delivery of ammonium glycyrrhizinate: In vitro percutaneous permeation through human skin and in vivo anti-inflammatory activity on human volunteers. *J. Control. Release* 106: 99–110.
30 Schäfer-Korting M., Mehnert W. and Korting H.C. (2007). Lipid nanoparticles for improved topical application of drugs for skin diseases. *Adv. Drug Deliv. Rev.* 59: 427–443.
31 Müller R.H., Radtke M. and Wissing S.A. (2002). Solid lipid nanoparticles (SLN) and nanostructured lipid carriers (NLC) in cosmetic and dermatological preparations. *Adv. Drug Deliv. Rev.* 54 (suppl.): S131.
32 Lopes L.B. (2014). Overcoming the cutaneous barrier with microemulsions. *Pharmaceutics* 6: 52–77.
33 Vogt A., Combadiere B., Hadam S., *et al.* (2006). 40 nm, but not 750 or 1,500 nm, Nanoparticles enter epidermal CD1a+ cells after transcutaneous application on human skin. *J. Invest. Dermatol.* 126: 1316–1322.
34 Alvarez-Román R., Naik A., Kalia Y.N., *et al.* (2004). Skin penetration and distribution of polymeric nanoparticles. *J. Control. Release* 99: 53.

35 Zhang Z., Tsai P.C., Ramezanli T. and Michniak-Kohn B. (2013). Polymeric nanoparticles-based topical delivery systems for the treatment of dermatological diseases. *Wiley Interdiscip. Rev. Nanomed. Nanobiotechnol.* 5: 205–218.
36 Hasanovic A., Zehl M., Reznicek G. and Valenta C. (2009). Chitosan-tripolyphosphate nanoparticles as a possible skin drug delivery system for aciclovir with enhanced stability. *J. Pharm. Pharmacol.* 61: 1609–1616.
37 Kim D.G., Jeong Y.I., Choi C., *et al.* (2006). Retinol-encapsulated low molecular water-soluble chitosan nanoparticles. *Int. J. Pharm.* 319: 130–138.
38 Tomoda K., Terashima H., Suzuki K., *et al.* (2011). Enhanced transdermal delivery of indomethacin-loaded PLGA nanoparticles by iontophoresis. *Colloids Surf B* 88: 706–710.
39 Shah P.P., Desai P.R., Patel A.R. and Singh M.S. (2012). Skin permeating nanogel for the cutaneous co-delivery of two anti-inflammatory drugs. *Biomaterials* 33: 1607–1617.
40 Kreilgaard M. (2002). Influence of microemulsions on cutaneous drug delivery. *Adv. Drug Deliv. Rev.* 54 (suppl.): S77–S98.
41 Wu H., Ramachandran C., Bielinska A.U., *et al.* (2001). Topical transfection using plasmid DNA in a water-in-oil nanoemulsion. *Int. J. Pharm.* 221: 23–34.
42 Lopes L.B., VanDeWall H., Li H.T., *et al.* (2010). Topical delivery of lycopene using microemulsions: enhanced skin penetration and tissue antioxidant activity. *J. Pharm. Sci.* 99: 1346–1357.
43 Pakpayat N., Nielloud F., Fortuné R., *et al.* (2009). Formulation of ascorbic acid microemulsions with alkyl polyglycosides. *Eur. J. Pharm. Biopharm.* 72: 444–452.
44 Baroli B., Ennas M.G., Loffredo F., *et al.* (2007). Penetration of metallic nanoparticles in human full-thickness skin. *J. Invest. Dermatol.* 127: 1701–1712.
45 Yong K.T. (2012). Quantum dots for biophotonics. *Theranostics* 2: 629–630.
46 Prow T.W., Monteiro-Riviere N.A., Inman A.O., *et al.* (2012). Quantum dot penetration into viable human skin. *Nanotoxicology* 6: 173–185.
47 Chu M.Q., Wu Q., Wang J.X., *et al.* (2007). In vitro and in vivo transdermal delivery capacity of quantum dots through mouse skin. *Nanotechnology* 18: 455103.
48 Zhang L.W., Yu W.W., Colvin V.L. and Monteiro-Riviere N.A. (2008). Biological interactions of quantum dot nanoparticles in skin and in human epidermal keratinocytes. *Toxicol. Appl. Pharmacol.* 228: 200–211.
49 Ryman-Rasmussen J.P., Riviere J.E. and Monteiro-Riviere N.A. (2006). Penetration of intact skin by quantum dots with diverse physicochemical properties. *Toxicol Sci.* 91: 159–165.
50 Zhang L.W. and Monteiro-Riviere N.A. (2008). Assessment of quantum dot penetration into intact, tape-stripped, abraded and flexed rat skin. *Skin Pharmacol. Phys.* 21: 166–180.
51 Nohynek G.J, Lademann J., Ribaud C. and Roberts M.S. (2007). Grey goo on the skin? Nanotechnology, cosmetic and sunscreen safety. *Crit. Rev. Toxicol.* 37: 251–277.
52 Kimura E., Kawano Y., Todo H., *et al.* (2012). Measurement of skin permeation/penetration of nanoparticles for their safety evaluation. *Biol. Pharm. Bull.* 35: 1476–186.
53 Byszewski P. and Klusek Z. (2001). Some properties of fullerenes and carbon nanotubes. *Opto-Electron. Rev.* 9: 203–210.

54 Rouse J.G., Yang J., Ryman-Rasmussen J., *et al.* (2007). Effects of mechanical flexion on the penetration of fullerene amino acid-derivatized peptide nanoparticles through skin. *Nano Lett.* 7: 155–160.

55 Siu K.S., Chen D., Zheng X., *et al.* (2014). Non-covalently functionalized single-walled carbon nanotube for topical siRNA delivery into melanoma. *Biomaterials* 35: 3435–3442.

56 Degim I.T., Burgess D.J. and Papadimitrakopoulos F. (2010). Carbon nanotubes for transdermal drug delivery. *J. Microencapsul.* 27: 669–681.

57 Bal S.M., Ding Z., Kersten G.F.A., *et al.* (2010). Microneedle-based transcutaneous Immunisation in mice with *N*-trimethyl chitosan adjuvanted diphtheria toxoid formulations. *Pharm. Res.* 27 (9): 1837–1847.

58 Prausnitz M.R. (2004). Microneedles for transdermal drug delivery. *Adv. Drug Deliv. Rev.* 56: 581–587.

59 Larrañeta E., McCrudden M.T.C., Courtenay A.J. and Donnelly R.F. (2016). Microneedles: a new frontier in nanomedicine delivery. *Pharm. Res.* 1–19.

60 McAllister D.V., Wang P.M., Davis S.P., *et al.* (2003). Microfabricated needles for transdermal delivery of macromolecules and nanoparticles: fabrication methods and transport studies. *Proc. Natl. Acad. Sci. U. S. A.* 100: 13755–13760.

61 Coulman S.A., Anstey A., Gateley C., *et al.* (2009). Microneedle mediated delivery of nanoparticles into human skin. *Int. J. Pharm.* 366: 190–200.

62 Zhang W., Gao J., Zhu Q., *et al.* (2010). Penetration and distribution of PLGA nanoparticles in the human skin treated with microneedles. *Int. J. Pharm.* 402: 205–212.

63 Gomaa Y.A., Garland M.J., McInnes F.J., *et al.* (2014). Microneedle/nanoencapsulation-mediated transdermal delivery: Mechanistic insights. *Eur. J. Pharm. Biopharm.* 86: 145–155.

64 Zhang D., Das D.B. and Rielly C.D. (2013). An experimental study of microneedle-assisted microparticle delivery. *J. Pharm. Sci.* 102: 3632–3644.

65 Zhang D., Das D.B. and Rielly C.D. (2014). Microneedle assisted micro-particle delivery from gene guns: Experiments using skin-mimicking agarose gel. *J. Pharm. Sci.* 103: 613–627.

66 Wang P.M., Cornwell M., Hill J. and Prausnitz M.R. (2006). Precise microinjection into skin using hollow microneedles. *J. Invest. Dermatol.* 126: 1080–1087.

67 Park J.H., Choi S.O., Kamath R., *et al.* (2007). Polymer particle-based micromolding to fabricate novel microstructures. *Biomed. Microdevices* 9: 223–234.

68 Häfeli U.O., Mokhtari A., Liepmann D. and Stoeber B. (2009). In vivo evaluation of a microneedle-based miniature syringe for intradermal drug delivery. *Biomed. Microdevices* 11: 943–950.

69 Jiang J., Moore J., Edelhauser H. and Prausnitz M. (2009). Intrascleral drug delivery to the eye using hollow microneedles. *Pharm. Res.* 26: 395–403.

70 Patel S.R., Lin A.S., Edelhauser H.F. and Prausnitz M.R. (2011). Suprachoroidal drug delivery to the back of the eye using hollow microneedles. *Pharm. Res.* 28: 166–176.

71 Kim Y.C., Edelhauser, H.F. and Prausnitz M.R. (2014). Particle-stabilized emulsion droplets for gravity-mediated targeting in the posterior segment of the eye. *Adv. Healthc. Mater.* 3: 1272–1282.

72 DeMuth P.C., Su X., Samuel R.E., *et al.* (2010). Nano-layered microneedles for transcutaneous delivery of polymer nanoparticles and plasmid DNA. *Adv. Mater.* 22: 4851.

73 Kennedy J., Larrañeta E., McCrudden M.T.C., *et al.* (2017). In vivo studies investigating biodistribution of nanoparticle-encapsulated rhodamine B delivered via dissolving microneedles. *J. Control. Release* 265: 57–65.

74 Chiappini C., Martinez J.O., De Rosa E., *et al.* (2015). Biodegradable nanoneedles for localized delivery of nanoparticles in vivo: Exploring the biointerface. *ACS Nano* 9: 5500–5509.

75 Ito Y., Hagiwara E., Saeki A., *et al.* (2007). Sustained-release self-dissolving micropiles for percutaneous absorption of insulin in mice. *J. Drug Target.*15 (5): 323–326.

76 Chen H., Zhu H., Zheng J., *et al.* (2009). Iontophoresis-driven penetration of nanovesicles through microneedle-induced skin microchannels for enhancing transdermal delivery of insulin. *J. Control. Release* 139 (1): 63–72.

77 Yu J., Zhang Y., Ye Y., *et al.* (2015). Microneedle-array patches loaded with hypoxia-sensitive vesicles provide fast glucose-responsive insulin delivery. *Proc. Natl. Acad. Sci. U. S. A.* 112 (27): 8260–8265.

78 Ma Y., Boese S., Luo Z., *et al.* (2015). Drug coated microneedles for minimally-invasive treatment of oral carcinomas: development and in vitro evaluation. *Biomed. Microdevices* 17 (2): 1.

79 Paleco R., Vučen S.R., Crean A.M., *et al.* (2014). Enhancement of the in vitro penetration of quercetin through pig skin by combined microneedles and lipid microparticles. *Int. J. Pharm.* 472 (1–2): 206–213.

80 Vučen S.R., Vuleta G., Crean A.M., *et al.* (2013). Improved percutaneous delivery of ketoprofen using combined application of nanocarriers and silicon microneedles. *J. Pharm. Pharmacol.* 65 (10): 1451–1462.

81 Donnelly R.F., Morrow D.I.J., Fay F., *et al.* (2010). Microneedle-mediated intradermal nanoparticle delivery: Potential for enhanced local administration of hydrophobic pre-formed photosensitisers. *Photodiagn. Photodyn Ther.* 7 (4): 222–231.

82 Gomaa Y.A., El-Khordagui L.K., Garland M.J., *et al.* (2012). Effect of microneedle treatment on the skin permeation of a nanoencapsulated dye. *J. Pharm. Pharmacol.* 64 (11): 1592–1602.

83 Ke C.J., Lin Y.J., Hu Y.C., *et al.* (2012). Multidrug release based on microneedle arrays filled with pH-responsive PLGA hollow microspheres. *Biomaterials* 33 (20): 5156–5165.

84 Lee S.G., Jeong J.H., Lee K.M., *et al.* (2014). Nanostructured lipid carrier-loaded hyaluronic acid microneedles for controlled dermal delivery of a lipophilic molecule. *Int. J. Nanomed.* 9: 289–299.

85 Vora L., Donnelly R., González-Vázquez P., *et al.* (2017). Novel bilayer dissolving microneedle arrays with concentrated PLGA nano-microparticles for targeted intradermal delivery: Proof of concept. *J. Control. Release* 265: 93–101.

86 Justin R., Román S., Chen D., *et al.* (2015). Biodegradable and conductive chitosan–graphene quantum dot nanocomposite microneedles for delivery of both small and large molecular weight therapeutics. *RSC Adv.* 5 (64): 51934–51946.

87 Hansen S. and Lehr, C.-M. (2012). Nanoparticles for transcutaneous vaccination. *Microb. Biotechnol.* 5 (2): 156–167.
88 Zaric M., Lyubomska O., Touzelet O., *et al.* (2013). Skin dendritic cell targeting via microneedle arrays laden with antigen-encapsulated poly-D,L-lactide-co-glycolide nanoparticles induces efficient antitumor and antiviral immune responses. *ACS Nano* 7 (3): 2042–2055.
89 Storni T., Kündig T.M., Senti G. and Johansen P. (2005). Immunity in response to particulate antigen-delivery systems. *Adv. Drug Deliv. Rev.* 57 (3): 333–355.
90 Gutierro I., Hernandez R.M., Igartua M., *et al.* (2002). Size dependent immune response after subcutaneous, oral and intranasal administration of BSA loaded nanospheres. *Vaccine* 21 (1-2): 67–77.
91 Jaganathan K.S. and Vyas S.P. (2006). Strong systemic and mucosal immune responses to surface-modified PLGA microspheres containing recombinant hepatitis B antigen administered intranasally. *Vaccine* 24 (19): 4201–4211.
92 Mahapatro A. and Singh D.K. (2011). Biodegradable nanoparticles are excellent vehicle for site directed in-vivo delivery of drugs and vaccines. *J. Nanobiotechnol.* 9 (1): 55.
93 Xueting W., Ning W., Ning L., *et al.* (2016). Multifunctional particle-constituted microneedle arrays as cutaneous or mucosal vaccine adjuvant-delivery systems. *Hum. Vaccines Immunother.* 8 (12): 2075–2089.
94 Bal S.M., Slütter B., Jiskoot W. and Bouwstra J.A. (2011). Small is beautiful: *N*-trimethyl chitosan–ovalbumin conjugates for microneedle-based transcutaneous immunisation. *Vaccine* 29 (23): 4025–4032.
95 Wang N., Zhen Y., Jin Y., *et al.* (2017). Combining different types of multifunctional liposomes loaded with ammonium bicarbonate to fabricate microneedle arrays as a vagina mucosal vaccine adjuvant-dual delivery system (VADDS). *J. Control. Release* 246: 12–29.
96 Zhen Y., Wang N., Gao Z, *et al.* (2015). Multifunctional liposomes constituting microneedles induced robust systemic and mucosal immunoresponses against the loaded antigens via oral mucosal vaccination. *Vaccine* 33 (35): 4330–4340.
97 DeMuth P.C., Garcia-Beltran W.F., Ai-Ling M.L., *et al.* (2013). Composite dissolving microneedles for coordinated control of antigen and adjuvant delivery kinetics in transcutaneous vaccination. *Adv. Funct. Mater.* 23 (2): 161–172.
98 Zaric M., Lyubomska O., Poux C., *et al.* (2015). Dissolving microneedle delivery of nanoparticle-encapsulated antigen elicits efficient cross-priming and Th1 immune responses by murine Langerhans cells. *J. Invest. Dermatol.* 135 (2): 425–434.
99 Kumar A., Li X.R., Sandoval M.A., *et al.* (2011). Permeation of antigen protein-conjugated nanoparticles and live bacteria through microneedle-treated mouse skin. *Int. J. Nanomed.* 6: 1253–1264.
100 Khan K.H. (2013). DNA vaccines: roles against diseases. *Germs* 3 (1): 26–35.
101 Kumar A., Wonganan P., Sandoval M.A., *et al.* (2012). Microneedle-mediated transcutaneous immunization with plasmid DNA coated on cationic PLGA nanoparticles. *J. Control. Release* 163 (2): 230–239.
102 McCaffrey J., McCrudden C.M., Ali A.A., *et al.* (2016). Transcending epithelial and intracellular biological barriers. A prototype DNA delivery device. *J. Control. Release* 226: 238–247.

103 McCarthy H.O., McCaffrey J., McCrudden C.M., *et al.* (2014). Development and characterization of self-assembling nanoparticles using a bio-inspired amphipathic peptide for gene delivery. *J. Control. Release* 189 (1): 141–149.

104 Bennett R., Yakkundi A., McKeen H.D., *et al.* (2015). RALA-mediated delivery of FKBPL nucleic acid therapeutics. *Nanomedicine* 10 (19): 2989–3001.

105 Yang, H.W., Ye, L., Guo, X.D., *et al.* (2017). Ebola vaccination using a DNA vaccine coated on PLGA- PLL/γPGA nanoparticles administered using a microneedle patch. *Adv. Healthc. Mater.* 6 (1).

106 Kim N.W., Lee M.S., Kim K.R., *et al.* (2014). Polyplex-releasing microneedles for enhanced cutaneous delivery of DNA vaccine. *J. Control. Release* 179: 11–17.

107 Hu Y., Xu B., Xu J., *et al.* (2015). Microneedle-assisted dendritic cell-targeted nanoparticles for transcutaneous DNA immunization. *Polym. Chem.* 6 (3): 373–379.

108 Kim C.S., Wilder-Smith P., Ahn Y.C., *et al.* (2009). Enhanced detection of early-stage oral cancer in vivo by optical coherence tomography using multimodal delivery of gold nanoparticles. *J. Biomed. Opt.* 14 (3): 034008.

109 Kim C.S., Ahn Y.C., Wilder-Smith P., *et al.* (2010). Efficient and facile delivery of gold nanoparticles in vivo using dissolvable microneedles for contrast-enhanced optical coherence tomography. *Biomed. Opt. Express* 1 (1): 106–113.

110 Gittard S.D., Miller P.R., Boehm R.D., *et al.* (2011). Multiphoton microscopy of transdermal quantum dot delivery using two photon polymerization-fabricated polymer microneedles. *Faraday Discuss.* 149 (1): 171–185.

111 Doraiswamy A., Ovsianikov A., Gittard S.D., *et al.* (2010). Fabrication of microneedles using two photon polymerization for transdermal delivery of nanomaterials. *J. Nanosci. Nanotechnol.* 10 (10): 6305–6312.

112 Raja W.K., Maccorkle S., Diwan I.M., *et al.* (2013). Transdermal delivery devices: fabrication, mechanics and drug release from silk. *Small* 9 (21): 3704–3713.

113 Yan L., Raphael A.P., Zhu X., *et al.* (2014). Nanocomposite-strengthened dissolving microneedles for improved transdermal delivery to human skin. *Adv. Healthc. Mater.* 3 (4): 555–564.

114 Kim M., Jung B. and Park J.H. (2012). Hydrogel swelling as a trigger to release biodegradable polymer microneedles in skin. *Biomaterials* 33 (2): 668–678.

115 Chen M.C., Ling M.H., Wang K.W., *et al.* (2015). Near-infrared light-responsive composite microneedles for on-demand transdermal drug delivery. *Biomacromolecules* 16 (5): 1598–1607.

116 Chen M.C., Wang K.W., Chen D.H., *et al.* (2015). Remotely triggered release of small molecules from LaB6@SiO2-loaded polycaprolactone microneedles. *Acta Biomater.* 13: 344–353.

7

Minimally-invasive Patient Monitoring and Diagnosis Using Microneedles

Aaron J. Courtenay, Marco T.A. Abbate, Maelíosa T.C. McCrudden and Ryan F. Donnelly

School of Pharmacy, Queen's University Belfast, Belfast BT9 7BL, UK

7.1 Introduction

7.1.1 What is Patient Monitoring?

Therapeutic monitoring in patients refers to a system of analyses aimed at the detection and quantification of drugs or endogenous markers primarily from serum or plasma matrices. Therapeutic monitoring is performed as a matter of routine over the course of continued treatment, in hospital settings, for example, or at the initial stages of a patient presenting in the healthcare facility. Clinically, therapeutic monitoring in patients is used extensively as a rapid diagnostic tool for many disease states and allows for the optimisation of drug therapies.

7.1.2 Why is Patient Monitoring Useful?

Throughout clinical settings, for example in hospitals and other community clinics such as GP surgeries and pharmacies, therapeutic monitoring plays a significant role in assisting prescribers and clinicians to provide a high standard of care to patients with a number of specific disease states [1]. Therapeutic drug monitoring can be used to diagnose acute and chronic conditions. Pathophysiological events taking place within the body can be monitored, often in real time. One such example is in instances of acute myocardial infarction. In this example, troponin levels, as determined by constant monitoring, can reflect the nature and significance of the blockage within the myocardial tissue. Similarly, in the event of a stroke, patient monitoring can shed light on the extent of tissue damage caused. This vital information directly influences clinical decision making. Chronic illnesses can also be monitored, most notably in cases of renal or hepatic insufficiency [2]. From a pharmacological standpoint, this information is critical when determining medication doses. If a patient has a reduced ability to metabolise or eliminate a particular medicine, they may be more susceptible to increased side-effects or ultimately fatal toxicity as a result of medication build-up within their system.

In many instances, patient monitoring is used for its ability to detect and quantify drugs with narrow therapeutic windows, providing information for the clinical team on occasions of suboptimal dosing or alternatively, toxicity. Drugs with a narrow therapeutic window include warfarin, lithium and some antibiotic therapies such as gentamicin.

Microneedles for Drug and Vaccine Delivery and Patient Monitoring, First Edition.
Edited by Ryan F. Donnelly, Thakur Raghu Raj Singh, Eneko Larrañeta, and Maelíosa T.C. McCrudden.

The plasma levels of these drugs are measured from whole blood samples and in the case of lithium or warfarin must be measured routinely for the duration of this particular therapy. Furthermore, the measurement of plasma concentrations of particular drugs can be used to monitor adherence to prescribed regimens. This can be a useful tool when assessing patients who are being treated for a long-term condition, and is particularly evident in patients suffering from HIV/AIDS, *Mycobacterium tuberculosis* infections and leukaemia. Similarly, plasma monitoring can provide insights into those patients suffering substance abuse or addiction, highlighting periods of relapse or extended "clean streaks." This is used in a clinical setting for therapeutic purposes but is also used by police to detect those driving or operating machinery when impaired by alcohol or an illicit substance. It is worth noting that patient monitoring may not simply refer to whole blood samples but can refer to alternative sample matrices such as stool or sputum samples. Non-invasive methods of acquiring patient samples, such as breath testing, are used routinely in a community setting to monitor patient carbon monoxide levels in nationwide smoking cessation campaigns.

Patient monitoring can also highlight pharmacokinetic and pharmacodynamic variations. Instances where patients show no obvious clinical response to medication can be as a result of pharmacodynamics variability, or how the drug interacts with the body. Here patient monitoring can play an important role. Pharmacokinetic variation between patients is common and as such each person will react differently to a drug substance, in terms of absorption, distribution, metabolism and elimination. As such, treatments often need to be individualised to ensure optimum treatment is provided. Two sub-groups of patients that often require significant dose adjustment are older people and neonates or premature neonates. These particular groups often have pharmacokinetic and pharmacodynamic profiles that result in under- or over-dosing. Up- and down-regulated metabolic pathways, protein binding, altered tissue volumes and subsequent apparent volumes of distribution can dramatically impact on circulating therapeutic levels and so patient monitoring becomes a vital aspect of patient care.

7.1.3 Limitations and Challenges of Therapeutic Monitoring

Although a variety of different sample types can be used for therapeutic monitoring in patients, including sputum, stool and breath samples, as is the case in *Helicobacter pylori* testing, the overwhelming majority of drug molecules and endogenous substances are detected through analysis of whole blood samples. This method of drug and substance detection is highly accurate, with clinically validated procedures in place for the handling of this type of patient sample; however, there are a number of drawbacks associated with this practice. The sample handling of whole blood is extremely costly and time consuming. Analytical equipment capable of detecting and quantifying drug substances from blood samples, such as high-performance liquid chromatography (HPLC) systems, cost in the region of £30 000. Highly skilled and appropriately trained staff are needed to operate such technology and the maintenance costs, including consumables, can approach thousands of pounds per year. In developed countries this type of system may be achievable, however, clearly such costs will not easily be translated to developing, low- and middle-income countries. Accompanying the expensive equipment, are highly complex and technically demanding sample extraction methods that must be carried out prior to sample analysis. Again, this adds to the cost and length of time required to provide an accurate read-out of individual samples.

In order to obtain the whole blood samples, the traditional approach of using a hypodermic needle and syringe is often utilised. It must be noted, however, that approximately 10% of the population suffer from needle phobias [3, 4]. One early review describes how, although varying degrees of anxiety are associated with needle phobias, in severe cases the significant drop in blood pressure as a result of vasovagal shock reflex can cause death [3]. The use of hypodermic needles has further drawbacks, most notably the increased risk of infection and transmission of blood borne diseases. Hepatitis C virus, hepatitis B virus and HIV contaminated sharps result in 3 million healthcare workers being injured annually [5]. Further considerations for police and law enforcement teams include the ability to gain blood samples in a timely manner. People detained for reasons of public safety or who have committed criminal acts and are thought to be under the influence of alcohol or other illicit substances are usually required to provide a blood sample for analysis. In the case of whole blood samples, suitably trained medical professionals are required to be present in order to perform the procedure and this is not always feasible.

Older patients are often concurrently taking multiple medicines to manage comorbidities and so monitoring these patients can require multiple blood samples to be taken. With repeated assault of the venous system and as the body ages, venous access can become more difficult, thus making the process of obtaining a blood sample more problematic [6]. A further complication encountered primarily in neonatal patients is the need to practice "minimal handling." Each time a medical intervention is carried out, there are a number of thermal and energetic losses to the patient. The need to keep infection risks low and to minimise the impact of monitoring results in compromises in the number and volume of blood samples that can be taken from a new-born baby [7].

These complex issues culminate in therapeutic monitoring not being used to its full potential. As a result of this, some drug substances in certain groups of patients, for example in the very young or very old, are poorly understood. Kinetic profiles have not been fully elucidated and, as such, monitoring in some patients is not conducted as often as it should be [8]. Disease diagnosis, particularly of infectious diseases where testing is validated and available, is a real concern in the developing world. Many low- and middle-income countries do not yet have the infrastructure required to routinely carry out such testing, yet these countries traditionally have a higher prevalence of infectious diseases. This paradox highlights the need for the development of minimally invasive and, ideally, non-invasive techniques that could permit rapid and repeated monitoring options for patients. Monitoring methodologies that do not require blood samples to be collected and which ultimately do not require the use of hypodermic needles at all may prove more acceptable to patients. Non- or minimally-invasive methods of patient monitoring could also improve disease control, sub-optimal drug dosing and provide information regarding prescribed regimen adherence, in a variety of different patient groups.

7.2 Sampling Techniques

7.2.1 Minimally and Non-invasive Sample Extraction

A small number of minimally and non-invasive sample extraction techniques have been developed in recent years. These methods primarily involve the extraction of interstitial fluid (ISF). It is important to note that ISF can, in many cases, be used as a substitute for whole blood samples as drug substance concentrations in whole blood can be accurately reflected in ISF samples [9, 10]. From a pharmacological standpoint, compounds that are free and unbound to plasma proteins are considered pharmacologically active. It has been reported that tissue concentrations can be more predictive of clinical response when compared with total drug concentrations when measured from plasma (including bound and free drug) [9, 10].

In order to access ISF, a suitable method to overcome the significant barrier properties afforded by the skin, and in particular the outer *stratum corneum* (SC) layer is required. This layer is primarily composed of dead, highly keratinised cells and provides highly efficient protection from the loss of body fluid. The need to maintain homeostasis within the skin is high and as such passive methods reliant on concentration gradients do not yield significant volumes of ISF. Two methods for ISF extraction from skin have been detailed in the literature. Firstly, clinical microdialysis has been described as minimally invasive. This method involves the insertion of a microcatheter probe into the skin whereby the principles of conventional dialysis result in the extraction of analytes. The method of analyte extraction is also associated with a number of challenges. Similar to intradermal injection, the positioning of the microcatheter is problematic and can easily be inserted incorrectly, even by a highly trained professional [10]. Secondly, reverse iontophoresis, has been described as non-invasive. This method relies on a small electric current applied to the skin via two small probes, which results in the extraction of neutral polar molecules and cations from skin ISF [9]. A schematic representation of iontophoretic analyte extraction from the skin is shown in Figure 7.1.

Reverse iontophoresis also requires highly trained personnel and costly iontophoretic equipment. In addition, as the skin itself carries an overall net negative charge, anionic species cannot be extracted in large quantities using this method. It has also been shown that patient perspiration can alter the accuracy of this method. These methods, although showing some promise, still require a blood sample to be collected in order to calibrate the respective clinical outputs. In order to overcome these issues a novel method for ISF extraction could significantly improve clinical monitoring capabilities in patients [8].

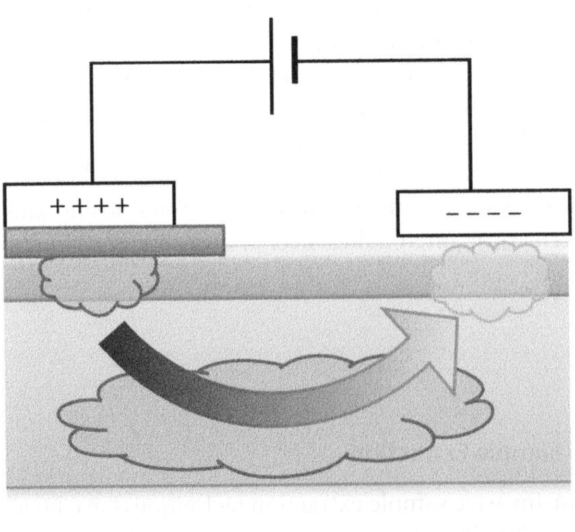

Figure 7.1 Schematic representation of an iontophoretic device for analyte extraction. Adapted from [11] Watkinson A.C., Kearney M.-C., Quinn H.L., *et al.* (2016). Future of the transdermal drug delivery market – have we barely touched the surface? *Expert Opin. Drug Delivery* 13 (4): 523–532.

7.2.2 Microneedles and Fluid Sampling Technology

Microneedle (MN) technology has recently emerged as a means to circumvent the barrier properties of the skin and access ISF. MN modalities vary significantly across research groups and company prototypes, with pyramidal, conical and cylindrical geometries being the most popular. Needle heights also vary considerably, between 50 and 900 μm, and needle densities of up to 100 MN/cm^2 have been reported. MN devices are, by definition, micron-scale structures, often fabricated from metal, silicon or polymeric materials, which penetrate the *SC* and have traditionally been used to deliver therapeutic compounds transdermally. Although there is a significant wealth of literature outlining the ability of MN technology to enhance transdermal delivery, over the past five years MN technology has also been used to access ISF and to act as a minimally invasive method of patient monitoring. MN technology has been shown to be favoured by patients, both young and old, for a number of reasons [12–14]. This is due to the fact that the technology does not cause pain (as the needles are too short to stimulate dermal nerve tissues) and generally the technology is considered convenient to use (both in terms of self-application and the potential for reduced dosing frequency). MN technology has shown promise in the ability to access ISF in a painless manner but also in the ability to manipulate the technology to access whole blood samples. It has been suggested that the technology could be modified through changes in geometry, needle length and material choice to switch between accessing ISF and whole blood [8]. As an example, consider the effect of MN penetration depth into the skin: it has been shown that depths of as little as 50–150 μm provide access to ISF sampling, whereas deeper penetration of up to 1000–2000 μm could facilitate whole blood sampling [15–17].

7.3 Microneedle Fluid Extraction Device Technical Considerations

In order for MN technology to provide a realistic alternative to current clinical practice, a number of physical, mechanical and chemical parameters must be considered. Manufacture of MN patches/devices constituting micron-scale needles that do not fracture upon application and that can successfully penetrate the *SC*, thus facilitating access to ISF or whole blood, can now be achieved routinely. The following section outlines some of the most basic principles governing the success or failure of an MN patch/device for ISF or whole blood sampling.

7.3.1 Mechanical Parameters

One of the most fundamental concepts involved in successful MN design relates to the mechanical strength of the MN structures. Are the MNs mechanically strong enough to penetrate the outer *SC* layer to provide sufficient access to the ISF or whole blood? This is a critical factor that must be considered from the initial fabrication stages. The MN tips need to be sharp and physically robust enough to overcome the resistive properties of the skin. MNs exhibiting properties such as a high resistance to buckling and fracturing are essential for any such MN product [18]. With respect to swelling or hydrogel-forming

MN arrays, these are polymeric networks that are hard in the dry state and swell upon fluid uptake. As the hydrogel system swells, continuous change in the physical properties of the hydrogel results from the uptake of fluid. Needles become less solid in nature and more gel-like. In order for hydrogel systems to be removed effectively from the skin, such materials need to maintain enough mechanical strength to allow the needle tips to remain attached to the MN array bed [19]. Solid MNs made from metal or silicon, particularly hollow MNs, often have sufficient mechanical strength to facilitate piercing of the skin, however an important point to note is that many materials previously used for solid MN preparation are not Food and Drug Administration (FDA) approved biocompatible materials, and fracture and deposition of such materials, such as silicon, have been listed as a potential concern within the literature [20].

7.3.2 Fluidics

In order to effectively access ISF or whole blood, there is a direct need to consider the forces involved in the movement of fluid through, for example, hollow MN devices. It has been reported that fluid flow rates vary between 1 and 100 μl/h [15]. A number of MN device prototypes use capillary action to stimulate fluid collection, with needle designs often taking inspiration from nature [17, 21, 22]. Although these needle shapes, such as snake-fang and volcano designs, can successfully penetrate the skin, the main drawback to hollow MNs remains the risk of breakage in and blockage of the skin. Regulatory bodies have already expressed concerns over the risk of materials such as silicon or metal fragments being left behind in the skin layers following removal of MN devices [23]. A number of research groups have expressed the desire for a higher degree of control when it comes to fluid collection, stating that passive diffusion alone may not provide adequate fluid collection at a rate acceptable to the end user or of a volume that would allow accurate detection of the analytes of interest. In order to overcome this potential limitation, the addition of a vacuum has been described, allowing aided withdrawal of ISF and whole blood – somewhat akin to a miniaturised hypodermic needle [24]. A number of alternative methods of fluid extraction have been identified. One particular example involved the use of poly(methyl methacrylate) to manufacture an MN system that used quadruplet grooves for blood storage following capillary force extraction [25]. Although this group was able to achieve suitable insertion, they were not able to demonstrate suitable fluid movement along the quadruplet grooves of the device. Mukerjee *et al.*, in 2004, were one of the first groups reporting a hollow/solid MN combined system fabricated from silicon. This design relied on the fluid flow concept for extraction of ISF, with specific care taken to design needle tips such that the risk of blockage by skin coring was reduced [17]. Confocal microscopy was used to image the insertion of this system *in vivo*, however some of the needle tips fractured and remained in the skin. This particular report highlights extraction of ISF from human ear lobes through MN induced channels, approximately 80 μm in width.

7.4 Microneedle Innovations

A number of research groups have explored the possibility of MN technology as a method of sample acquisition, possibility due to the many advantage they offer in terms of patient acceptability and ease-of-use. Through the use of MN technology these researchers have taken significant strides forward in developing novel systems for ISF and whole blood sampling in conjunction with methods of specific analyte detection. A large portion of the available literature describes systems involved in the detection and quantification of glucose. As such, it has been deemed practical to highlight this work specifically, whilst understanding that many of the principles involved in the studies described are applicable to the detection and quantitation of other analytes of interest.

7.4.1 Glucose Monitoring

Specific analyte monitoring has grown in popularity with many publications focusing on minimally and non-invasive glucose monitoring. Testing of blood glucose concentration is commonplace in patients suffering from diabetes. Consequently, achieving a method for assessing blood glucose concentrations could have a significant positive impact on the day-to-day lives of patients and provide an alternate income stream for pharmaceutical companies. The intense interest from academic and industrial research communities in this area is not surprising, considering the blood glucose testing market is estimated to exceed US$12 billion by 2020 [26]. Although the majority of methods used in MN monitoring have not dramatically changed since their first development, the field has matured, now striving to provide highly accurate and specific systems for patient monitoring. A number of novel techniques have been developed in order to access blood and ISF glucose and these will now be discussed in more detail.

One of the first devices pertaining to ISF glucose monitoring, reported in patent literature, refers to a reverse iontophoresis device. Gartstein *et al.* patented a reverse iontophoresis device based on an array of hollow MNs, a reservoir, electrodes (both anode and cathode), an electrical power source and a control system including a central processing unit and a visual indicator. A third electrode was designed to act as a bio-electrochemical sensor [27]. The authors describe a system whereby the reservoir could contain a hydrogel loaded with glucose oxidase for detection and quantification of glucose using optical detection methods. Glass MN have been used in conjunction with vibrational techniques allowing for glucose monitoring following successful skin penetration [28]. In this experimental set-up a vacuum was induced for 5–10 min, allowing volumes of 1–10 μl ISF to be collected from a 1 cm^2 area. As with many MN products to date, mild erythema was reported and when subjects were asked they reported the procedure was "relatively painless." It should be noted that transient mild erythema is not necessarily considered a limiting factor, however, if a device were to cause sustained or prolonged moderate or severe erythema, this would be unacceptable. This study reported that the concentrations of glucose found in the ISF were proportional to whole blood sample concentrations. Again, with this particular set-up, a whole blood sample was required for calibration purposes.

These glucose extraction methods are somewhat analogous to strategies employed in the GlucoWatch Biographer. The impressive nature of this work is revealed in the group's ability to demonstrate detection of the dynamic blood glucose levels without any significant lag in response, an ideal for minimally invasive patient monitoring [28, 29]. This research group further explored the accuracy of electrochemical monitors for glucose measurement, comparing them with GC/MS (gas chromatography/mass spectrometry). Whole blood and ISF yielded a concentration bias towards ISF that should be acknowledged [30]. Also noteworthy is the fact that the GlucoWatch Biographer has since failed to be accepted into the market with reports of inconvenient use and a number of false readings, resulting in a response of no confidence in the product. As such, the manufacturers have, since 2012, halted production and stopped further development of the product.

Since this initial technology, there have been a number of publications where minimally invasive monitoring using MN technology has been implemented to detect and quantify glucose in ISF and whole blood. In 2011 poly(carbonate) MN arrays were used alongside a sponge to collect ISF. This style of device relies on passive diffusion and osmotic pressure gradients to draw out ISF, however, this presented its own set of limitations. Although glucose concentrations were shown to accurately represent blood concentrations, this was only tested in healthy volunteers [31]. This concept was brought forward by Sakaguchi *et al.* who assessed the technology through a process of comparative validations with oral glucose tolerance tests in healthy volunteers and diabetic patients [32]. The two-step process involved MNs being applied to the arm and removed, followed by the application of a poly(vinyl alcohol) hydrogel, which had been previously immersed in a potassium chloride solution, at the site of the MN application. This resulted in the extraction of ISF such that the glucose and sodium concentrations could be measured 2 h post glucose ingestion. Plasma concentrations of these analytes were taken at 30 min intervals and it was demonstrated that ISF glucose concentrations closely matched true blood values. Although the two step application process is a notable limitation of this technology, the fact that the process was a good indicator of post-prandial plasma glucose levels signified the possibility of this technology to significantly benefit patients.

In a 2014 study, dissolving MNs, at a density of 300 MN/cm^2, were prepared from sodium chondroitin sulfate and dextran and used to monitor dermal ISF of rats. In this instance the dissolving MNs were applied to the rats and removed, facilitating the formation of pores in the skin. Following this initial application step, a wet unwoven cloth containing 10–30 μl of water was placed on the site of application. This research group showed that increasing the water content and contact time of the unwoven cloth with the site of application improved glucose extraction [33]. Again, although this work demonstrates the capacity for MNs to be used in this way, a two-step process may not be accepted by end users unless this has been streamlined in some way.

Valdés-Ramírez *et al.* have shown the potential for glucose/oxygen MN biofuel cells as biosensors for subdermal glucose monitoring. Here carbon-paste bioanode and cathodes were integrated into hollow MN arrays. Based on a system of glucose dehydrogenase, these MN arrays are able to harvest sustainable power signals proportional to the glucose fuel concentration in the skin. Importantly, these signals were independent of the background electroactive signals that usually affect amperometric sensors. The system is described as "on-body" and "pain-free" allowing for continuous transdermal monitoring of glucose [34]. The authors state that further studies focusing on the effects of proteins and biofouling need to be fully assessed to ensure this device can provide continued efficacy. Many implantable devices suffer from the limitation of colonisation and bio-fouling, and it is anticipated that long-term use of such a device may lead to increased risk of infection, or a progressive deterioration in function.

Hydrogel-forming MN arrays provide a unique mechanism to obtain ISF or whole blood samples. MN arrays fabricated from polymeric materials that have been crosslinked are hard in the dry state, allowing them to successfully penetrate the *SC*. Following insertion into the skin, the MN tips come into contact with the ISF, which then begins to diffuse into the MN array. The resultant swollen hydrogel has been used extensively for drug delivery purposes, however, recently it has been exploited for its ability to absorb ISF. In a publication by Caffarel-Salvador *et al.* in 2015 [35], the Donnelly research group showed detection, via uptake into swollen hydrogel MNs, of a number of important biological analytes and drugs *in vitro* and *in vivo*. The extraction process involved the use of a crosslinked matrix of poly(methyl vinyl ether-*co*-maleic acid) and poly(ethylene glycol) to form the MN arrays. Following a wear-time of up to 1 h, the MN arrays were removed and immersed in a wash buffer solution and the analytes of interest were analysed using validated HPLC methods (Figure 7.2).

The group assessed ISF glucose concentrations in human volunteers following administration of 75 g of oral glucose powder. The blood glucose levels were cross referenced with whole blood glucose measurements taken using the traditional finger prick lancet and glucometer combination [35]. This novel polymeric methodology provides self-disabling MN technology that allows for convenient wear, ease-of-use for patients and accurate detection of specific analytes in ISF.

In the same year, an integrated system of hollow MNs in combination with a paper-based sensor was described. The one-touch-activated blood multi-diagnostic system required finger pressure to activate, with glucose and cholesterol concentrations then successfully evaluated from whole blood. The miniaturised all-in-one-point of care diagnostic system aimed to provide a low cost and simplified operation method. An outline of the device is shown in Figure 7.3.

In this study, hollow MNs manufactured from nickel and coated in parylene were used, with an outer diameter of 120 μm and needle bevel angle of 15°. The volume of

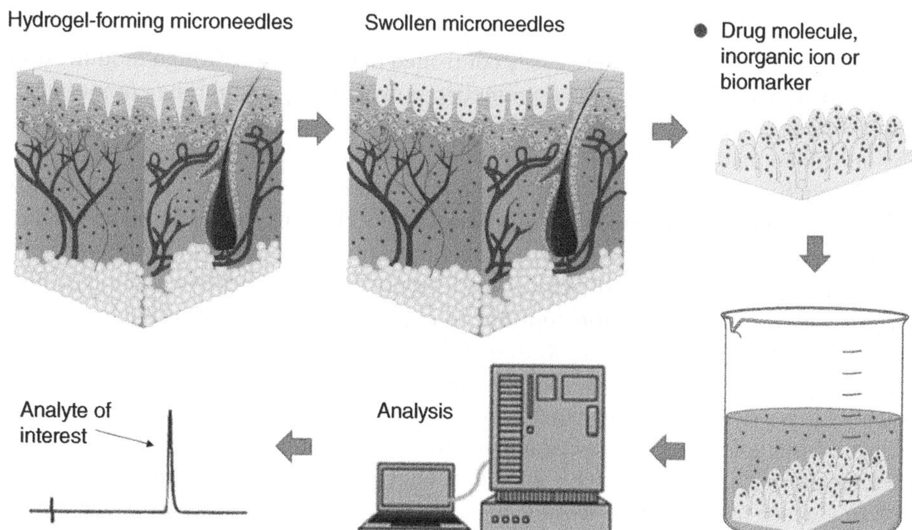

Figure 7.2 Schematic representation of hydrogel-forming MN arrays used for therapeutic drug monitoring. Reproduced with permission from [35] Caffarel-Salvador E., Brady A.J., Eltayib E., *et al.* (2015). Hydrogel-forming microneedle arrays allow detection of drugs and glucose *in vivo*: Potential for use in diagnosis and therapeutic drug monitoring. *PLoS One*. 10 (12): 1–21.

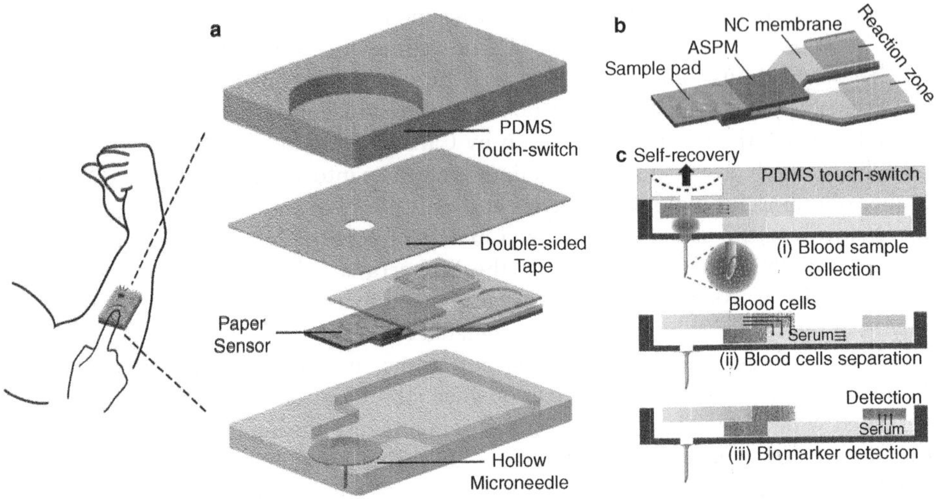

Figure 7.3 Schematic representation of the one-touch-activated blood multi-diagnostic system (OBMS). (a) Structure of the OBMS. (b) Diagram of the paper-based multiplex sensor consisting of the sample pad, asymmetric poly(sulfone) membrane (ASPM), nitrocellulose (NC) membrane and reaction zones. (c) Operating principle of the paper-based sensor for blood sample multi-diagnosis: (i) blood sample collection, removing the applied finger force causes the self-recovery of the poly(dimethylsiloxane) (PDMS) touch-switch, inducing a negative pressure to collect a blood sample and induce its absorption by the sample pad; (ii) blood cell separation, filtering of the blood cells based on size exclusion by the ASPM; (iii) biomarker detection, detection of the multiplex biomarkers by colorimetric assay. Reproduced with permission from [36] Li C.G., Joung H.-A., Noh H., *et al.* (2015). One-touch-activated blood multidiagnostic system using a minimally invasive hollow microneedle integrated with a paper-based sensor. *Lab Chip* 15 (16): 3286–3292.

blood extracted and flow rate were easily controlled in this device through modification of the poly(dimethylsiloxane) and asymmetric poly(sulfone) membrane components. A rabbit model was used for this initial study but the authors conceded that real progression will be made when their device can be validated in human clinical trials [36]. The one-touch-activated blood multi-diagnostic system described by Li *et al.* was also used to detect cholesterol concentrations in whole blood. Again, paper based enzymatic reaction pads were used to assess analyte concentration. This system allowed for both glucose and cholesterol to be detected within 5 min of device application to the skin [36].

As methods for MN fabrication continue to improve, the need to develop lower cost devices has become more important. In one study, a low power consuming and easy to use continuous glucose monitoring system was developed [37]. The research group that pioneered this work claimed that such a device is of low cost, however with MNs manufactured from gold, mass production using this methodology will likely incur significant financial costs that would have knock-on effects in high retail pricing. The system involves glucose oxidase, immobilised on self-assembled monolayers with mercaptopropionic acid onto gold MNs. Cyclical voltametric studies showed that the transdermal sensing unit could detect glucose between 30 and 400 mg/dl in ISF. Several challenges will be faced by this technology prior to full commercialisation, such as long-term stability of the detection enzyme, oxygen deficiency, calibration and any potential inflammatory response. The low power consumption of this particular device is a design feature of increasing importance. As point-of-care sensing devices become more complex and probably, more digitised, low power consumption will provide the end user with a longer battery life and, ultimately, less disruption to the monitoring process and increased convenience for the user [37]. Incorporation of wireless technology into this device allows the potential for full integration into a mobile network of data, which could provide insight for future therapies and prescribing decisions. This "connected" technology introduces concerns of confidentiality and patient safety however, an area familiar to the digital manufacturing industry. This highlights the need for an integrated approach to this research, drawing on developments in cyber security and expertise from digital technology experts to provide crucial input into design and manufacture so that monitoring devices are safe for patient use and capable of withstanding modern day cyber security pressures.

Further work carried out in 2015 demonstrated that the combination of MN designs, such as solid state MN arrays, with a small quantity of hydrogel material allowed for the detection of ISF and subsequent glucose and sodium ion detection within 1 min [38]. This particular all-solid-state sodium ion-selective electrode system is in its infancy with initial *in vitro* tests proving promising. Additional *in vivo* testing is planned, however, for successful commercialisation to be achieved it is anticipated that extraction processes lasting "several hours" will not be acceptable. Such lengthy processes render the final glucose measurements redundant, yet in this work the authors propose the measurement of glucose directly within the hydrogel without the need to extract in an immersion fluid – unfortunately, the authors do not provide insight as to how this could be done [38].

Continuous tissue monitoring systems involving MN array electrodes manufactured from epoxy-based negative photoresist have been described [39]. Similar to the previously described research by Hwa *et al.* in 2015 [37], this biosensor relies on MNs functionalised with glucose oxidase. Aluminium master templates were manufactured using electrical discharge machining, from which poly(dimethoxy siloxane) moulds were created. Using vacuum and centrifugation techniques, the group created epoxy MN arrays at a rate of >80 per week. One main limitation of this work appears to be regarding scalability of the manufacturing technology however. Manufacturing collaborators will undoubtedly be required to industrialise this process. That notwithstanding, the research group were able to demonstrate the ability to detect changing glucose concentrations in the range of 0–30 mM, with no significant cross-talk between the MN array electrodes. The lack of cross-talk between arrays suggests the technology could be used for single analyte detection and more complex multiplex measurements. The MN electrodes were sterilised using gamma irradiation prior to use *in vivo* in humans. The group reported a slight reduction in performance post-sterilisation however, stating that the device "performed adequately post sterilisation."

The need for sterility is an issue that has been widely discussed in the area of drug delivery via MN devices. Despite this, regulatory authorities have yet to confirm the need for sterility of such products and devices [40]. As MN products move closer to licensure, regulatory hurdles such as the need for sterility or low bioburden manufacture simply must be defined. Only when this has occurred can these standards be fully considered and met [23]. The number of recent publications focussing on glucose detection and monitoring, using minimally invasive devices, serves to highlight that this is an attractive sector for researchers. If the recent developments can produce alternatives to finger prick/glucometer testing, they could, in time, provide significant income and revenue in sales but more importantly, they could also provide tremendous patient benefit.

7.5 Microneedle Innovations in Analyte Monitoring

7.5.1 Therapeutic Drug and Biomarker Detection

A number of drug compounds and biomarkers have been targeted within the literature in MN-mediated patient monitoring. A notable paper, from our own research team, describing *in vivo* monitoring of caffeine and theophylline was published in 2015 by Caffarel-Salvador *et al.* [35]. In this work, hydrogel-forming MN arrays were used to sample ISF and subsequently determine concentrations of the methylxanthine compounds, theophylline and caffeine. Theophylline and caffeine are often used clinically to alleviate neonatal apnoea and have narrow therapeutic windows. Therefore, it is important for clinicians to be able to rapidly detect and quantify circulating levels within the infant's body, ensuring toxicity and the subsequent manifestation of side-effects can be monitored. Unfortunately, in this study, we could not directly correlate ISF to blood concentrations of theophylline and caffeine. ISF concentrations of the analytes did, however, show trends indicative of blood concentrations. It was therefore suggested in the paper that, upon further development, this technology could be of significant benefit to neonatal and infant populations in critical care situations, removing the need for extensive blood sampling and reducing discomfort and pain to the patients.

Further to this work, the same research group have shown hydrogel-forming MN arrays to be useful in the detection of compounds with narrow therapeutic windows, such as lithium [41]. Treatment of bi-polar disorder in patients using lithium can be difficult, as the therapeutic concentrations in the body fall within a limited range. Also, serum half-life of lithium is variable, increasing on chronic administration. As such, regular lithium monitoring has to be carried out to avoid sub-optimal dosing and toxicity. Sprague-Dawley rats were used as an *in vivo* model for human subjects in this paper, and although differences in pharmacokinetic profiles were noted, the technology successfully extracted ISF. Subsequent analysis led to the determination of lithium ion concentrations in the rats. This work highlighted the potential for outpatient monitoring, with patients administering the MN patches themselves following instruction from a pharmacist and reading a Patient Information Leaflet (PIL), or viewing a website showing simple application instructions [41]. The concept of hydrogel-forming MN arrays has proven popular with research groups fabricating a variety of MN patches using poly(methyl vinyl ether-*co*-maleic acid) and poly(ethylene glycol). Romanyuk *et al.* described this exact process and proposed a simple method for analyte extraction from the MN arrays post-insertion, using a conventional laboratory microtube [42]. Figure 7.4 shows a schematic representation of the casting method of micromoulding fabrication used to manufacture these MN arrays. This particular method was first described by Donnelly and his group in 2011.

In this study, the group were able to clarify that if a low viscosity casting solution was used, then excess solution could be removed from the mould surface while leaving the mould cavities filled with casting solution. In contrast, if a high viscosity solution is used, then the process of removing the excess solution from the mould surface also removes

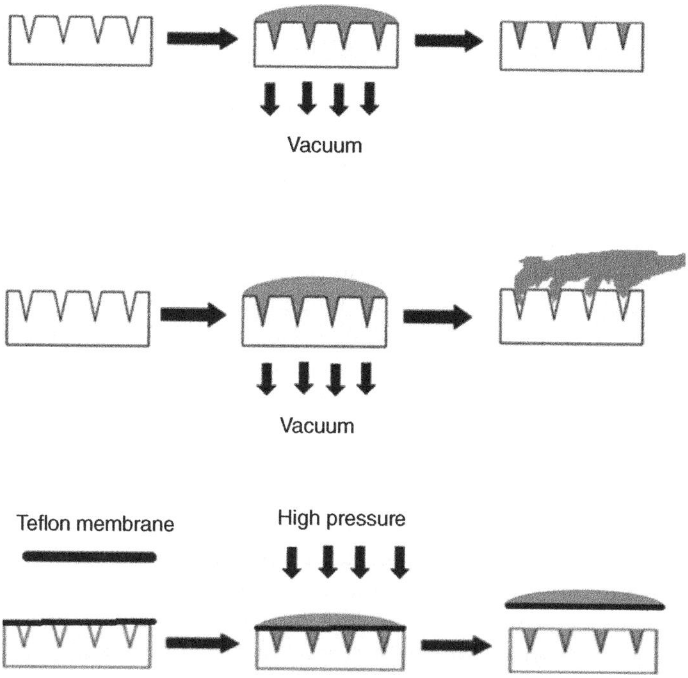

Figure 7.4 Schematic representation of the moulding of MN patches. In the fabrication process, a polymeric casting solution is cast onto a micromould, vacuum is applied to pull the solution into the cavities of the mould and then excess solution is removed from the mould surface using a razor blade. Reproduced with permission from Romanyuk *et al.* supporting information (2015) [43] Romanyuk A.V., Zvezdin V.N., Samant P., *et al.* (2015). Collection of analytes from microneedle patches. Supporting Information. *Sci. Rep.* 12: 1–9.

solution from the mould cavities. The addition of a hydrophobic (i.e. Teflon) mask with holes aligned with the mould cavities allowed the casting solution to remain in the mould cavities even after removing the excess solution from the mould surface. When casting was done with a highly viscous solution (15% poly(methyl vinyl ether-*co*-maleic acid) and 7.5% poly(ethylene glycol)) containing a pink dye (1 mM sulforhodamine), very little dye remained in the micromould cavities. This research group opted for pyramidal needle shapes with heights and widths of 600 μm and 300 μm, respectively. Application of the MNs to rat skin *in vivo* resulted in ISF fluid extraction into the hydrogel network. The MN array was then affixed to the inner lid of a conventional laboratory microtube, treated with an extraction fluid and centrifuged. This allowed the group to assess concentrations of the model analyte, sulforhodamine, providing a simple method of analyte extraction from the array post MN removal.

7.6 Microneedle Electrode Technology

7.6.1 Electro-biochemical Monitoring

Worth consideration is the fact that, although the vast majority of research to date has focused on sampling of ISF or whole blood for the purposes of analyte detection and quantification, a small number of publications have been published describing MNs as electrodes for monitoring of physiological electrical signals. Electrode–skin interface impedance, electromyography and electrocardiography are routinely monitored in clinical settings, providing healthcare professionals with detailed information regarding a patient's electro-physiological profile. To this end, Chen *et al.* fabricated MN arrays by initially casting an iron-epoxy resin and through the application of a magnetic field, they were able to produce sharp needle tips [44]. The MN tips were coated with titanium/gold and manufactured into an MN electrode. When compared with commercially available silver/silver chloride electrodes, the MN electrodes performed well, being able to detect the periodical electrocardiographic signal fluctuations in humans. The system was also sensitive enough to detect the electrical profile following rhythmical contraction of the bicep brachii muscle, without the need for prior skin preparation [44]. This paper highlights the possibility that MN technology could be used as a minimally invasive method of monitoring bio-electro signalling, with at least the same degree of accuracy as commercially available conventional electrodes. In 2015 Arai *et al.* demonstrated the capacity for silicon-based MN to be converted into electrodes allowing for the detection and monitoring of contact impedance with a view to electroencephalographic monitoring [45]. As an extension of MN electrodes for monitoring, Wang *et al.* 2017 designed a flexible parylene-based MN electrode array. Successful penetration and insertion studies demonstrated that the silicon MN electrodes could be worn and, further, the device detected competitive impedance densities, which ultimately correlated with traditional wet electrode electroencephalographic monitoring [46].

In one particular study, researchers have used hollow MN technology to allow detection of organophosphorus compounds. These nerve agents are some of the most toxic chemicals used in chemical warfare and impact significantly on soldiers, farmers and civilians in conflict areas [47]. The hollow MN arrays were packed with a carbon paste derived from a homogenous mixture of graphite powder and mineral oil. With a total of 3×3 MN arrays used in each case, two MNs were employed as working and counter electrodes and one was modified using Ag/AgCl ink as a reference standard [48]. Up till now, MN electrode bio-monitoring has relied on amperometric detection, however this system provides voltammetric transduction, affording increased selectivity. The low detection limits were achieved as a result of the ability of the system to discriminate between common electrochemical interferents, high stability and speed throughout the course of long ISF exposure. The "easy-to-operate" bio-sensor claims to cause no burden to the wearer and that future iterations will provide wireless transmission for ease of data processing.

As previously discussed, there is a significant need to provide monitoring for patients who have become dependent on substances such as illicit drugs and alcohol. Mishra *et al.*, in 2017, used an electrochemical detection system and combined this with modified hollow MNs fabricated using injection moulding from a master template [48]. The bore of each MN was filled with either platinum or silver wire to generate a solid MN with a tip containing a microcavity. Alcohol oxidase was immobilised onto the platinum wire microcavity using an electro-polymerising method and following interaction with alcohol provided a chrono-amperometric output. When used in an *ex vivo* mouse skin model the bio-sensor was able to detect alcohol from simulated ISF following MN application. Similar amperometric detection systems have been described previously, but have focused on nitric oxide (NO) detection following tumour development. MN coated with poly(3,4-ethylenedioxythiophene) and functionalised with hemin molecules were mounted on the end of an endoscopy scope and used to detect NO from cancer tumours, allowing both visualisation and initial sensing of colon cancer [49]. The idea of protein and bio-marker mapping has become a popular concept promising significant information on compounds present within a particular microenvironment. A series of monoplex and multiplex immunodiagnostic MN arrays were fabricated using poly(lactic acid) [50]. This system provided cytokine extraction from skin tissues and when coupled with densitometric computational analysis successfully captured specific antigens at the MN tips. Through the use of an MN coating map, each individual MN could specifically be mapped to detect a particular antigen, for example human tumour necrosis factor-α (TNF-α) and mouse interleukin-6 (IL-6). An additional study in this report highlighted the relative stability of the system for up to 30 days post-manufacture. Understandably, product lifetime will become a significant factor throughout the industrialisation process and so it is important to consider this from the initial stages of microfabrication.

7.7 Sampling and Analytical Systems Integration

7.7.1 Limitations and Challenges Associated with Systems Integration

System integration, whereby analyte detection or quantification happens on-site – or within the device – has in some instances streamlined the patient monitoring process. Rapid point-of-care testing kits provide unique advantages over traditional methods of monitoring, including rapid detection allowing for immediate feedback to the patient or clinician. Such systems, although promising significant benefits, are not without challenges. The literature states a number of concerns regarding sensor and MN spoilage due to contamination with competing endogenous biomolecules or even microbial contamination [15, 51]. Biofouling may only prove challenging with devices that have a long wear time of, for example, days or weeks. It is possible that devices designed for shorter wear times may not suffer significantly from deleterious biofouling events [52]. Disposable or single use sampling platforms may provide the answer to this particular issue, however, this may undoubtedly increase the cost of production. A notable advantage of these integrated detection systems relates to their ability to connect to treatment options. Consider a closed-loop drug delivery system, where ISF or equivalent fluid is continuously monitored (e.g. the case of glucose, discussed previously) by an MN device. Using microprocessors, the data received from the extraction stage could be suitable interpreted, an appropriate dose of treatment medicine (in this case insulin) could then be delivered using another controlled release method. In theory this process sounds simple, however, caution is needed with regards to the design of such a system. Numerous problems relating to electronic faults, errors in programming, physical and electronic damage, both accidental and malicious, and patient data protection need to be considered in detail ensuring patient safety is maintained [8].

The concept of combining "lab-on-chip" devices with MN technology may provide a useful method of continuous monitoring in patients without the need for complex dynamic chemical processing. Through a combination of pyramidal MNs (1000 μm in height and 500 μm base width) and a microfluidic chip, Miller *et al.* produced an interesting proof-of-concept patient monitoring device. Through integration of these two technologies, a series of microfluidic channels and electrode transducers were used to detect myoglobin (a skeletal muscle injury marker) and troponin (a cardiac injury marker) following immunoassay detection [53]. Next generation "on body" monitoring devices have shown significant commercial success, for example, Fitbit® and Apple Watch®. Although an MN based detection device is still some distance from commercial markets, there appears to be real promise for this technology to disrupt current markets.

A further example of highly specific and accurate integrated systems is demonstrated by a group using hollow MNs. Ranamukhaarachchi *et al.*, in 2016, used hollow MNs to extract extremely small volumes of ISF (<1 nl) with the inner lumen of each tip functionalised to capture drug candidates without the need for sample transfer (Figure 7.5) [54].

This MN-optofluidic biosensor follows an enzyme based assay process for the detection and quantitation of vancomycin, requiring only a diode laser and photodetector as external analytical equipment. The hollow MN were gold coated and subsequently functionalised with vancomycin-specific capture antibodies, facilitating detection limits of 100 nM concentration in volumes as low as 0.6 nl.

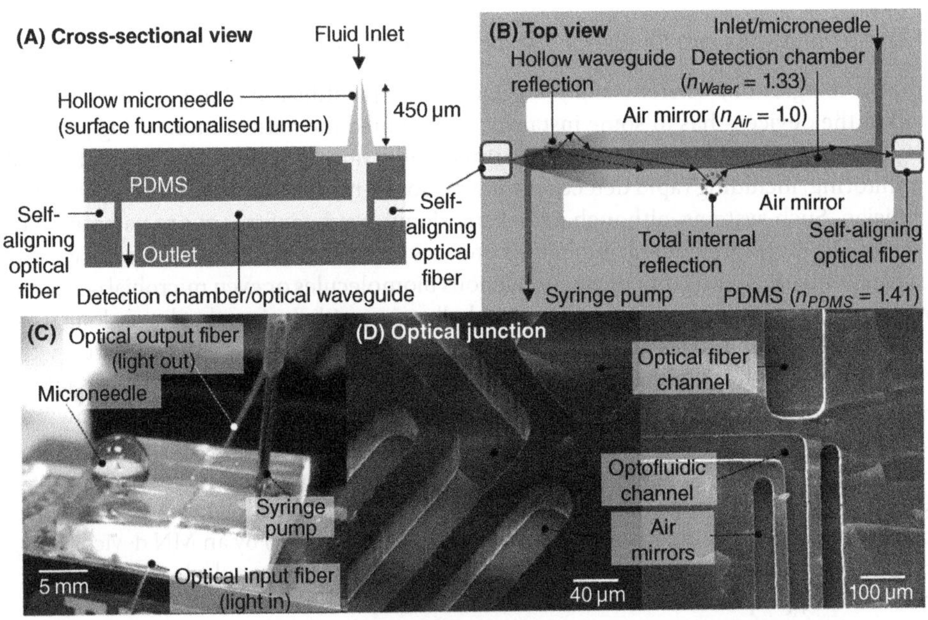

Figure 7.5 Design and images of the integrated MN-optofluidic biosensor. (A) Cross-sectional schematic view and (B) top view design of the devices. A surface-functionalised MN is integrated to a PDMS optofluidic device equipped with optical fibres. The mechanism of the optical waveguide equipped with air mirrors to guide the incident light from the input fibre through the TMB end-product in the detection chamber to the output fibre using total internal reflection and hollow waveguide reflection is shown in (B). (C) An image of the integrated MN-optofluidic device during the TMB assay; and (D) scanning electron micrographs of PDMS optofluidic devices at the junction of air mirrors, microfluidic channel carrying the TMB end-product and optical fibre self-alignment channels. Reproduced with permission from [54] Ranamukhaarachchi SA, Padeste C, Dübner M, Häfeli UO, Stoeber B, Cadarso VJ. (2016). Integrated hollow microneedle-optofluidic biosensor for therapeutic drug monitoring in sub-nanoliter volumes. *Sci Rep.* 6(1):29075.

Several other designs consisting of hollow or other suitable MNs for sampling patient's blood, lymph or ISF, in order to analyse the collected sample simultaneously using inbuilt sensing systems, have been proposed and patented. However, while a small number of studies demonstrating the proof of concept of using MNs for extraction of interstitial fluid or blood from the human volunteers without significant pain have been reported, a workable, fully integrated, device has not to date been demonstrated *in vitro* or *in vivo* or brought to the market [55, 56]. Indeed, according to the Banga Group, who recently reviewed such systems, success will depend on several factors such as: MN design and geometry, the capabilities of the sensing device, efficient integration of the two components into a single device, proper contact between the device and the skin and patient adherence factors [57]. For devices designed to sense the analyte immediately following collection of bodily fluids, the sensor would be a critical component. It would be beneficial if the sensing device had a reusable sensor, unless it was specifically designed for single use, that would allow rapid, accurate, analysis of the analyte(s) of interest. Sensors could be based upon a wide range of analytical technologies amenable to miniaturisation, including potentiometric, physiochemical, thermal, spectroscopic, calorimetric, optical, light scattering, gravimetric, amperometric, electronics or electrochemical systems [58].

7.8 Interstitial Fluid and Blood Sampling

7.8.1 Devices and Patents

Over the past 10 years, a number of devices and device concepts have been submitted for patent filing and review. This section provides a brief outline of some of the important patent filings and devices developed in MN technology, pertaining to fluid extraction and drug analysis.

Gartstein *et al.* patented a reverse iontophoresis device based on an array of hollow MNs, a reservoir, electrodes (both anode and cathode), an electrical power source and a control system including a central processing unit and a visual indicator. A third electrode was designed to act as a bio-electrochemical sensor [27]. Prausnitz *et al.* 2004 further claimed that such a device could be incorporated into a wrist band and conveniently worn by a patient for drug delivery, sampling of biological fluid or both [59]. The inventors further described how hollow MNs could be filled with a gel that has a sensing functionality associated with it. In an application for sensing, based on binding to a substrate or reaction mediated by an enzyme, the substrate or enzyme could be immobilised in the needle interior. Wave guides could also be incorporated into the MN device to direct light to a specific location, or for detection, for example, using means such as a pH-sensitive dye for colour evaluation. Similarly, light could be transmitted through the MNs for measurement of blood glucose based on infrared spectra. Alternatively, a colour change could be measured in the presence of immobilised glucose oxidase.

An alternative device was then described by Sherman *et al.* [60]. Instead of an electrical current extracting fluid from skin, it functioned by means of a user-activated diaphragm pump attached to an array of disposable hollow MNs. Electrical or optical methods were claimed as methods of detection. Alternatively, the MNs could be coated with an appropriate material, so that they themselves became the sensor.

Uniquely, this patent described methods by which a disposable MN array could be hygienically destroyed. Methods included heating/melting and chemical destruction. Ackley described how optical fibres could be placed in the central lumen of hollow MNs and used for delivery of light into tissue and subsequent sensing activities [61]. Whitson patented a device based on hollow MN arrays that puncture the skin and can, in contrast to other patents in this area, draw blood [62]. The concentration of a blood analyte of interest could then be determined directly, either calorimetrically or potentiometrically. The device consisted of a disposable MN patch with 200 hollow MNs, each 127 μm in height arranged in a 20×20 array, which was coupled to a test chamber. The MNs were designed to withdraw 1 μl of blood from the outermost dermal microcirculation by capillary action in a painless fashion. The sample would then be communicated to the test chamber containing a reagent which would, in turn, react with the blood analyte of interest to produce a detectable signal. Results obtained would be displayed on a screen. Since the blood would not come into contact with the patient's skin or the outside of the device, cross-contamination between patients or between the patient and clinician would be minimised.

Gonnelli patented a MN (single or array) membrane device consisting of hollow MNs (10–500 μm in height), the openings of which were covered by a membrane composed of an analyte-selective material [63]. Frazier and colleagues explain about "active" MNs with a hollow shaft having single or multiple outlet ports and integrated with a suitable detection device for real-time analysis of extracted biological samples [64]. Similarly, Cho describes a simple handheld "integrated" MN device for diagnostic application [65]. However, Cormier *et al.* patented a simple patch-based design comprising thin microblades for sampling body fluids with minimal tissue damage during insertion. The microblades, which could be solid or hollow, were 25–500 μm in length and arranged in arrays of 10–1000 blades/cm^2. Some microblades had additional barb structures to ensure better anchoring in the skin. These inventors also disclosed a self-sustained reverse iontophoresis-based sampling device [66].

Zimmermann *et al.* reported a continuous, self-calibrating, disposable and hollow MN-based glucose monitor. Hollow MNs in this system were integrated with a poly-Si dialysis membrane and a flow-through sensor, which was enzyme-based [67]. A clinical diagnostic system, for body fluid sampling, consisting of multiple stacks of disposable MN-based "functional plastic biochips" was developed by Ahn *et al.* [68]. Recently, the Kendall Group have shown that surface-modified MNs can be used for specific capture of intradermal biomarkers [69]. The group vaccinated mice with FluVax® by intramuscular injection and measured the resultant anti-FluVax IgG (AF-IgG) 21 days later. To do this, they grafted hetero-bifunctional PEG to gold-coated MNs. Surface modification was completed and AF-IgG capture proteins were grafted onto the PEG. Following insertion into mouse skin, MNs were washed to remove loosely bound species and then an appropriately labelled detection antibody was added to bind specifically to captured AF-IgG and then quantified fluorometrically. This technology certainly cuts down on the time required for analysis and obviates the requirement for blood samples. However, further development will undoubtedly be required to increase applicability and quantitative capabilities.

Although the majority of MN systems have been primarily designed for transdermal and intradermal drug delivery, a number of these systems involving hollow MNs have been proposed as potentially useful for fluid extraction from the body [70–75]. The

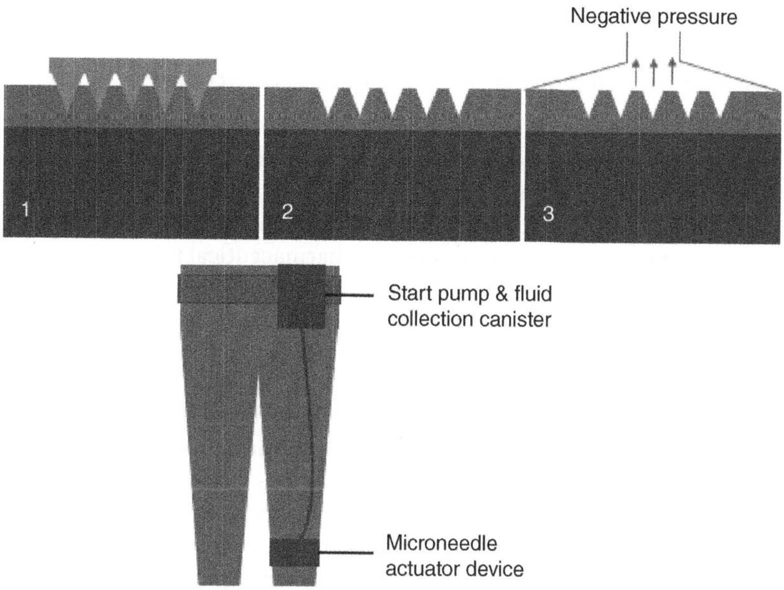

Figure 7.6 Schematic representation of (1) Renephra MN inserted into skin, (2) Renephra MN removed from skin yielding microchannels and (3) negative pressure applied removing excess ISF. The fluid is then collected in a canister of the Renephra device. Adapted from Dawidowska I., Taylor A., Ebah L. Renephra website (2017). http://www.renephra.com/technology.html (accessed 28 July 2017).

Renephra device (Figure 7.6) uses MN to penetrate the skin at oedematous sites, such as the ankles, and implements a vacuum to remove excess fluid, which is then stored in an absorbent material (www.renephra.com/technology.html (accessed 28 July, 2017)).

This company focuses on fluid extraction to reduce the osmotic and oncotic pressures exerted on the body during various disease states. As such the company tries to treat chronic oedema, claiming to have a potential impact for oncology patients and indeed patients in other disease areas. Such a device shares technical similarities to other MN products designed for fluid extraction and subsequent compound separation and analysis.

In 2014, Seventh Sense Biosystems first received CE mark approval for a Touch Activated Phlebotomy blood collection system – TAP™. The company continued, in 2015, to file three separate patents based on the TAP system. This small device adheres to the forearm and using a single push button action to draw whole blood, samples up to 100 μl. Seventh Sense Biosystems have submitted this for FDA approval, to allow them to begin selling their device, however, as yet this has not been realised (www.7sbio.com (accessed 28 July 2017). A start-up company based in Seattle, USA called Tasso Inc., have also launched a blood collection system. The HemoLink™ design incorporates microfluidic technology and relies primarily on vacuum suction to withdraw blood into a collection tube without the need for needles. The device has yet to be FDA approved and thus is not yet available for commercial use (www.tassoinc.com (accessed 28 July 2017). Despite the promise of these systems, neither product has reached the market or begun clinical trials.

7.9 Developments Moving Forwards

7.9.1 Industrialisation and Commercialisation: Hurdles to Overcome

Industrialisation of MN technology has been driven forward primarily as a result of the collaboration between key stakeholders within the clinical, academic and industrial settings. The need for alternative monitoring systems has focused research towards minimally and non-invasive sample acquisition techniques that, when combined with novel analytical processing, now show real promise as pharmaceutical patient monitoring products. As with all novel medical device systems, there will be a number of barriers to overcome, although some have already been addressed in the academic literature, as discussed previously in this chapter. It is important, in order for such patient monitoring systems to be accepted by patients and clinicians, that these important stakeholders are included in the design process from the outset. In a qualitative study carried out by Mooney *et al.* published in 2015, the opinions of paediatricians on MN mediated monitoring were collated and analysed across common themes, providing information on the translation of MN devices from laboratory to clinical practice [77]. It has been widely discussed that for MN technology to be accepted by the end-user it is important to engage stakeholders from the outset: from concept and design to manufacture and production. A total of 145 responses were collated in this research study and collectively they indicated that MN technology was gaining favour with clinicians. This was primarily due to the perceived ease of use of these devices, compared with the routine and significant problems associated with traditional hypodermic needle usage in the clinical setting. Respondents believed an alternative method of sample collection was needed for children, with 83% of paediatricians stating there was a particular need in premature neonates. These assertions, together with the ongoing progress being made in this field of diagnostics, suggest that an MN device for monitoring purposes is worthy of the ongoing expenditure in research and development departments.

Studies have shown that patients can apply MN devices successfully following counselling from a pharmacist and in conjunction with a PIL [78]. In order to gain acceptance with regulatory bodies, and in parallel with this work, the FDA and the UK's Medicines and Healthcare Products Regulatory Agency (MHRA) have indicated there may be a need for some method of MN insertion confirmation. An example of how this issue could be addressed has been described in one study which showed that a visual mechanism of feedback could be included to ensure patients had correctly inserted an MN array [79]. Here the study described a pressure sensitive film being used to inform volunteers whether or not they had applied the appropriate amount of pressure required to achieve successful MN insertion. Despite this reassurance from both academic and clinical settings, significant marketing may be required to educate the general population with regards to both use and safety of such novel technology. Consideration must also be focused on the packaging of MN-based patient monitoring products, not only to ensure the items are suitably protected in transport, but also that the product as a whole is patient friendly and easy to use. The packaging should not only provide mechanical protection but may also need to provide moisture and light protection if the monitoring device involves labile detection agents. Tried and tested systems, particularly for cholesterol and glucose monitoring, although cumbersome, are well established – a truly easy to use, reliable, accurate product will be the only way to break into this market.

Patient monitoring, by its very nature, requires long-term or repeated ISF or whole blood extraction. It is, as yet, unknown what the physiological effects of continual application of MN devices will have on the skin. Only recently, one report highlighted the effects of repeat application of polymeric MN arrays to immunocompetent, hairless mice [80]. The MN were repeatedly applied to the same application site on the mice over a series of weeks and, encouragingly, no detrimental effects to barrier or protective function of the skin over the course of the study timeline were observed. This study provides robust evidence to support assessment of the repeat application of polymeric MN arrays in higher order animal models and, ultimately, in human volunteers.

Another concern relates to the development of local skin and systemic infection as a result of microbial infiltration. Previously, infection risk as a result of microbial infiltration into the skin following MN application has been assessed as minimal [81, 82]. It should be noted that, to date, there have been no reported cases of local or systemic infection caused as a result of MN application. It has been suggested that microbial infiltration through the micro-channels caused by MN application are unlikely to result in significant contamination of the skin. The micro-channels caused by MN application are filled with aqueous components of ISF and so microorganisms colonised on the skin surface may be more likely to remain on the hydrophobic *SC*. Associated with this, the skin is known to be a highly immunocompetent organ, containing vast numbers of immune cells able to capture and detect foreign microbes prior to the development of an infection [83]. Further, the skin hosts non-immune enzymes that can act to defend against intruding microbes. Current best practice for hypodermic injections involves skin cleansing prior to puncture of the skins outer barrier, however it is known, colloquially, that in many instances (such as insulin for patients with diabetes) this guidance is not strictly adhered to [84]. Ideally skin cleansing prior to application of any MN based device would not need to be performed, to avoid this inconvenient step and reduce the complexity of the process further. It will be for regulatory authorities to consider the available evidence base and decide as to the necessity for this cleansing stage.

Although infiltration of microorganism's resident on the surface of the skin has been shown not to be problematic, another concern that has been raised is the need for product sterility and endotoxin loading. A number of sterilisation methods have been proposed for solid MN products, such as e-beam, gamma irradiation, microwave radiation, ethylene oxide and wet/dry heat sterilisation; however, the destructive nature of these methods can lead to detrimental effects on MN mechanical strength. Gamma-radiation at doses required to ensure sterility have been shown not to cause significant changes to the mechanical strength or swelling capacity of polymeric MN arrays [40]. This work also assessed the endotoxin loading of polymeric MN arrays, fabricated using industry scalable techniques. The reported endotoxin levels were below and FDA's maximum allowable endotoxin concentration for medical devices in contact with lymphatic tissue. This not only highlighted the antimicrobial nature of poly(methyl vinyl ether-*co*-maleic acid), but also showed that fabrication under normal laboratory conditions did not result in excessive endotoxin loadings in each MN array.

Discussions between academics, industry and regulators have already begun, with the initial indications suggesting that low-bioburden manufacturing to cGMP standards may be sufficient to ensure patient safety, negating the need for sterile manufacturing [85]. In turn, this decision, either way, will have a significant impact on the manufacturing techniques employed in the fabrication of MN products and ultimately this will affect the cost of manufacture and hence the cost of sale. Currently, MNs can be manufactured through many microfabrication methods, often requiring multiple stages. Coated and solid MN arrays, for example, use micromoulding techniques from silicon or metal templates. Silicon manufacture and handling usually requires clean room facilities. Therefore, companies wishing to move forward with this type of product may need to invest significant initial capital to facilitate the optimisation and design of appropriate and reproducible mass production methods. Types of MN that can be terminally sterilised (if required) may be more attractive to industrial collaborators. Further regulatory guidance is needed in order to fully classify MN based monitoring products, however, it has been suggested that they will likely fall under the medical device category [8]. Once this distinction is clarified, this may allow for the adaptation of current quality control methods that will be needed for MNs. Traditional methods of quality control may not be entirely suited to MN products as they are inherently different to both transdermal patches and hypodermic needles.

Patient monitoring of ISF, or biomolecules found within skin ISF, may be favoured over whole blood sampling, due to the numerous well-documented blood-related complications. However, when used for quantitative analysis of analyte concentration whole blood can often allow a fuller picture to be elucidated. Many drug compounds bind to proteins within whole blood with only a fraction of the drug being "free" within the vascular system. Understanding the relationship of this drug–protein binding can be a key parameter when making clinical decisions on dosing. Skin ISF concentrations of some drugs can exhibit lag times as the drug distributes and reaches equilibrium between the various compartments within the body.

7.10 Conclusion

Patient monitoring is well established in the clinical setting, providing valuable information on drug and bio-marker concentrations within various compartments of the body. Stool samples, oral salivary fluid, ocular lacrimal fluid, whole blood and ISF have all been used clinically to elucidate analyte concentrations. Knowledge of specific analyte concentrations is vital to guide clinicians on prescribing and dosing choices, allowing optimisation of drug therapies, minimising toxicity and unwanted side-effects. Patient monitoring can also be used for diagnostic purposes, as well as proving the vaccination status of patients. Access to whole blood and ISF often provides the most complete information following suitable processing of samples. However, gaining access to whole blood and ISF often involves invasive procedures such as hypodermic needle venepuncture – well known to be unpleasant for patients, requiring highly skilled medical practitioners, sharps disposal protocols and ultimately conferring significant cost. This chapter has highlighted some of the methods reported in the literature used to access ISF and whole blood using minimally- and non-invasive sampling techniques. Reverse iontophoresis and clinical microdialysis have been demonstrated as providing an alternative pathway to achieving whole blood and ISF samples, however, these processes are not without significant drawbacks in terms of patient acceptability, reproducibility and accuracy – for example, in many cases, whole blood calibration samples are required, regardless of the sampling methodology utilised.

MN technology serves as one alternative as a minimally invasive method of extracting ISF. This refers to a diverse range of systems, involving microstructures usually less than 1 mm in height, that can be used to puncture the outer surface of the skin, the *SC*. The resultant production of aqueous pores in the skin surface allows extraction of skin fluids that can then be analysed. The analysis can either occur on-board within the system or removed for external evaluation using laboratory based analytical techniques. Solid, hollow and hydrogel-forming MN arrays have been used to extract samples from the body. Metals, ceramics and a number of polymers have all been used to fabricate these structures using a range of micro-fabrication techniques. As the functionalities of these novel devices fall outside the current standard guidance for transdermal patches and hypodermic needles, alternative standards for fabrication and quality control remain to be documented and detailed. For this to occur, regulatory authorities must agree on a number of outstanding concerns. Such concerns include the need for clarity on the sterility of MN devices, acceptable endotoxin loading limits, the need, or not, for skin cleansing prior to use and packaging requirements. As the evidence base for each of these concerns grows through academic and industrial research, it is anticipated that clarity on these regulatory concerns will soon be addressed. The accompaniment of MN technology with portable, wearable point-of-care testing technology shows significant promise in being able to reduce the monitoring/sampling burden on patients and provide rapid and accurate data on bio-marker, drug or analyte concentrations. This can only help improve therapeutic outcomes for patients. Industrial collaborators within the medical device and bio-sensing sectors are likely to press forward with development so that both the significant financial incentive and the ability to enhance patient care can be realised using MN technology.

References

1 Sorrentino D., Nguyen V., Henderson C. and Bankole A. (2016). Therapeutic drug monitoring and clinical outcomes in immune mediated diseases. *Inflamm. Bowel Dis.* 22 (10): 2527–2537.

2 Sedman A.J., Molitoris B.A., Nakata L.M. and Gal J. (1986). Therapeutic drug monitoring in patients with chronic renal failure: evaluation of the Abbott TDx drug assay system. *Am. J. Nephrol.* 6(2): 132–134.

3 Hamilton J. (1995) Needle phobia: a neglected diagnosis. *J. Fam. Pr.* 41: 169–175.

4 Kersten, G. and Hirschberg H. (2007). Needle-free vaccine delivery. *Expert Opin. Drug Delivery* 4: 459–474.

5 Giudice E.L. and Campbell J.D. (2006). Needle-free vaccine delivery. *Adv. Drug Delivery Rev.* 58: 68–89.

6 Lomonte C., Forneris G., Gallieni M., *et al.* (2016). The vascular access in the elderly: a position statement of the Vascular Access Working Group of the Italian Society of Nephrology. *J. Nephrol.* 29 (2):175–184.

7 Sekkat N., Naik A. and Kalia Y. (2002). Reverse iontophoretic monitoring in premature neonates: feasability and potential. *J. Control. Release* 81: 83–89.

8 Donnelly R.F., Mooney K., Caffarel-Salvador E., *et al.* (2014). Microneedle-mediated minimally invasive patient monitoring. *Ther. Drug Monit.* 36 (1):10–17. Available from: http://www.ncbi.nlm.nih.gov/pubmed/24365984 (accessed 25 July 2017).

9 Leboulanger B., Guy R.H. and Delgado-Charro M.B. (2004). Reverse iontophoresis for non-invasive transdermal monitoring. *Physiol. Meas.* 25: 35–50.
10 Brunner M. and Derendorf H. (2006). Clinical microdialysis: current applications and potential use in drug development. *Trends Anal. Chem.* 25: 674–680.
11 Watkinson A.C., Kearney M.-C., Quinn H.L., *et al.* (2016). Future of the transdermal drug delivery market – have we barely touched the surface? *Expert Opin. Drug Delivery* 13 (4): 523–532.
12 Marshall S., Sahm L.J. and Moore A.C. (2016). Microneedle technology for immunisation: Perception, acceptability and suitability for paediatric use. *Vaccine* 34 (6): 723–734. Available from: https://doi.org/10.1016/j.vaccine.2015.12.002 (accessed 24 July 2017).
13 Mooney K., McElnay J.C. and Donnelly R.F. (2014). Children's views on microneedle use as an alternative to blood sampling for patient monitoring. *Int. J. Pharm. Pract.* 22 (5): 335–344.
14 Quinn H.L., Hughes C.M. and Donnelly R.F. (2017). *In vivo* and qualitative studies investigating the tranlational potential of microneedles for use in the older population. *Drug Delivery Transl. Res.* doi: 10.1007/s13346-017-0393-4.
15 Gardeniers H.J.G.E., Luttge R. and Berenschot E.J.W. (2003). Silicon micromachined hollow microneeldes for transdermal liquid transport. *J. Microelectromech. Syst.* 12: 855–862.
16 Chaudhri B., Ceyssens F. and De Moor P. (2010). A high aspect ratio SU-8 fabrication technique for hollow microneedles for transdermal drug delivery and blood extraction. *J. Micromech. Microeng.* (20): 064006.
17 Mukerjee E.V., Collins S.D., Isseroff R. and Smith R.L. (2004). Microneedle array for transdermal biological fluid extraction and *in situ* analysis. *Sens. Actuators, A Phys.* 114 (2–3): 267–275.
18 Lutton R.E.M., Moore J., Larrañeta E., *et al.* (2015). Microneedle characterisation: the need for universal acceptance criteria and GMP specifications when moving towards commercialisation. *Drug Delivery Transl. Res.* 5 (4): 313–331. Available from: http://link.springer.com/10.1007/s13346-015-0237-z (accessed 27 July 2017).
19 Lutton R.E.M., Larrañeta E., Kearney M.C., *et al.* (2015). A novel scalable manufacturing process for the production of hydrogel-forming microneedle arrays. *Int. J. Pharm.* 494 (1): 417–429.
20 Donnelly R.F., Majithiya R., Singh T.R.R., *et al.* (2011). Design, optimization and characterisation of polymeric microneedle arrays prepared by a novel laser-based micromoulding technique. *Pharm Res.* 28 (1): 41–57.
21 Zhang P. and Jullien G. (2003). Micromachined needles for microbiological sample and drug delivery system. *Proceedings of International Conference on MEMS, NANO and Smart Systems,* Banff, Canada (20–23 July 2003).
22 Liu R., Wang X. and Feng Y. (2006) Theoretical analytical flow model in hollow microneedles for non-forced fluid extraction. *Proceedings of the First IEEE International Conference on Nano/Micro Engineered and Molecular Systems,* Zhihai, China (18–21 January 2006). New York, NY: IEEE, 1039–1042.
23 Donnelly R.F. and Woolfson A.D. (2014). Patient safety and beyond: what should we expect from microneedle arrays in the transdermal delivery arena? *Ther. Delivery* 5 (2): 1–10.

24 Tsuchiya K., Jinnin S. and Yamamoto H. (2010). Design and development of a biocompatible painless micorneedle bu the ion sputtering deposition method. *Precis. Eng.* 34: 461–466.

25 Khumpuang S., Kawaguchi G. and Sugiyama S. (2004). Quadruplets-microneedle array for blood extraction. *Proceedings of the Nanotechnology Conference and Trade Show*, Boston, MA (7–11 March 2004). Cambridge, UK: Nano Science Technology Inst., 205–208.

26 GIA Inc. (2015). The Global Blood Glucose Monitoring Devices Market Trends, Drivers & Projections. *Strategyr.* 2015. Available from: http://www.strategyr.com/MarketResearch/Blood_Glucose_Blood_Sugar_Monitoring_Devices_Market_Trends.asp (accessed 24 July 2017).

27 Gartstein V., Yuzhakov V.V., Owens G.D., Sherman F.F. (1999) Apparatus and method for using an intracutaneous microneedle array. US Patent 6,256,533, filed 9 June 1999 and issued 3 July 2001.

28 Sieg A., Guy R. and Delgado-Charro M. (2004). Electroosmosis in transdermal iontophoresis: implications for noninvasive and calibration-free glucose monitoring. *Biophys. J.* 8: 76–80.

29 Potts R., Tamada J. and Tierney M. (2002). Glucose monitoring by reverse iontophoresis. *Metab. Res. Rev.* 18 (1): 49–53.

30 Vesper H., Wang P. and Archibold E. (2006). Assessment of truness of a glucose monitor using intersitial fluid and whole blood speicimen matrix. *Diabetes Technol. Ther.* 13: 1194–1200.

31 Sato T., Okada S. and Hagino K. (2011). Measurement of glucose area under the curve using minimally invasive interstitial fluid extrction technology: evaluation of glucose monitoring concepts without blood sampling. *Diabetes Technol. Ther.* 14: 485–491.

32 Sakaguchi K., Hirota Y. and Hashimoto N. (2012). A minimally invasive system for glucose area under the curvemeasurement using interstitial fluid extraction technology: evaluation of the accuracy and udefulness with oral glucose tolerance tests in subjects with and without diabetes. *Diabetes Technol. Ther.* 14: 485–491.

33 Ito Y., Taniguchi M., Hayashi A., *et al.* (2014). Application of dissolving microneedles to glucose monitoring through dermal interstitial fluid. *Biol. Pharm. Bull.* 37 (11): 1776–1781.

34 Valdés-Ramírez G., Li Y.C., Kim J., *et al.* (2014). Microneedle-based self-powered glucose sensor. *Electrochem. Commun.* 47: 58–62.

35 Caffarel-Salvador E., Brady A.J., Eltayib E., *et al.* (2015). Hydrogel-forming microneedle arrays allow detection of drugs and glucose *in vivo*: Potential for use in diagnosis and therapeutic drug monitoring. *PLoS One.* 10 (12): 1–21.

36 Li C.G., Joung H.-A., Noh H., *et al.* (2015). One-touch-activated blood multidiagnostic system using a minimally invasive hollow microneedle integrated with a paper-based sensor. *Lab Chip* 15 (16): 3286–3292.

37 Hwa K.Y., Subramani B., Chang P.W., *et al.* (2015). Transdermal microneedle array-based sensor for real time continuous glucose monitoring. *Int. J. Electrochem. Sci.* 10 (3): 2455–2466.

38 Kojima J., Hosoya S., Suminaka C., *et al.* (2015). An integrated glucose sensor with an all-solid-state sodium. *Micromachines* 6: 831–841.

39 Sharma S., Huang Z., Rogers M., *et al.* (2016). Evaluation of a minimally invasive glucose biosensor for continuous tissue monitoring. *Anal. Bioanal. Chem.* 408 (29): 8427–8435.

40 McCrudden M.T.C., Alkilani A.Z., Courtenay A.J., *et al.* (2015). Considerations in the sterile manufacture of polymeric microneedle arrays. *Drug Delivery Transl. Res.* 5 (1): 3–14.

41 Eltayib E., Brady A.J., Caffarel-Salvador E., *et al.* (2016). Hydrogel-forming microneedle arrays: Potential for use in minimally-invasive lithium monitoring. *Eur. J. Pharm. Biopharm.* 102: 123–131.

42 Romanyuk A.V., Zvezdin V.N., Samant P., *et al.* (2014). Collection of analytes from microneedle patches. *Anal. Chem.* 86 (21): 10520–10523.

43 Romanyuk A.V., Zvezdin V.N., Samant P., *et al.* (2015). Collection of analytes from microneedle patches. Supporting Information. *Sci. Rep.* 12: 1–9.

44 Chen K.Y., Ren L., Chen Z.P., *et al.* (2016). Fabrication of micro-needle electrodes for bio-signal recording by a magnetization-induced self-assembly method. *Sensors* 16 (9): 1–15.

45 Arai M., Nichinaka Y. and Miki N. (2015). Electroencephalogram measurement using polymer-based dry microneedle electrode. *Jpn. J. Appl. Phys.* 54: 06FP14.

46 Wang R., Jiang X., Wang W. and Li Z. (2017). A microneedle electrode array on flexible substrate for long-term EEG monitoring. *Sens. Actuators, B Chem.* 244: 750–758.

47 Noort D., Benschop H.P. and Black R.M. (2002). Biomonitoring of exposure to chemical warfare agents: a review. *Toxicol. Appl. Pharmacol.* 184 (2): 116–126.

48 Mishra RK, Vinu Mohan, A.M., Soto F., *et al.* (2017). Microneedle biosensor for minimally-invasive transdermal detection of nerve agents. *Analyst* 142: 918–924.

49 Keum D.H., Jung H.S., Wang T., *et al.* (2015). Microneedle biosensor for real-time electrical detection of nitric oxide for *in situ* cancer diagnosis during endomicroscopy. *Adv. Healthc. Mater.* 4: 1153–1158.

50 Ng K.W., Lau W.M. and Williams A.C. (2015). Towards pain-free diagnosis of skin diseases through multiplexed microneedles: biomarker extraction and detection using a highly sensitive blotting method. *Drug Delivery Transl. Res.* 5 (4): 387–396.

51 Wilke N., Morrissey A. and Hilbert C. (2005). Silicon microneedle electrode arrays with temperature monitoring for electroporation. *Sens. Actuators A Phys.* 123: 319–325.

52 Harding J.L. and Reynolds M.M. (2014). Combating medical device fouling. *Trends Biotechnol.* 32 (3): 140–146.

53 Miller P., Moorman M., Manginell R., *et al.* (2016). Towards an integrated microneedle total analysis chip for protein detection. *Electroanalysis* 28 (6): 1305–1310.

54 Ranamukhaarachchi S.A., Padeste C., Dübner M., *et al.* (2016). Integrated hollow microneedle-optofluidic biosensor for therapeutic drug monitoring in sub-nanoliter volumes. *Sci. Rep.* 6 (1): 29075.

55 Smart W.H. and Subramanian K. (2000). The use of silicon microfabrication technology in painless blood glucose monitoring. *Diabetes Technol. Ther.* 2: 549–559.

56 Wang P.M., Cornwell M. and Prausnitz M.R. (2005). Minimally invasive extraction of dermal interstitial fluid for glucose monitoring using microneedles. *Diabetes Technol. Ther.* 7: 131–141.

57 Banga A.K. (2011). Microneedles and their applications. *Recent Pat. Drug Delivery Formulation* 5 (2): 95–132.
58 Yuzhakov V.V., Gardstein V., Owens G.D. (2000) Microneedle apparatus used for marking skin and for dispensing semi-permanent subcutaneous markeup. US Patent 6,565,532 B1, filed 12 July 2000 and issued 20 May 2003.
59 Prausnitz M.R., Allen M.G., Henry S., McAllister D.V., Ackley D.E., Jackson T. (1999) Devices and methods for enhanced microneedle penetration of biological barriers. US Patent 6,743,211, filed 23 November 1999 and issued 1 June 2004.
60 Sherman F.F., Arias F., Gartstein V., et al. (2001). Portable intersitital fluid monitoring system. US Patent 6,591,124 B2, filed 11 May 2001 and issued 8 July 2003.
61 Ackley D.E. (2001). Microneedle array systems. US Patent 7,027,478 B2 filed 20 December 2001 and issued 11 April 2006.
62 Whiston R.C. (2001). Hollow microneedle patch. US Patent 6,603,987 B2, filed 11 June 11, 2001 and issued 5 August 5, 2003.
63 Gonnelli R.R. (2008). Microneedle with membrane. US Patent 20,090,043,250 A1, filed May 12, 2008 and issued 12 February 2009.
64 Frazier A.B., Andrade D.A., Bartholomeusz F. and Brazzle J.D. (2008). Active microneedles and microneedle arrays. US Patent 20,090,069,697 A1, filed 18 November 2008 and issued 12 March 2009.
65 Cho S.T. (2004). Microneedles for minimally invasive drug delivery. US Patent 20,040,186,419 A1, filed 29 January 2004 and issued 23 September 2004.
66 Cormier M.J.N. and Theeuwes F.T. (2001). Device and method for enhancing transdermal sampling. US Patent 6,219,574 B1, filed 17 June 1997 and issued 17 April 2001.
67 Zimmermann S., Feinbork D., Stoeber B., *et al.* (2003). A microneedle-based glucose monitor: fabricated on a wafer-level using in-device enzyme immobilisation. *Proceedings of the 12th International Conference on Solid State Sensors and Actuators,* Boston, MA (8–12 June 2003). New York, NY: IEEE, 99–102.
68 Ahn C., Choi J., Ceaucage G., *et al.* (2004). Disposable smart lab-on-a-chip for point-of-care clinical diagnosis. *Proc. IEEE.* 154–173.
69 Corrie S.R., Fernando G.J.P., Crichton M.L., *et al.* (2010). Surface-modified microprojection arrays for intradermal biomarker capture with low non-specific protein binding. *Lab Chip* (10): 2655–2658.
70 Lastovich A.G., Evans J.D. and Pettis R.J. (2002). Microdevice and method of manufacturing a microdevice. WO Patent 2,002,005,890, filed 11 July 2001 and issued 25 April 2002.
71 Palmer P.J. (2001). Method and apparatus for enhancing penetration of a puncturing member into the intradermal space. WO Patent 2,001,093,931, filed 6 June 2001 and issued 13 December 2001.
72 Wilkinson B.J. and Newby M. (2003). Method and device for intradermally delivery a substance. WO Patent 2,003,084,598, filed 31 March 2003 and issued 16 October 2003.
73 Tokumoto S., Matsudo T. and Kuwahara T. (2007). Transdermal drug administration apparatus having microneedles. WO Patent 2,007,091,608, filed 7 February 2007 and issued 16 August 2007.
74 Yeshrun Y., Levin Y., Almagor Y., *et al.* (2008). Microneedle device. WO Patent 2,008,047,360, filed 17 October 17 2006 and issued 24 April 2008.

75 Inou A., Takigawa M., Sekiguchi T. and Nakahara K. (2010). Transdermal administration device. WO Patent 2,010,010,974, filed 27 July 2009 and issued 28 January 2010.

76 Dauncey M. (2014). Microneedle-needle-based devices and methods for the removal of fluid from a body, WO Patent 2,014,016,579, filed 23 July 2013 and issued 30 January 2014.

77 Mooney K., McElnay J.C. and Donnelly R.F. (2015). Paediatricians' opinions of microneedle-mediated monitoring : a key stage in the translation of microneedle technology from laboratory into clinical practice. *Drug Delivery Transl. Res.* 5 (4): 346–359.

78 Donnelly R.F., Moffatt K., Alkilani A.Z., *et al.* (2014). Hydrogel-forming microneedle arrays can be effectively inserted in skin by self-application: a pilot study centred on Pharmacist intervention and a patient information leaflet. *Pharm. Res.* 31(8): 1989–1999.

79 Vicente-Pérez E.M., Quinn H.L., Mcalister E., *et al.* (2016). The use of a pressure-indicating sensor film to provide feedback upon hydrogel-forming microneedle array self-application *in vivo*. *Pharm Res.* 3072–3080.

80 Vicente-Perez E.M., Larrañeta E., Mccrudden M.T.C, *et al.* (2017). Repeat application of microneedles does not alter skin appearance or barrier function and causes no measurable disturbance of serum biomarkers of infection , inflammation or immunity in mice *in vivo*. *Eur. J. Pharm. Biopharm.* 117: 400–407.

81 Zahn J., Trebotich D. and Liepmann D. (2005). Microdialysis microneedles for continuous medical monitoring. *Biomed. Microdevices* 7: 56–59.

82 Donnelly R.F., Singh T.R.R., Alkilani A.Z., *et al.* (2013). Hydrogel-forming microneedle arrays exhibit antimicrobial properties: potential for enhanced patient safety. *Int. J. Pharm.* 451 (1–2): 76–91.

83 Zaric M., Lyubomska O., Touzelet O., *et al.* (2013). Skin dendritic cell targeting *via* microneedle arrays laden with co-glycolide nanoparticles induces efficient antitumor and antiviral immune responses. *ACS Nano* 7 (3): 2042–2055.

84 World Health Organization. (2010). WHO best practices for injections and related procedures toolkit. II. Safe Injections Global Network. Geneva: World Health Organization, 1–51.

85 Norman J. and Strasinger C. (2016). Scientific considerations for product development, manufacturing and quality control. *Proceedings of the 4th International Conference on Microneedles*, London, UK (23–25 May 2016).

8

Delivery of Photosensitisers and Precursors Using Microneedles

Mary-Carmel Kearney[1], Sarah Brown[1], Iman Hamdan[1,2] and Ryan F. Donnelly[1]

[1] *Queen's University Belfast, School of Pharmacy, Belfast BT9 7BL, UK*
[2] *Middle East University, School of Pharmacy, P.O. Box. 383 Amman 11831, Jordan*

8.1 Introduction

8.1.1 Photodynamic Therapy

Photodynamic therapy (PDT) is a clinical procedure used as either a primary or adjunctive treatment to induce cell death in premalignant and malignant lesions through the interaction of a photosensitising agent with light and oxygen [1]. Most commonly, the lesions treated using this therapy are located on the skin, but PDT can also be adapted to treat lesions occurring on additional sites, such as lungs, the oesophagus and eyes [2]. Exposure of photosensitiser molecules to visible light, in the presence of oxygen, initiates a sequence of photochemical reactions [3]. When energy, in the form of light, is transferred to a photosensitiser the molecule is excited to a higher energy state. This energy can be lost by fluorescence light emission, heat emission, formation of free radicals via type I photochemical reactions or by generation of singlet oxygen in type II reactions, as depicted in Figure 8.1. Both type I and type II reactions contribute to cytotoxicity, however, the production of singlet oxygen is considered the central pathway for clinical PDT [4]. Interaction of singlet oxygen and other reactive oxygen species, such as free radicals, with cell organelles induces either cellular apoptosis or necrosis. The complex PDT-induced cellular ablation processes are not fully understood [5], however, it is thought that the apoptosis mechanism occurs at lower light intensities, with cellular necrosis more prominent when higher light intensities are employed. Modification of light and photosensitiser treatment parameters facilitates control of these antitumour effects. It is currently known that the antitumour effects are brought about by three main mechanisms, which are not independent but somewhat inter-related. The generated toxic oxygen species may directly kill tumour cells through interaction with cellular components, they may damage the tumour vasculature depriving the tumour of oxygen and nutrients and/or they can stimulate an immune response due to release of inflammatory mediators from treated cells [6]. A combination of these mechanisms all contribute to cell death.

Photosensitising agents are most commonly administered intravenously but can also be orally and topically delivered depending on the intended site of action [8]. Following administration, the drug can be absorbed by all cells, but it tends to accumulate in

Microneedles for Drug and Vaccine Delivery and Patient Monitoring, First Edition.
Edited by Ryan F. Donnelly, Thakur Raghu Raj Singh, Eneko Larrañeta, and Maelíosa T.C. McCrudden.

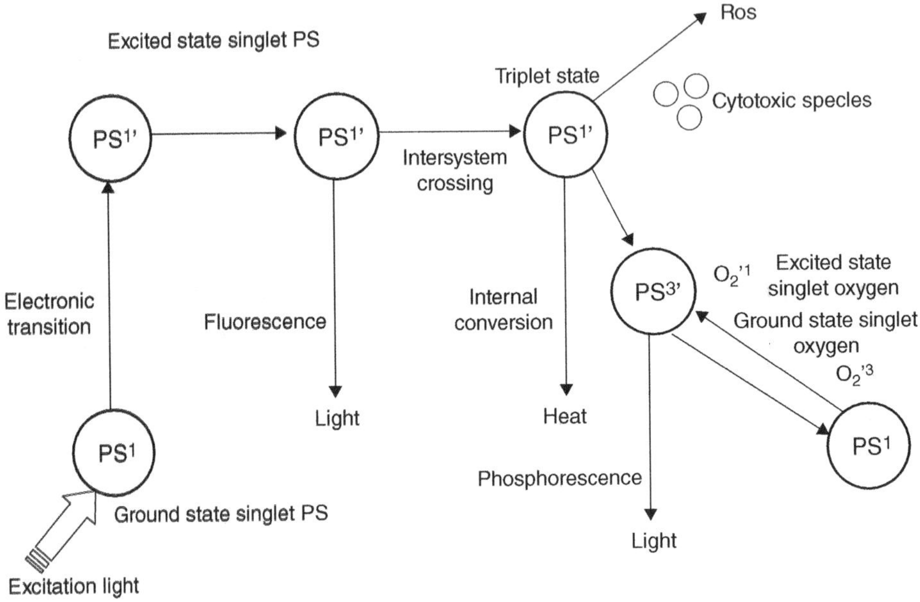

Figure 8.1 Photophysical and photochemical mechanisms of PDT. Reproduced with permission from [7] C.A. Robertson, D.H. Evans and H. Abrahamse (2009). Photodynamic therapy (PDT): A short review on cellular mechanisms and cancer research applications for PDT. *J. Photochem. Photobiol. B Biol.* 96: 1–8 .

rapidly dividing cancerous cells, offering targeted drug delivery [9]. It is thought that tumour accumulation occurs due to a combination of factors, including reduced lymphatic drainage of the tumour, high vascular permeability of photosensitisers and affinity of these agents for dividing endothelium [3]. Irradiation of the tumour site, with light of a specific wavelength and fluence rate for a defined period, will propagate the photochemical reactions discussed, producing singlet oxygen as well as other reactive oxygen species. Interaction of these species with cell components leads to irreversible cell damage. These photochemical reactions are localised to the irradiated area, thereby minimising damage to surrounding healthy cells. Singlet oxygen has a very short half-life of <40 ns, with a correspondingly low radius of action of <20 nm [10]. The total therapy treatment area is dependent on multiple, interdependent factors including the photosensitiser, light dose, tissue site, oxygen availability as well as time between drug administration and light exposure [8].

8.1.2 Photosensitisers

PDT has progressed considerably from the early application of sunlight and haematoporphyrin derivative, to the use of Photofrin® and to second generation preformed photosensitisers and topical application of the prodrug 5-aminolevulinic acid (ALA), which leads to *in situ* synthesis of the endogenous photosensitiser, protoporphyrin IX (PpIX) [11]. Initial reports of PDT demonstrated the benefit of this treatment for early stage, inoperable cancers, such as breast and cervical cancers [2], however, these photosensitisers were associated with a number of limitations. For example, haematoporphyrin is a crude mixture of partially unidentified porphyrins and shows poor target tissue selectivity. In addition, the low extinction coefficients of older agents require high doses to be administered and the relatively short wavelength of absorption (<640 nm) results in poor light penetration deep into the tissue. Systemic delivery of such agents typically causes prolonged and severe cutaneous photosensitivity, with the consequence that patients must remain indoors up to several weeks after treatment. Finally, photosensitisers can take 48–72 h to accumulate in tumours before irradiation can take place. Many of these limitations have been at least partially overcome by the development of second generation photosensitisers, such as porphyrin derivatives, chlorins and phthalocyanines [12]. These molecules are monomeric, well characterised, effective generators of singlet oxygen and absorb at longer wavelengths, thereby, facilitating deeper tissue penetration of light. However, many of these compounds are extremely hydrophobic in nature, making their formulation and delivery challenging. In addition, prolonged photosensitivity following intravenous administration can still be a significant issue. To circumvent some of these issues, novel delivery methods and treatment methodologies have been explored.

8.2 Topical Application of Photodynamic Therapy

ALA and its methyl ester, methyl aminolaevulinate (MAL) are two photosensitising agents currently licenced, in various forms, for topical application [6]. They are not specifically photosensitisers but rather precursors of the endogenously produced photosensitising molecule, PpIX. Both compounds can be administered intravenously and orally, however, topical delivery is often considered superior as it localises photosensitisation to the site of application and reduces the systemic side-effects frequently encountered following oral or intravenous administration [13–16]. Topical application of photosensitisers is not associated with cutaneous photosensitivity, it avoids damage to surrounding tissues and minimises scarring [15, 17]. Clinically, PDT based on topical application of ALA has been widely reported to give high clearance rates when used in the treatment of superficial skin lesions, such as actinic keratosis, and can be easily applied to various sites of the body [18]. However, deep lesions, such as nodular basal cell carcinomas, or lesions with overlying keratinous debris, are reported as being resistant. Such treatment failures have been attributed to light scatter and the shallow penetration of water-soluble drugs such as ALA, as these compounds do not possess the typical physiochemical characteristics necessary for passive skin diffusion (detailed previously). Intratumoural injection of ALA in animal models and human patients has yielded comparable or better intralesional PpIX production to that induced upon topical application. However, comparing the results of separate studies is difficult, due to considerable variability between clinicians, with little measurable control over depth of injection or localisation of injected ALA [19, 20].

ALA hydrochloride has a low molecular weight of 167.6 Da, which confers suitability in terms of size for permeation across the *stratum corneum*. However, permeation is limited to a depth of approximately 1 mm due to its hydrophilicity (log P −1.5) and its propensity to bind to tissue components [13, 15, 21, 22]. These physicochemical characteristics make clearance of highly keratotic lesions and carcinomas with depths extending up to 5 mm particularly difficult to achieve with ALA. MAL has increased lipophilicity in comparison with the parent molecule with improved permeation across cellular membranes. Interestingly, however, *in vitro* studies comparing PpIX production did not find a significant enhancement as a result of delivering the esterified compound [22, 23]. Morrow *et al.* reported that ALA-containing creams elicited a greater PpIX production in comparison with MAL, possibly attributable to the fact that the methyl ester is still relatively hydrophilic with a log P of −0.9 [22]. Transport across the *stratum corneum* is most likely via the aqueous shunt route rather than the intracellular route and as a result the increased lipophilicity is unlikely to be of consequence in terms of topical delivery. Reported benefits of MAL include a reduction in systemic absorption in comparison with ALA [24] as well as a reduction in pain when used in the treatment of facial actinic keratosis [25].

To date, there has been minimal success in the topical delivery of preformed photosensitisers, such as the phthalocyanines and porphyrins [12]. These photosensitisers are thought to offer a potential means of increasing PpIX concentration in dermal regions within nodular basal cell carcinomas, as endogenous synthesis in these regions is typically poor. Preformed photosensitisers absorb light at longer wavelengths, which are capable of deeper tissue penetration and also offer a less painful form of PDT as there is reduced photosensitiser accumulation in the nerve cells [9, 26]. In terms of physicochemical properties, preformed photosensitisers tend to be relatively large structures that have high partition coefficients. These characteristics do not render the compounds suitable for passive permeation across the skin. Compounds require a balance of lipophilicity for successful topical drug delivery, however, extreme hydrophobicity limits delivery beyond the outer lipid-rich layers of the skin [27]. Typically, preformed photosensitisers are positively charged with a resultant propensity to accumulate within the skin due to electrostatic interactions [26]. This cationic nature can prove challenging for formulation, as electrostatic interactions with negatively charged components can hamper drug release. Unsurprisingly, topical delivery of preformed photosensitisers has been limited, with the compounds conventionally administered intravenously. One of the leading drawbacks of this, however, is the inducement of prolonged cutaneous photosensitivity, with patients having to avoid light for up to 3 months [16, 26, 28, 29].

8.3 Methods to Enhance Topical Photodynamic Therapy

To broaden the range of photosensitisers amenable to intradermal photosensitiser delivery and improve the efficacy of compounds already formulated for topical application, various enhancement methods have been explored. These methods have included: the formulation of ALA-ester derivatives [14, 30]; curettage/debulking of lesions [31]; use of penetration enhancers [32]; tape stripping [33]; iontophoresis [34]; sonophoresis [35]; and jet injection [28, 36]. Whilst these methods have demonstrated the ability to improve photosensitiser delivery, they are not without limitations. For example, the use of ALA-ester derivatives, involves a waiting time of several hours between administration and light application. In addition, these molecules do not possess the desirable characteristics of preformed photosensitisers, for example, long absorption wavelengths [37]. Jet injectors have not been shown to enhance the delivery of photosensitisers to the deeper layers of the skin and may result in loss of photosensitiser [28], in addition to causing pain on administration [38]. In a novel approach, Donnelly *et al.* have reported the use of PDT in combination with the emerging drug delivery technology of microneedles (MNs) [13]. Since the first description of this unique combination, considerable further work has been conducted in the field [26, 39, 40]. This chapter discusses the various MN approaches that have been explored to enhance PDT and charts the advancements that have been made in the area.

8.3.1 Microneedle-mediated Photodynamic Therapy

MNs have shown promise as a method for enhancing photosensitiser delivery and efficacy [41]. Administration is painless and their use can overcome some of the challenges of conventional photosensitiser administration, such as extended waiting times between photosensitiser application and light treatment and improvement of PDT efficacy, without causing prolonged disruption of skin barrier function [40, 42, 43]. MN-mediated PDT also offers the potential of increased local administration of the photosensitiser and addresses many of the limitations of conventional PDT for treating skin lesions. There are various approaches to the use of MN technology for drug delivery, as outlined in previous chapters. The most commonly used approaches for MN-mediated PDT are, skin pre-treatment using MNs, and, photosensitiser delivery using MNS containing the photosensitising agent itself, with each discussed in the following sections.

8.3.2 Photodynamic Therapy and Skin Pre-treatment Using Microneedles

The first report of MNs for enhanced topical delivery of a photosensitising agent, ALA, was by Donnelly *et al.* [13]. In this study, silicon MN arrays were shown to enhance skin permeation of ALA *in vitro* and *in vivo* (Figure 8.2). Excised murine skin and full thickness porcine skin were punctured using arrays measuring 270 μm in height in a 6×7 configuration. MNs were applied for 30 s, after which a bioadhesive patch, containing ALA, was applied for a number of hours. Permeation studies, using murine skin, showed a significant increase in ALA delivery in comparison with the control. In porcine skin, there was increased ALA permeation in the upper dermal layers, but at a mean depth of 1.8755 mm there was no enhancement of ALA concentrations as a result of MN application. *In vivo* studies in a mouse model demonstrated that MN application could facilitate reduced photosensitiser application times and dose to achieve target PpIX concentrations. The authors also noted that ALA concentrations were only at photosensitising levels at the target site, minimising adverse effects often associated with this treatment. Another group have adopted a similar approach, comparing PpIX production in pigs following application of a 20% w/w ALA-containing cream (35 mg/cm^2), with or without MN pre-treatment [44]. An MT-Microneedle Therapy roller system was used to treat the skin prior to application of the photosensitising cream. MNs of three heights were investigated: 0.5, 1 and 1.5 mm. Similar conclusions were drawn from this study in terms of the potential benefit of reducing therapy application time and dose required for clinical effect, therefore, increasing both patient and clinician convenience. It has been noted, however, that application of semi-solid products such as creams, particularly to moist regions of the body, is not the most aesthetically acceptable procedures and a solid, one-step application may be preferred.

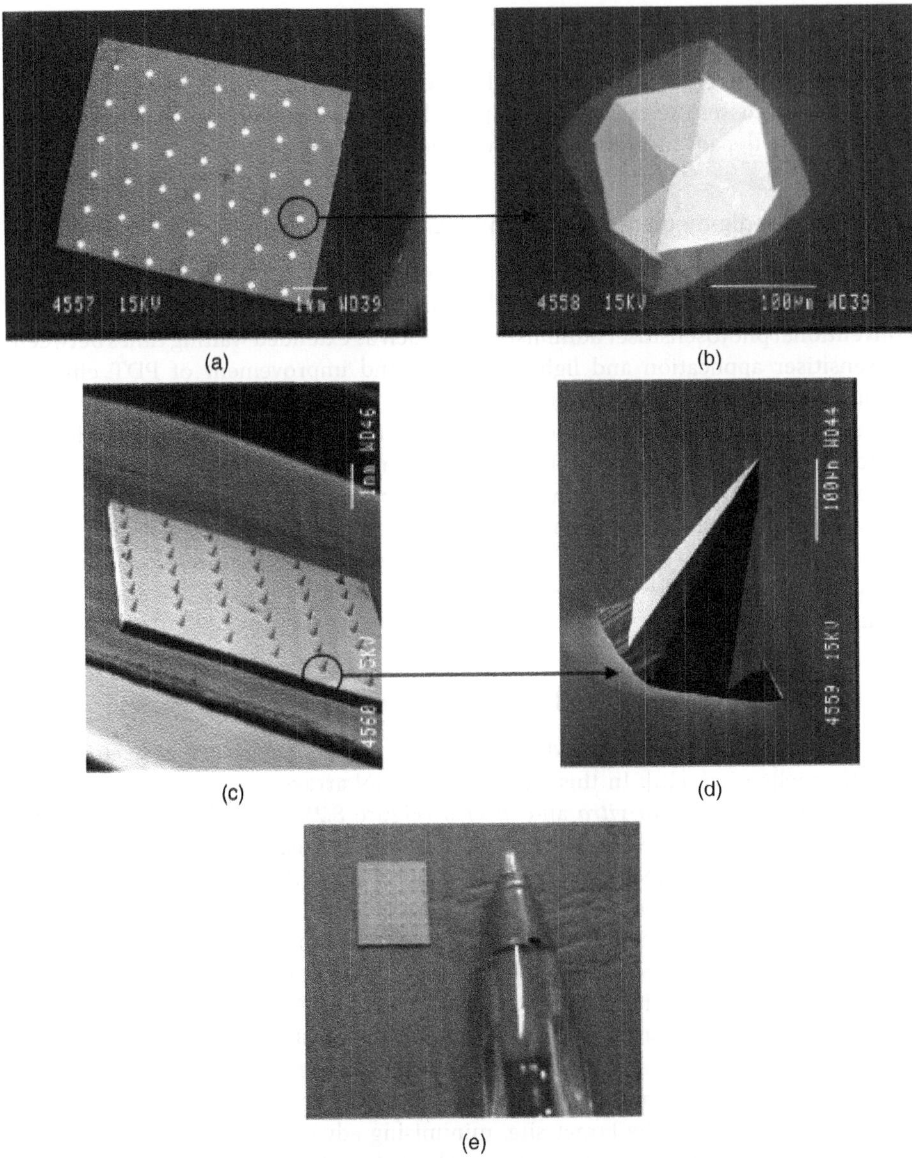

Figure 8.2 Scanning electron micrograph (SEM) images taken of: (a) a typical silicon MN array from directly above, (b) of a single MN from above, (c) of an array from the side and (d) of an individual MN from the side. (e) Digital photograph of a typical silicon MN array. For SEM, individual MN arrays were mounted onto aluminium stubs using double-sided adhesive tape and coated in gold (Polaron® E5150 sputter coater, Quorum Technologies, Ringmer, UK). Specimens were then visualised using a JEOL JSM840 scanning electron microscope (Jeol, Tokyo, Japan) and images captured on Ilford FP4 black and white roll film (Jessops, Leicester, UK), which was then developed and digitally scanned. Reproduced with permission from [13] R.F. Donnelly, D.I.J. Morrow, P.A. McCarron, *et al.* (2008). Microneedle-mediated intradermal delivery of 5-aminolevulinic acid: potential for enhanced topical photodynamic therapy. *J. Control. Release* 129: 154–162

The impact of skin pre-treatment with MNs on ALA- and MAL-induced PpIX production, as well as MN impact on pain sensations during light exposure and erythema after PDT, were investigated by Mikolajewska *et al.* [42]. Arrays of 11×11 MNs (height approximately 600 μm, base diameter 300 μm and interspacing 300 μm) were prepared from aqueous blends of 20% w/w poly(methylvinylether/maleic anhydride) (PMVE/MA). The skin of 14 healthy volunteers was pre-treated with MNs. Equal amounts of creams containing 2, 8 and 16% w/w ALA and MAL were applied on 1 cm^2 areas for 4 h. In addition, 16% w/w ALA and MAL creams were applied for 24 h. Subsequently, PpIX fluorescence spectra were measured. Skin spots where 16% w/w ALA and MAL creams were applied were then exposed to red light (632 nm, 77 mW/cm^2). The time for pain to occur was measured in seconds, and the erythemal response was monitored for up to 6 h after the end of light exposure. Use of MNs increased PpIX fluorescence after a 4 h incubation time with 2 and 8% w/w ALA or MAL creams, but not with the 16% w/w ALA or MAL creams. Pre-treatment with MNs did not increase pain sensations during light exposure, nor did it influence erythema occurrence. The authors concluded that MNs are a promising tool for increasing the efficiency of topical PDT by improving cutaneous delivery of ALA and MAL, without increased incidence of side-effects.

MN-mediated PDT has advanced to clinical studies investigating treatment of skin lesions, primarily using MNs for pre-treatment purposes [39, 45, 46]. One example is described by Torezan and colleagues who have conducted a pilot human study comparing the effect of PDT in treatment of actinically damaged skin using MAL cream applied in combination with MN treatment, with MAL applied following conventional lesion curettage [47]. Ten patients with a minimum of three facial actinic keratosis lesions were recruited to the study. Patients were treated with Metvix® cream (16% w/w MAL hydrochloride), applied to each side of the face, with gentle curettage prior to application on one side and MN treatment on the other. MN treatment involved application of cream to the face followed by passing of a Dermaroller® (192 stainless-steel needles measuring 1.5 mm in height) across the treated area 7–8 times in multiple directions. An occlusive dressing was applied over the treated site for 90 min, after which the area was cleansed and illuminated with a light dose of 37 J/cm^2. Patients were assessed for a number of outcomes including pain (using a visual analogue scale (VAS)), lesion clearance and multiple clinical parameters.

The results showed that MN-assisted PDT was more painful than conventional treatment (VAS 6±0.4 versus 4±0.6) and that one patient developed a secondary infection on the MN-treated site, 7 days post-therapy. Increased pain sensation can be explained by the relatively large height of the MNs used in the study, which have the propensity to stimulate dermal nerves. In comparison with studies that found MN-assisted PDT to be no more painful than conventional treatment, the MNs used in this study were considerably longer. Infection due to MN-assisted PDT has not been reported in other clinical studies and may have occurred on this occasion due to the application procedure, which involved MN treatment after cream application. Metvix cream is licenced for topical application, not intradermal delivery and is, therefore, unlikely to be a sterile product. In terms of clinical outcomes, it was found that 90-days post-treatment, both treatment protocols showed improved lesion clearance, with an average clearance rate of 83.3%. MN-assisted PDT resulted in more pronounced cosmetic improvements in comparison with the control group and the clearance rate was also greater (90.5 versus 86%), confirming the benefit of the combined MN and PDT treatment.

The application of metal MNs to pierce the skin prior to application of an ALA-containing nanoemulsion has been shown to be more effective in the treatment of keratotic actinic keratosis and Bowen's disease than conventional PDT [45]. Lesions on three patients, which displayed no significant response to curettage treatment prior to the engagement of PDT, displayed clearance following two to four courses of treatment combining MN pre-treatment and PDT [45]. Pinpoint bleeding, erythema and an increase in pain during irradiation were observed after MN application. However, similar to the study conducted by Torezan *et al.*, this study used MNs measuring 2 mm in length. In contrast to this, another PDT study using metal MNs with lengths of 300 μm demonstrated improvement in ALA delivery with no bleeding and relatively low pain and erythema scores [39, 45]. The application of MNs prior to topical ALA application also resulted in more uniform ALA penetration and activation [39, 42]. As such, when translated to a clinical environment, this MN-mediated enhancement offers the potential for PDT, which could more consistently provide complete clearance of skin lesions in comparison with the current topical application alone.

Lev-Tov *et al.* have recently reported a randomised controlled evaluator-blind human trial evaluating the effect of MN-assisted ALA PDT for the treatment of actinic keratosis [48]. A total of 51 participants were enrolled in the trial, with each randomly allocated into either an "MN" or "sham" treatment group. Participants were further subdivided according to ALA incubation time (20, 40 or 60 min). Those in the MN group were treated with a Microchannel Skin System® (3M™Company) MN device followed by ALA cream application, while the "sham" group were treated with an applicator without any MNs, prior to cream application. Following cream incubation, lesions were exposed to blue light at a wavelength of 417±5 nm, for 8 min (total fluence of 4.8 J/cm^2). Actinic keratosis lesions were mapped prior to and four weeks after treatment. Results from the study found that there was no statistical difference in terms of complete lesion response rate between the MN and sham treatment groups. MNs were found to possibly reduce the incubation time required for ALA, which in a clinical setting would facilitate a greater number of patients to be treated per day. A secondary outcome measured was pain caused by MN exposure prior to PDT treatment. The researchers found that MN treatment stimulated significantly greater pain in comparison with the control group, but concluded that this difference was relatively minimal. A follow-up study by this same group aimed to further reduce photosensitiser incubation time [49]. A total of 32 participants completed the study, where they were randomised to receive topically applied photosensitiser with an incubation time of either 10 or 20 min following MN pre-treatment. Results found that a 20 min incubation time was more effective than 10 min, yet equally as effective as the 1 h incubation periods previously investigated [48], with a lesion clearance rate of 76%. Interestingly, participants reported little or no pain when incubation times of 10 and 20 min were used, which is encouraging, when comparing with results from the earlier study. As recognised by the authors, a larger-scale randomised clinical study is needed to validate the efficacy and side-effect profile of this treatment option.

With a view to comparing the efficacy of different skin preparatory methods prior to PDT on PpIX uptake, Bay *et al.* conducted a small-scale, randomised clinical trial [50]. Twelve participants were recruited to the study with each exposed to a standardised skin preparation with curettage, microdermabrasion using abrasive pads, microneedling using dermarollers, ablative fractional laser (AFXL), non-AFXL and no pre-treatment, followed by application of Metvix cream for 3 h and subsequent red-light illumination (total dose 37 J/cm^2). PpIX fluorescence following the treatment schedule was assessed, where it was found that AFXL pre-treatment induced the highest fluorescence intensities in comparison with other tested options. Similarly, while all pre-treatment methods caused some mild to moderate clinical local skin reactions (erythema and oedema), the most severe were induced by AFXL pre-treatment. Adverse effects caused by AFXL persisted from time of treatment to 7 days post-treatment, with skin barrier function, as determined by transepidermal water loss measurements, also impaired for at least 3 days post-treatment. This contrasts with non-AFXL, microdermabrasion and MN treatment, where skin barrier function had recovered to baseline immediately following treatment. While AFXL treatment may have induced the highest PpIX fluorescence, it is acknowledged by the authors that this does not necessarily translate into enhanced efficacy of PDT. The superior PpIX production and local reactions due to AFXL are thought to be because of the greater volume of tissue ablated during the laser treatment in comparison with the other treatment options. The MNs used in this work were relatively short (only 200 μm in height) and the microchannels created in the skin considerably smaller than those induced by laser treatment. Modification of the MN geometry and configuration would enable generation of larger channels, with potential for further enhancement of PpIX production, yet it has been demonstrated that skin barrier function recovers relatively quickly following MN application with minimal local skin reactions [51].

Recognising that preformed photosensitisers offer certain advantages over ALA-induced PDT, such as absorption of light at longer wavelengths and increased photosensitising potency in comparison with ALA, Donnelly *et al.* evaluated the effect of MN pre-treatment on intradermal delivery of meso-tetra (*N*-methyl-4-pyridyl) porphine tetra tosylate (TMP) from a bioadhesive patch [26]. *In vitro* studies involved application of silicon MNs (6×7 configuration, 270 μm height) to excised murine or porcine skin for 30 s, followed by application of the TMP-containing patch for up to 6 h. At 1.8755 mm, the typical thickness of basal cell skin carcinomas, MN pre-treatment facilitated a significantly greater TMP concentration in comparison with the patch alone (1004.2 versus 87.48 μg/cm^3). *In vitro* studies also showed that there was some transdermal permeation of TMP, however, the total concentration was still relatively low. *In vivo* studies in nude mice measured TMP-induced fluorescence with results showing that TMP fluorescence was both higher (double) and more widespread in MN treated skin in comparison with patch application alone. Importantly, fluorescence was isolated to the treated skin region, and was not detected at distant skin sites, minimising the risk of widespread photosensitivity, a major issue following intravenous administration of photosensitisers [52]. This study demonstrated that MNs could be used to facilitate intradermal delivery of preformed photosensitisers that are normally precluded from topical application due to their relatively large sizes, and often high hydrophobicity. It was concluded by the authors, however, that some degree of water solubility is necessary for this type of MN application as the created channels in the skin are aqueous in nature.

A commonly cited limitation of MN-assisted PDT is penetration depth of the applied light. Typically, the penetration depth of optical light (400–600 nm) is 0.1–1 mm for human skin, yet many lesions extend beyond these depths. In a bid to address this issue, Kosoglu *et al.* have developed silica-based fibre optic MNs [53]. These MNs are intended for insertion directly into the target lesion, enabling direct tumour illumination, overcoming the light-impeding barrier often presented by skin tissue. Initial studies by this group were focused on MN design, with the aim of developing sharp MNs that would cause minimal pain and tumour vasculature damage. Subsequent work has focused on MN design to ensure optimum light transmission along with photothermal heat generation [54], and concluded that thinner MNs enabled more favourable light delivery. More recently, Kim *et al.* have developed a novel integrated optical MN array to address the same problem [55]. The prototype device described by this group consists of cured poly(lactic acid), transparent MNs (11×11 configuration in 1 cm^2, 1.6 mm in height) and a microlens array (1 mm spacing and 2 mm thick) fabricated from the same polymer (Figure 8.3). The lens acts to focus incident light into each MN and to propagate the intensity of this incoming light. The light intensity transmitted through the MNs was characterised with and without the microlens, with the lens providing a fourfold increase in light transmission. Light penetration in bovine and porcine tissue was also evaluated using the MN device. Only 7.6% of incident light (490 nm) from a laser beam was transmitted through 2.7 mm of porcine muscle when directly illuminated in comparison with 32% when using the optical MN array. The authors concluded that their device could enhance current phototherapies by extending the penetration depth and intensity of incident light. Further work is necessary to demonstrate the therapeutic efficacy of this device, but it presents a novel and exciting approach to overcoming the issue of light penetration depth.

The majority of MN-mediated PDT studies have focused on the ablation of pre-cancerous or cancerous cells, however, interestingly PDT is also known to improve the cosmetic appearance of photo-aged and/or damaged skin [39] as well as being effective in the treatment of acne [56]. This has been attributed to enhanced collagen production, increased levels of transforming growth factor beta and transforming growth factor beta type II receptors in the epidermis and reduced matrix metalloprotease activity post-PDT [57]. In addition to this, skin treatment with MNs has also been found to stimulate a cosmetically enhancing effect [58]. It is hypothesised that collagen and elastin production may be stimulated by MN puncture, as well as causing an up-regulation in the expression of genes that promote extracellular matrix

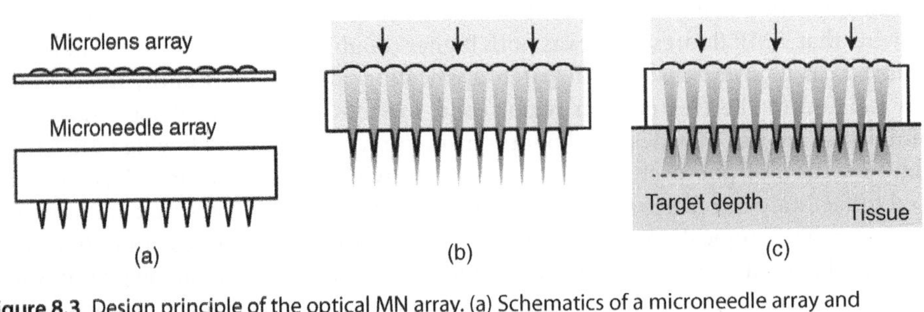

Figure 8.3 Design principle of the optical MN array. (a) Schematics of a microneedle array and microlens array. (b) Illustration of an assembled optical MN array. The microlens array focuses incident light through the MNs. (c) Illustration of light delivery into a tissue. The design has been optimised to achieve uniform and maximal light intensity at a target depth. Reproduced from [55] M. Kim, J. An, K.S. Kim, *et al.* (2016). Optical lens-microneedle array for percutaneous light delivery. *Biomed. Opt. Express* 7: 4220–4227.

remodelling. This results in increased skin elasticity and moisture content with an overall improvement in skin barrier function and aesthetic presentation [59]. This has led to the application of combined MN and PDT as a potential treatment modality for scars and photodamaged skin [39]. Clementoni *et al.* investigated photodynamic photorejuvenation of the face with a combination of microneedling, red light and broadband pulsed light. A full-face treatment of 21 patients was performed with 630 nm light and broadband pulsed light after multiple passes with an MN roller and 1 h ALA incubation. The primary endpoint was clinical improvement, scored during two separate live assessments by three physicians blinded to previous scores, using a five-point standardised photo-aging scale. The secondary endpoint was evaluation of patient satisfaction based on a quartile scale comparing baseline with 6-month post-treatment photography. Statistically significant improvement was seen in the global photo-aging scores, as well as sub-components of the scale (fine lines, mottled pigmentation, sallowness, tactile roughness and telangiectasias) at 3 months, as compared with baseline live assessment, and at 6-month live assessment compared with 3 months. In addition, 90% of patients judged clinical improvement to be greater than 50% at 6 months compared with baseline photography. The authors concluded that use of an MN roller to "pre-treat" prior to application of ALA appears to be well tolerated and allows for even absorption and perhaps deeper penetration of ALA following a defined incubation period. Use of red light and broadband pulsed light allowed for deeper activation of ALA, potentially accounting for marked clinical improvement in photo-aging.

In a different cosmetic approach, solid MNs were trialled for enhancement of PDT to treat alopecia areata [46, 60]. Yoo *et al.* reported the application of MAL to the right side of eight patients' scalps, following pre-treatment with 70% v/v alcohol and an MN roller. The left sides acted as controls, with no MN rolling and were only cleansed with 70% v/v alcohol. Immediately after MN preparation, MAL was applied on the right scalp area under occlusion for 3 h. Both sides were then illuminated with a red light (average wavelength 630 nm; light dose 37 J/cm^2) for 7.5 min. Each patient received three treatments at 4-week intervals. A photograph of each patient was taken before and after treatment. Biopsy was performed on the samples that were taken from both sides of the alopecia totalis lesions after 16 weeks (4 weeks after the final treatment). After the three treatments, none of the patients achieved hair growth in the MN-rolled lesion or in the unrolled lesion. Furthermore, there was no increment in the density of the anagen hair follicles and no difference in the histologic findings of the groups. As PDT alone has not been shown to improve treatment of alopecia areata, it is not surprising that its use in combination with MNs did not elicit the hypothesised response. However, it is not known whether the lack of effect is due to poor photosensitiser absorption or lack of effect of PDT in the treatment of alopecia areata [46, 60]. It is possible that MN pre-treatment may not have enhanced photosensitiser absorption as the 1 mm MNs used induced bleeding, which may have prompted removal of the photosensitiser from the site of action, thus hindering absorption. This highlights the importance of experimental considerations such as appropriate MN length and design for specific applications.

The majority of the studies investigating solid MNs for enhancing delivery and efficacy of photosensitising agents have been composed of metal or silicon. These materials, however, are not known to be biocompatible, thus giving rise to safety concerns, including adverse effects as a result of broken MN fragments being deposited in the skin [61]. Pioneering studies combining MNs with PDT used silicon MNs and, while the limitations of this design are recognised, it should be acknowledged that this work provided proof-of-concept for the potential of this novel approach. Reservations regarding safety, however, will need to be addressed before MN arrays composed of these materials are acceptable for clinical use [17, 40, 43, 62].

8.3.3 Delivery of Photosensitisers Using Microneedles Containing the Active Agent

To enhance both the convenience and efficacy of MN-mediated PDT, research has progressed to the development of MN arrays with the photosensitising agent fabricated in the device itself. Typically, the arrays are composed of biocompatible polymer(s) and encapsulated drug. The photosensitising agent may be directly loaded into polymeric MNs (single encapsulation), or loaded into polymeric nanoparticles that are then incorporated into the soluble MN matrix (double encapsulation) [43]. Benefits of incorporating the photosensitising agent into nanoparticles, rather than directly into the MN matrix, include minimisation of solubility issues, particularly with hydrophobic pre-formed photosensitisers and maintenance of the photosensitiser in its monomeric form, which can enhance bioavailability and light absorption [17]. Owing to the viscoelasticity of the polymers used for MN fabrication, potential fragmentation of MNs is much less likely than with those composed of metal or silicon. In addition, it is expected that any polymer deposited in the skin will be safely degraded by the conventional metabolic pathways of the body [42, 43, 62]. This approach could help simplify the administration of a photosensitising agent from a two- to a one-step process, which is significant as previous studies have reported ease-of-use as paramount to achieving acceptance of MN delivery devices by both practitioners and patients [17, 40, 43, 61]. To achieve this, carbohydrates, such as galactose, have been considered as materials for a dissolving MN system containing ALA. However, these MNs displayed poor release of ALA across a model membrane and also failed to improve the penetration of the ALA through excised porcine skin, potentially due to partially dissolved carbohydrate remaining in the MN-created holes and minimising potential drug release [63]. Alternatively, MNs composed of polymers that dissolve in the skin, such as PMVE/MA [40], have been successfully fabricated with better stability profiles than the latter MNs [62, 64, 65].

As the majority of commercially available and clinically used pre-formed photosensitisers are hydrophobic, the same group aimed to deliver a model hydrophobic dye into the skin using dissolving polymeric MNs [17]. Nile red was first incorporated into 150 nm diameter poly-lactide-*co*-glycolic acid (PLGA) nanoparticles using an emulsion and salting-out process yielding loadings of 3.87 μg Nile red/mg of PLGA. Polymeric MN arrays (11×11 configuration, height 600 μm, base diameter 300 μm, interspacing 300 μm), designed to dissolve in the skin, were prepared from aqueous blends of 20% w/w PMVE/MA and tailored to contain 1.0 mg of Nile red-loaded PLGA nanoparticles. Intradermal delivery of Nile red was determined *in vitro*. Tissue penetration studies using excised porcine skin revealed that high tissue concentrations of Nile red were observed at 1.125 mm (382.63 ng/cm^3) following MN-mediated delivery. The authors stated that this one-step delivery strategy for the local delivery of highly hydrophobic agents overcomes many of the disadvantages of current delivery strategies for hydrophobic preformed photosensitisers. Importantly, local delivery should overcome the potentially debilitating prolonged photosensitivity, which is frequently seen following systemic administration of such agents. As photosensitisers do not exert their effect through direct cellular contact but rather by transfer of energy to another molecule, such as oxygen, so that the molecule reaches an excited state and causes damage to surrounding cells [66], the success of PDT is not exclusively dependent on photosensitiser release from nanoparticles. However, the molecule that accepts energy from the photosensitiser to reach a high-energy state (e.g. oxygen) must be able to diffuse in and out of the nanoparticles to make contact with the photosensitiser. Therefore, PLGA, which dissolves slowly at normal physiological pH, with subsequent slow release of photosensitiser, is still a suitable material for nanoparticle formulation in this system, as it allows oxygen diffusion through its matrix [17]. Furthermore, approaches based on the conjugation of PLGA nanoparticle surfaces with various targeting ligands, such as aptamers, folic acid, peptides or antibodies have demonstrated improved cell specific drug targeting to various cancer cell lines, including those derived from skin tumours [42]. One feature of a dissolving MN system that must be borne in mind during development and characterisation, is the effect of photosensitiser loading on MN mechanical strength and insertion capabilities. As drug loading is increased, there can be an associated adverse impact on MN mechanical strength [43]. The high potency of many photosensitising agents, however, ensures that this is not a major issue.

While most studies investigating MNs incorporating the active agent within the device are dissolving systems, other MN forms have also been explored for MN-assisted PDT. For example, Jain *et al.* (2016) have investigated ALA coated MNs for PDT of skin tumours [67]. They adopted this treatment approach to overcome the frequently reported criticism of rapid pore closure following application of a "poke and patch" or "skin pre-treatment" methodology. MNs (57 per array, 700 μm height, 200 μm width), etched from stainless-steel sheets, were coated with approximately 350 μg ALA per array by repeated insertion of MNs into a 25% w/v ALA solution. Coated MNs were inserted into Balb/c mice for 5 min and the dermal pharmacokinetics of ALA conversion to PpIX assessed at defined time points, up to 48 h post-treatment. PpIX production was compared with two other groups, one which had ALA-containing cream (25 mg of 20% w/w ALA) applied for the duration of the time interval and a second which involved MN pre-treatment followed by application of ALA-containing cream (25 mg of 20% w/w ALA). It was found that PpIX production was significantly greater for both MN groups in comparison with cream alone. PpIX production was similar for both MN groups, however, the dose required for the coated MN group was much lower in comparison with the MN pre-treatment group (Figure 8.4).

The authors also investigated the distribution profile of PpIX within the skin, and through this found that MN-assisted delivery of ALA induced PpIX production to depths of up to 480 μm in comparison with approximate depths of 170 μm for cream application alone. While there was no significant difference in the average depths of PpIX signals from both MN groups, the coated system produced considerably higher PpIX fluorescence intensity. To determine the anti-tumour effect of the various ALA delivery mechanisms, nodular subcutaneous tumours were induced in Balb/c mice. Following ALA incubation (5 min for coated MN, 4 h for cream application), tumours were irradiated with red light (633 nm; 118 J/cm^2) and tumour size determined on alternate days, up to 11 days post-illumination. Results showed that tumour regression

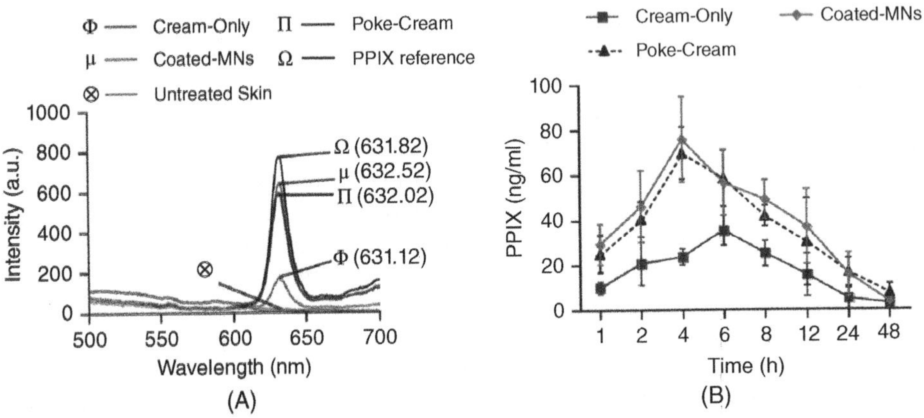

Figure 8.4 PPIX produced *in vivo* in murine skin after application of 5-ALA in Balb/c mice. (A) Emission spectra of a reference PPIX standard and of murine skin extracts after delivery of 5-ALA in "Cream-Only" group (topical cream with 5 mg 5-ALA applied on intact skin), "Poke-Cream" group (topical cream with 5 mg 5-ALA applied on skin poked with uncoated microneedles), and "Coated-MNs" group (microneedles coated with 350 μg 5-ALA per patch). (B) Pharmacokinetics of PPIX in murine skin after delivery of 5-ALA in the different groups. Reproduced with permission from [67] K. Berg (2001). Basic principles of 5-aminolevulinic acid-based photodynamic therapy. In: *Photodynamic Therapy and Fluorescence Diagnosis in Dermatology* (115–143). Elsevier Science B.V.

was significantly greater in mice treated with combined MN and PDT in comparison with ALA-containing cream application alone. Of particular promise was the low dose of ALA required for this tumour suppressing effect using coated MNs. Only 1.75 mg of ALA were delivered using these MNs, with 5 mg applied using the skin pre-treatment approach. In addition, the latter approach required cream occlusion and incubation times of 4 h, whereas the coated MNs were only applied for 5 min. The merits of this coated MN system are apparent for the potential treatment of shallow, surface lesions, however, longer needles, or possibly a different photosensitiser would be required to target deeper tumours.

Su *et al.* also recognised the time-limited nature of the permeability enhancing effect offered by MN pre-treatment due to the inherent elasticity of the skin and the rapid rate at which it repairs [68]. To overcome this, they have developed dissolving MNs with an "extended-length" design containing nanoparticles for application in PDT. Pyramidal dissolving MNs were prepared from poly(vinyl alcohol) (PVA) and poly(vinylpyrrolidone) (PVP) polymers, with poly(D,L-lactide-*co*-glycolide) nanoparticles encapsulated within the MN matrix. Nanoparticles containing coumarin-6 (as a model compound) were prepared, formulated with the MN matrix and filled into custom-made moulds. Two MN designs were prepared as depicted in Figure 8.5, with a commercially available 3M™ stainless-steel MN device also used as a comparator. Dissolving MNs were inserted into the dorsal skin of male Sprague-Dawley rats for 3 min with stainless-steel MNs inserted for the same time prior to topical application of an aqueous solution containing nanoparticles for 3 min. Nanoparticle distribution in the skin was determined by measuring coumarin-6 fluorescence intensity. Results showed that the fluorescence signal following application of extended length MNs was approximately three times greater than the 3M™ design PVA/PVP MNs, with no or minimal signal from the stainless-steel MN treatment site. Histological sections

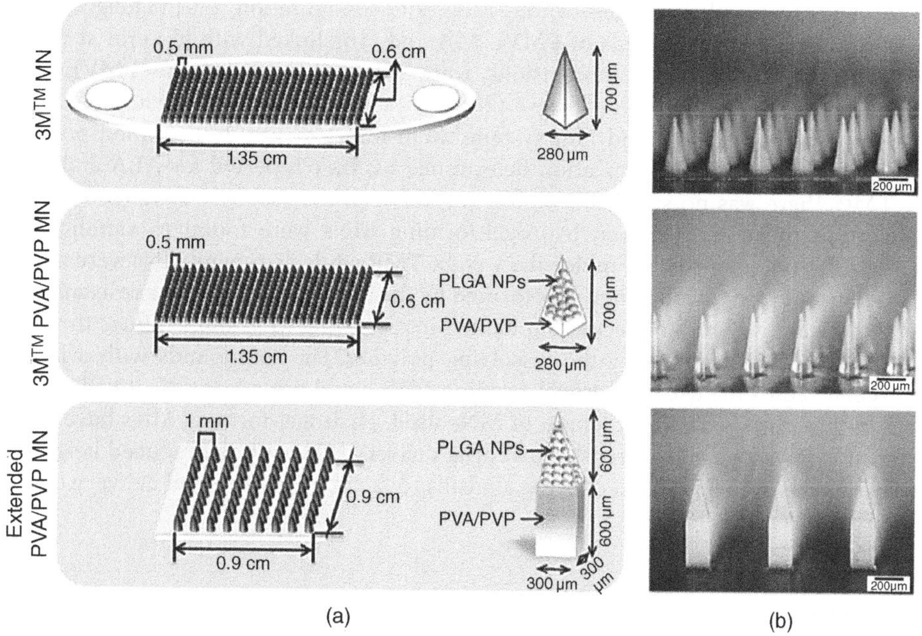

Figure 8.5 (a) Detailed specifications and (b) brightfield micrographs of the commercial 3M™ MN, the 3M™ PVA/PVP MNs and the extended PVA/PVP MNs. For the extended MN, coumarin-6 loaded nanoparticles (yellow colour) were encapsulated within the whole pyramidal structure at a length of 600 µm. Reproduced with permission from [68] L.-C. Su and M.-C. Chen (2017). Efficient delivery of nanoparticles to deep skin layers using dissolvable microneedles with an extended-length design. *J. Mater. Chem. B* 5: 3355–3363 .

and 3D confocal reconstruction images of the skin confirmed that approximately 90% of nanoparticles from the extended length MNs were distributed in the viable epidermis and dermis and remained in the skin for at least 5 days. This contrasts with approximately 70% of nanoparticles from the comparator MN group, demonstrating the enhanced efficiency of the novel extended MN design. The benefit of this system is the depth of MN application and system ease-of-use. Once inserted in the skin, the MN matrix rapidly dissolves, exposing the contained NPs. Contained drug is subsequently released in a controlled and sustained manner over a number of days. Further studies with a photosensitising agent encapsulated within the nanoparticles are needed to ascertain the efficacy of this MN system for PDT.

Another MN design tested for photosensitiser delivery is hydrogel-forming MNs. As discussed in previous chapters, these MNs swell when inserted into the skin, due to imbibition of interstitial fluid, forming continuous hydrogel pathways for intradermal drug delivery [69]. These MN arrays are removed intact from the skin with no polymeric material remaining at the application site. In terms of PDT, this is particularly beneficial as hydrogel fragments have been known to cause scattering of light, thereby complicating or reducing the efficacy of treatment [40, 69].

Both dissolving and hydrogel-forming MNs, pre-combined with photosensitiser-loaded films for a one-step application, have been shown to improve the intradermal delivery of ALA, with the latter also proving effective for enhanced delivery of TMP *in vitro* [40]. In this work, hydrogel-forming MNs (7×7 configuration, 280 μm height) were prepared from aqueous blends of PMVE/MA and crosslinked with glycerol at 60°C. Dissolving MNs, of the same dimensions, were prepared from 20% w/w PMVE/MA. MNs were combined with bioadhesive films containing 19 mg/cm^2 of either ALA or TMP. The combined MN and film systems were inserted into dermatomed porcine skin and photosensitiser permeation determined over a 6 h period for ALA and 24 h for TMP. There was no significant difference between the two MN systems in terms of ALA permeation. However, hydrogel-forming MNs were found to significantly increase delivery of the larger molecular weight TMP, while dissolving MNs were not as efficacious [40]. The swollen network formed by the hydrogel reduced the resistance to TMP diffusion, whereas molecules contained in dissolving MNs must diffuse through the thick gel layer formed by the dissolving polymer. For compounds with a larger molecular radius the use of hydrogel-forming MNs could prove very effective [40]. Of note in this work is the short length of MNs used. Hydrogel-forming MNs have been extensively investigated for transdermal drug delivery [61], with the reduced height of projections used in this work, offering a means to facilitate intradermal delivery of the photosensitiser.

8.4 Microneedles and Photothermal Therapy

A related, yet distinct cancer treatment to PDT is photothermal therapy (PTT). PTT involves the use of plasmonic nanoparticles to induce destruction of target hyperproliferative cells [70]. Nanoparticles are administered either intravenously or intratumourally followed by irradiation of the tumour, using a laser of specific energy. This energy is converted into heat by the nanoparticles, resulting in tumour ablation. A limitation of this treatment, however, is the rapid clearance of nanoparticles from the tumour site, necessitating repeated administration, often leading to adverse effects. By using MNs as a means of delivering nanoparticles directly to the tumour and also incorporating a chemotherapeutic agent, Chen *et al.* sought to develop an effective, targeted, minimally-invasive and low-toxicity treatment regimen for superficial tumours [71]. They described the fabrication of polycaprolactone MNs containing photosensitive lanthanum hexabromide nanoparticles and doxorubicin, with a rapidly dissolving baseplate composed of PVA and PVP (Figure 8.6). MNs were applied to mice bearing a subcutaneous tumour on their back, with the MNs then exposed to a laser (5 W/cm^2) until their temperature increased to 50°C for 3 min, with repeated light exposure at day 3 and day 6. Tumour volume and mouse body weight were compared for this treatment group to tumours either not treated, treated with intratumoural doxorubicin or nanoparticle-containing MNs and light but not containing the doxorubicin. Results from these experiments were very encouraging with a synergistic effect produced by near-infrared PTT and chemotherapy in comparison with each individual treatment. The combined delivery of PTT and chemotherapy using MNs enabled triggered drug release. When MNs were exposed to the laser and underwent a phase transition, doxorubicin was released. Similarly, when the laser was "switched off" drug release was discontinued. This may help explain the reduced toxicity of this MN-mediated delivery of doxorubicin in comparison with intratumoural delivery, as the chemotherapy was directly targeted to tumour cells and drug release occurred in a step-wise and triggered

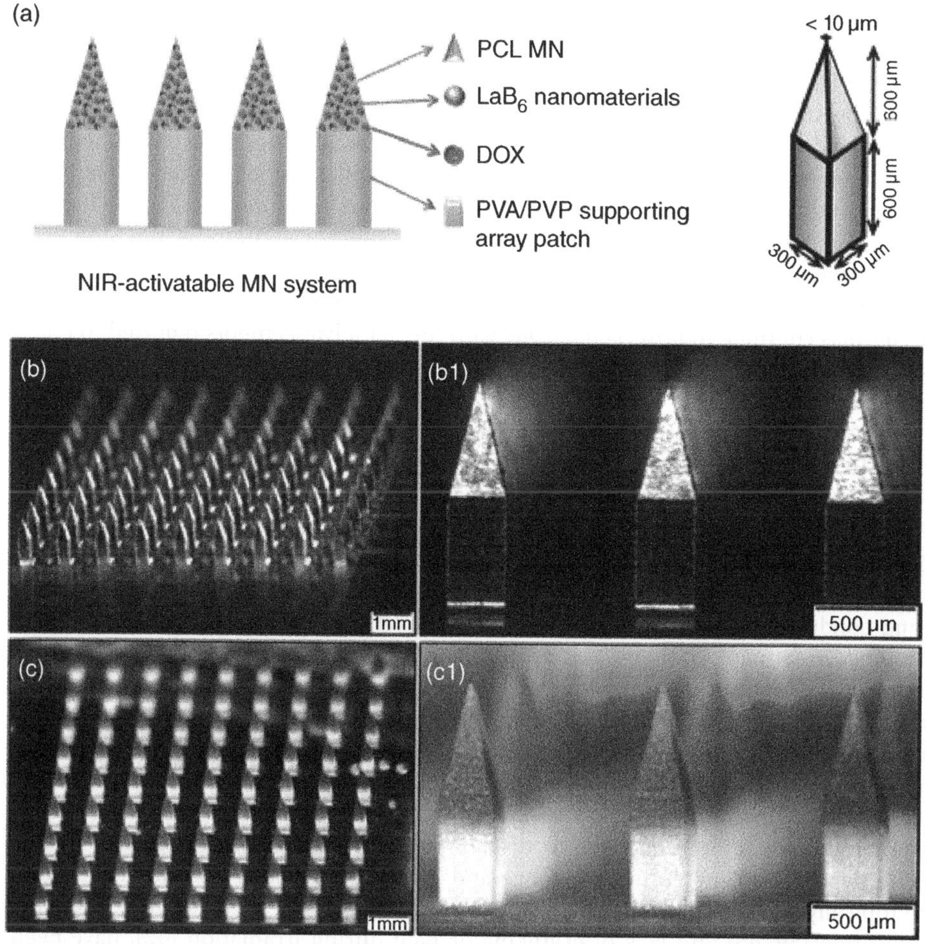

Figure 8.6 Characterisation of near-infrared light-activatable MN system. This system consists of embeddable polycaprolactone MNs containing photosensitive nanoparticles (lanthanum hexaboride; LaB_6) and an anticancer drug (doxorubicin), and a dissolvable PVA/PVP supporting array patch. (a) Schematic illustrations of the MN system. (b and b1) Brightfield micrographs of stainless-steel MN master structure and (c and c1) DOX-loaded PCL MN system. (b) and (c) Low magnification; and (b1) and (c1) high magnification. The inset in (a) shows the MN specifications. Reproduced with permission from [71] M.C. Chen, Z.W. Lin and M.H. Ling (2016). Near-infrared light-activatable microneedle system for treating superficial tumors by combination of chemotherapy and photothermal therapy. *ACS Nano* 10: 93–101 .

manner. The same group have adapted this approach for the delivery of other agents such as lidocaine, to enable patient-controlled analgesia [72]. Similar to the previous study, polycaprolactone MNs containing lanthanum hexabromide and lidocaine were prepared. MNs were tested in male Sprague-Dawley rats and following insertion were exposed to laser light, inducing a reversible phase-transition within the MNs and subsequent drug release.

Based on the positive results of using MNs for delivery of combined PTT and chemotherapy, it is not surprising that there are several groups working in this area.

Hao *et al.* (2017) described the development of a near-infrared responsive PEGylated gold nanorod coated poly(L-lactide) MN system containing docetaxel-loaded micelles for treatment of an A431 epidermoid tumour [73]. The system proved effective in eradicating the tested tumour in mice through generation of heat and controlled delivery of docetaxel to the cancerous cells. Advantages of this system include reduced dose requirement of chemotherapy agent, ease of application of the device, targeted delivery of therapy and the ablative effect on the tumour, resulting in complete tumour eradication, without recurrence. While effective for superficial tumours, the MNs used in this study were only 480 μm in height and are, therefore, limited to the treatment of shallow, surface lesions. The effective targeting of deeper lesions such as basal cell carcinomas continue to pose a challenge, with further advancements required to reach these depths.

8.5 Conclusion

A delivery system that could overcome the penetration-limiting properties of photosensitising agents is highly desirable. Primarily, it would enable successful treatment of deeper skin lesions, without occurrence of adverse effects. The large volume of work being conducted in MN-assisted PDT provides verification for the potential of this combined treatment strategy. The most common application is in the management of cancerous or precancerous lesions, but, as discussed in this chapter, it can be adapted to treat an increasing number of conditions. In some cases of relatively large or unusually contoured lesions, conventional topical photosensitiser application may be the most appropriate option. However, it is evident that MN technology can offer considerable enhancements to current treatment protocols. MNs obviate the requirement for photosensitisers to display physicochemical properties that enable tissue penetration, increasing the range and efficacy of photosensitising agents that are topically administered. Clinically desirable traits of MNs, such as simple, painless application [39, 45] and lack of erythema or pain during irradiation [42], have been apparent in clinical studies. Most importantly, MNs have been shown to enhance intradermal delivery of photosensitisers, thereby offering the potential to increase PpIX generation and the efficacy of PDT. This offers several benefits, including increase in photosensitiser concentrations at the target site [13, 26], enhanced cosmetic results [45] and, avoidance of prolonged cutaneous photosensitisation [13, 26].

MN-mediated enhancement of photosensitiser delivery, using the MN "pre-treatment" method limits the increased tissue ablative effects to the upper layers of the skin. While potentially not applicable for treating deep lesions such as basal cell carcinomas, this treatment approach could be useful in increasing photosensitiser concentrations in superficial skin lesions [13, 26]. Several studies have described the enhanced intradermal photosensitiser delivery using dissolving MNs. While most effective for small molecules, such as ALA, large molecular weight photosensitisers have also been administered to upper layers of the skin using a dissolving system [40]. Dissolvable MNs share the benefits of solid MNs, such as avoidance of prolonged cutaneous photosensitivity and short application times yet enable a more simplified administration protocol and reproducible delivery via hand application [42]. One potential drawback of the dissolving system is the fact that dissolved polymer will be deposited in the skin and may subsequently scatter applied light and reduce the absorption by the photosensitiser molecules, with a potentially negative impact on PDT success [40]. Regarding biocompatibility concerns of polymeric MNs, there have been no reports of non-degradable polymer deposited in the skin following application of MNs fabricated from FDA-approved, biocompatible materials. Results from initial small-scale *in vitro* and *in vivo* studies are reasons to be highly optimistic. Continued work is required to determine the clinical effect of dissolving MNs in human studies. The advancement of MN technology in tandem with PDT offers considerable promise as a superior treatment option to current therapy.

Delivery of photosensitisers to deeper lesions is challenging, however, the novel approaches being adopted to addresses this issue are promising. Combining MNs with PTT, as well as modification of MN design, could prove effective in remedying aggressive and deep tumours. Hydrogel-forming MNs are likely to have good clinician and patient acceptability due to their simple, one-step application and reproducible delivery. As they progress towards regulatory approval [40] this may, in time, allow MN patches to be applied by patients themselves, prior to clinic visits, thereby eliminating the waiting time between photosensitiser and light application [42]. The next steps required in progression of this approach include *in vivo* and clinical studies to assess the enhancement of intradermal preformed photosensitiser delivery, possibly using hydrogel-forming MNs. In addition, the clinical efficacy of the treatment with regards to tumour regression and clearance need to be explored.

This niche area of MN technology and PDT is continuing to expand, and it could be envisaged, with the positive clinical results to date, that MNs may become a mainstay approach in the treatment of skin lesions. Aligning the advancements in MN formulation and design with light-delivery strategies will help to further propel this technology towards routine clinical use. Numerous small-scale clinical studies have been conducted to date, with a unanimous conclusion that MNs enable enhanced photosensitiser delivery, PpIX production and, in many cases, offer superior lesion clearance in comparison with conventional treatment. MNs using biocompatible polymers and simple application procedures may prove to be the most logical MN design, but the various methods currently being employed in the field will allow a greater range of lesions to be treated using this therapy. It would be hoped that clinical studies investigating preformed photosensitisers are in the pipeline, as they too may help in the battle against deeper, more resistant tumours. Owing to the different protocols adopted in the various clinical trials thus far, it is often difficult to draw true comparisons and determine the optimum mode of MN-assisted PDT. As the portfolio of novel MN approaches continues to evolve, it is expected that there will be a correlating increase in the number and extent of clinical studies being conducted, all of which will help to validate the effect of using MNs to enhance PDT. Ultimately, if the technology can be safely and effectively advanced from a clinical trial stage to regular use by oncologists and dermatologists, there should be a respective improvement in overall therapy outcomes.

References

1 R.R. Allison and K. Moghissi (2013). Photodynamic therapy (PDT): PDT mechanisms. *Clin. Endosc.* 46: 24–29.

2 C. Hopper (2000). Photodynamic therapy: a clinical reality in the treatment of cancer. *Lancet Oncol.* 1: 212–219.
3 D.E.J.G.J. Dolmans, D. Fukumura and R.K. Jain (2003). Photodynamic therapy for cancer. *Nat. Rev. Cancer* 3: 375–379.
4 Y.N. Konan, R. Gurny and E. Allémann (2002). State of the art in the delivery of photosensitizers for photodynamic therapy. *J. Photochem. Photobiol. B Biol.* 66: 89–106.
5 A. Juarranz, P. Jaen, F. Sanz-Rodriguez, *et al.* (2008). Photodynamic therapy of cancer. Basic principles and applications. *Clin. Transl. Oncol.* 10: 148–154.
6 T.J. Dougherty, C.J. Gomer, B.W. Henderson, *et al.* (1998). Photodynamic therapy. *J Natl. Cancer Inst.* 90: 889–905.
7 C.A. Robertson, D.H. Evans and H. Abrahamse (2009). Photodynamic therapy (PDT): A short review on cellular mechanisms and cancer research applications for PDT. *J. Photochem. Photobiol. B Biol.* 96: 1–8.
8 A.P. Castano, T.N. Demidova and M.R. Hamblin (2004). Mechanisms in photodynamic therapy: part one—photosensitizers, photochemistry and cellular localization. *Photodiagn. Photodyn. Ther.* 1: 279–293.
9 A. Juzeniene, Q. Peng and J. Moan (2007). Milestones in the development of photodynamic therapy and fluorescence diagnosis. *Photochem. Photobiol. Sci.* 6: 1234–1245.
10 J. Moan and K. Berg (1991). The photodegradation of porphyrins in cells can be used to estimate the lifetime of singlet oxygen. *Photochem. Photobiol.* 53: 549–553.
11 Q. Peng, K. Berg, J. Moan, *et al.* (1997). 5-Aminolevulinic acid-based photodynamic therapy: principles and experimental research. *Photochem. Photobiol.* 65: 235–251.
12 M.J. Garland, C.M. Cassidy, D. Woolfson and R.F. Donnelly (2009). Designing photosensitizers for photodynamic therapy: strategies, challenges and promising developments. *Future Med. Chem.* 1: 667–691.
13 R.F. Donnelly, D.I.J. Morrow, P.A. McCarron, *et al.* (2008). Microneedle-mediated intradermal delivery of 5-aminolevulinic acid: potential for enhanced topical photodynamic therapy. *J. Control. Release* 129: 154–162.
14 L.-W. Zhang, Y.-P. Fang and J.-Y. Fang (2011). Enhancement techniques for improving 5-aminolevulinic acid delivery through the skin. *Dermatologica Sin.* 29: 1–7.
15 J.C. Kennedy (2006). Introduction. In: Photodynamic Therapy with ALA: A Clinical Hand*book* (1–14). Cambridge: The Royal Society of Chemistry.
16 B. Krammer and K. Plaetzer (2008). ALA and its clinical impact, from bench to bedside. *Photochem. Photobiol. Sci.* 7: 283–289.
17 R.F. Donnelly, D.I.J. Morrow, F. Fay, *et al.* (2010). Microneedle-mediated intradermal nanoparticle delivery: Potential for enhanced local administration of hydrophobic pre-formed photosensitisers. *Photodiagn. Photodyn. Ther.* 7: 222–231.
18 C.A. Morton, R.-M. Szeimies, A. Sidoroff and L.R. Braathen (2013). European guidelines for topical photodynamic therapy part 1: treatment delivery and current indications - actinic keratoses, Bowen's disease, basal cell carcinoma. *J. Eur. Acad. Dermatol. Venereol.* 27: 536–544.
19 C.A. Morton, R.M. MacKie, C. Whitehurst, *et al.* (1998). Photodynamic therapy for basal cell carcinoma: Effect of tumour thickness and duration of photosensitiser application on response. *Arch. Dermatol.* 134: 248–249.

20 R.F. Donnelly, P. Juzenas, P.A. McCarron, *et al.* (2006). Influence of formulation factors on methyl-ALA-induced protoporphyrin IX accumulation in vivo. *Photodiagn. Photodyn. Ther.* 3: 190–201.

21 D.I.J. Morrow, P.A. McCarron, A.D. Woolfson, et al. (2010). Influence of penetration enhancers on topical delivery of 5-aminolevulinic acid from bioadhesive patches. *J. Pharm. Pharmacol.* 62: 685–695.

22 D.I.J. Morrow, P.A. Mccarron, A.D. Woolfson, *et al.* (2010). Hexyl aminolaevulinate is a more effective topical photosensitiser precursor than methyl aminolaevulinate and 5-aminolaevulinic acids when applied in equimolar doses. *J. Pharm. Sci.* 99: 3486–3498.

23 A. Juzeniene, P. Juzenas, L.-W. Ma, *et al.* (2006). Topical application of 5-aminolaevulinic acid, methyl 5-aminolaevulinate and hexyl 5-aminolaevulinate on normal human skin. *Br. J. Dermatol.* 155: 791–799.

24 J. Moan, L.-W. Ma, A. Juzeniene, *et al.* (2003) Pharmacology of protoporphyrin IX in nude mice after application of ALA and ALA esters. *Int. J. Cancer* 103: 132–135.

25 D.-Y. Ko, K.-H. Kim and K.-H. Song (2014). Comparative study of photodynamic therapy with topical methyl aminolevulinate versus 5-aminolevulinic acid for facial actinic keratosis with long-term follow-up. *Ann. Dermatol.* 26: 321–331.

26 R.F. Donnelly, D.I.J. Morrow, P.A. McCarron, *et al.* (2009). Microneedle arrays permit enhanced intradermal delivery of a preformed photosensitizer. *Photochem. Photobiol.* 85: 195–204.

27 D.I.J. Morrow, P.A. McCarron, A.D. Woolfson, *et al.* (2010). Novel patch-based systems for the localised delivery of ALA-esters. *J. Photochem. Photobiol. B* 101: 59–69.

28 R.F. Donnelly, D.I.J. Morrow, P.A. McCarron, *et al.* (2007). Influence of solution viscosity and injection protocol on distribution patterns of jet injectors: application to photodynamic tumour targeting. *J. Photochem. Photobiol. B* 89: 98–109.

29 S. Wang, W. Fan, G. Kim, *et al.* (2011). Novel methods to incorporate photosensitizers into nanocarriers for cancer treatment by photodynamic therapy. *Lasers Surg. Med.* 43: 686–695.

30 F.S. De Rosa, A.C. Tedesco, R.F.V. Lopez, *et al.* (2003). In vitro skin permeation and retention of 5-aminolevulinic acid ester derivatives for photodynamic therapy. *J. Control. Release* 89: 261–269.

31 A.M. Soler, T. Warloe, A. Berner and K.E. Giercksky (2008). A follow-up study of recurrence and cosmesis in completely responding superficial and nodular basal cell carcinomas treated with methyl 5-aminolaevulinate-based photodynamic therapy alone and with prior curettage. *Br. J. Dermatol.* 145: 467–471.

32 M.B.R. Pierre, E. Ricci, A.C. Tedesco and M.V.L.B. Bentley (2006). Oleic acid as optimizer of the skin delivery of 5-aminolevulinic acid in photodynamic therapy. *Pharm. Res.* 23: 360–366.

33 S.R. Wiegell, I.-M. Stender, R. Na and H.C. Wulf (2003). Pain associated with photodynamic therapy using 5-aminolevulinic acid or 5-aminolevulinic acid methylester on tape-stripped normal skin. *Arch. Dermatol.* 139: 1173–1177.

34 S.J. Fallows, M.J. Garland, C.M. Cassidy, *et al.* (2012). Electrically-responsive anti-adherent hydrogels for photodynamic antimicrobial chemotherapy. *J. Photochem. Photobiol. B.* 114: 61–72.

35 G. Krishnan, J.E. Grice, M.S. Roberts, *et al.* (2013). Enhanced sonophoretic delivery of 5-aminolevulinic acid: preliminary human ex vivo permeation data. *Skin Res. Technol.* 19: e283–289.

36 X. Li, X. Wang, J. Gu, *et al.* (2013). Needle-free injection of 5-aminolevulinic acid in photodynamic therapy for the treatment of condylomata acuminata. *Exp. Ther. Med.* 6: 236–240.

37 R.F. Donnelly, P.A. McCarron and A.D. Woolfson (2005). Drug delivery of aminolevulinic acid from topical formulations intended for photodynamic therapy. *Photochem. Photobiol.* 81: 750–767.

38 T. Gratieri, I. Alberti, M. Lapteva and Y.N. Kalia (2013). Next generation intra- and transdermal therapeutic systems: using non- and minimally-invasive technologies to increase drug delivery into and across the skin. *Eur. J. Pharm. Sci.* 50: 609–622.

39 M.T. Clementoni, M. B-Roscher and G.S. Munavalli (2010). Photodynamic photorejuvenation of the face with a combination of microneedling, red light, and broadband pulsed light. *Lasers Surg. Med.* 42: 150–159.

40 R.F. Donnelly, D.I.J. Morrow, M.T.C. McCrudden, *et al.* (2014). Hydrogel-forming and dissolving microneedles for enhanced delivery of photosensitizers and precursors. *Photochem. Photobiol.* 90: 641–647.

41 M.-C. Kearney, S. Brown, M.T.C. McCrudden, *et al.* (2014). Potential of microneedles in enhancing delivery of photosensitising agents for photodynamic therapy. *Photodiagn. Photodyn. Ther.* 11: 459–466.

42 P. Mikolajewska, R.F. Donnelly, M.J. Garland, *et al.* (2010). Microneedle pre-treatment of human skin improves 5-aminolevulininc acid (ALA)- and 5-aminolevulinic acid methyl ester (MAL)-induced PpIX production for topical photodynamic therapy without increase in pain or erythema. *Pharm. Res.* 27: 2213–2220.

43 J.-H. Park, M.G. Allen and M.R. Prausnitz (2006). Polymer microneedles for controlled-release drug delivery. *Pharm. Res.* 23: 1008–1019.

44 P.G.S. Rodrigues, P.F.C. de Menezes, A.K.L. Fujita, *et al.* (2015). Assessment of ALA-induced PpIX production in porcine skin pretreated with microneedles. *J. Biophotonics* 8: 723–729.

45 G. Kolde, E. Rowe and H. Meffert (2013). Effective photodynamic therapy of actinic keratoses and Bowen's disease using microneedle perforation. *Br. J. Dermatol.* 168: 450–452.

46 J.W. Lee, K.H. Yoo, B.J. Kim and M.N. Kim (2010). Photodynamic therapy with methyl 5-aminolevulinate acid combined with microneedle treatment in patients with extensive alopecia areata. *Clin. Exp. Dermatol.* 35: 548–549.

47 L. Torezan, Y. Chaves, A. Niwa, J.A. Sanches, *et al.* (2013). A pilot split-face study comparing conventional methyl aminolevulinate- photodynamic therapy (PDT) with microneedling-assisted PDT on actinically damaged skin. *Dermatol. Surg.* 39: 1197–1201.

48 H. Lev-Tov, L. Larsen, R. Zackria, *et al.* (2017). Microneedles assisted incubation during aminolevulinic acid phototdynamic therapy of actinic keratoses - a randomized controlled evaluator blind trial. *Br. J. Dermatol.* 176: 543–545c.

49 T.A. Petukhova, L.A. Hassoun, N. Foolad, *et al.* (2017). Effect of expedited microneedle-assisted photodynamic therapy for field treatment of actinic keratoses. *JAMA Dermatol.* 140: 41–46.

50 C. Bay, C.M. Lerche, B. Ferrick, *et al.* (2017). Comparison of physical pretreatment regimens to enhance protoporphyrin IX uptake in photodynamic therapy. *JAMA Dermatol.* 153: 270–278.
51 Y.A. Gomaa, D.I.J. Morrow, M.J. Garland, *et al.* (2010). Effects of microneedle length, density, insertion time and multiple applications on human skin barrier function: Assessments by transepidermal water loss. *Toxicol. Vitr.* 24: 1971–1978.
52 W.E. Grant, A. MacRobert, S.G. Bown, *et al.* (1993). Photodynamic therapy of oral cancer: photosensitisation with systemic aminolaevulinic acid. *Lancet* 342: 147–148.
53 M.A. Kosoglu, R.L. Hood, Y. Chen, *et al.* (2010). Fiber optic microneedles for transdermal light delivery: Ex vivo porcine skin penetration. *J. Biomech. Eng.* 132: 910141–910147.
54 M.A. Kosoglu, R.L. Hood, J.H. Rossmeisl, *et al.* (2011). Fiberoptic microneedles: Novel optical diffusers for interstitial delivery of therapeutic light. *Lasers Surg. Med.* 43: 914–920.
55 M. Kim, J. An, K.S. Kim, *et al.* (2016). Optical lens-microneedle array for percutaneous light delivery. *Biomed. Opt. Express* 7: 4220–4227.
56 M. Boen, J. Brownell, P. Patel and M.M. Tsoukas (2017). The role of photodynamic therapy in acne: An evidence-based review. *Am. J. Clin. Dermatol.* 18: 311–321.
57 M.Y. Park, S. Sohn, E.-S. Lee and Y.C. Kim (2010). Photorejuvenation induced by 5-aminolevulinic acid photodynamic therapy in patients with actinic keratosis: A histologic analysis. *J. Am. Acad. Dermatol.* 62: 85–95.
58 M.T.C. McCrudden, E. McAlister, A.J. Courtenay, *et al.* (2015). Microneedle applications in improving skin appearance. *Exp. Dermatol.* 24: 561–566.
59 K.Y. Park, H.K. Kim, S.E. Kim, *et al.* (2012). Treatment of striae distensae using needling therapy: a pilot study. *Dermatol. Surg.* 38: 1823–1828.
60 K.H. Yoo, J.W. Lee, K. Li, *et al.* (2010). Photodynamic therapy with methyl 5-aminolevulinate acid might be ineffective in recalcitrant alopecia totalis regardless of using a microneedle roller to increase skin penetration. *Dermatol. Surg.* 36: 618–622.
61 H.L. Quinn, M.-C. Kearney, A.J. Courtenay, *et al.* (2014). The role of microneedles for drug and vaccine delivery. *Expert Opin. Drug Deliv.* 11: 1769–1780.
62 J.-H. Park, M.G. Allen and M.R. Prausnitz (2005). Biodegradable polymer microneedles: fabrication, mechanics and transdermal drug delivery. *J. Control. Release* 104: 51–66.
63 R.F. Donnelly, D.I.J. Morrow, T.R.R. Singh, *et al.* (2009). Processing difficulties and instability of carbohydrate microneedle arrays. *Drug Dev. Ind. Pharm.* 35: 1242–1254.
64 Y.K. Demir, Z. Akan and O. Kerimoglu (2013). Characterization of polymeric microneedle arrays for transdermal drug delivery. *PLoS One* 8: e77289.
65 T.-M. Tuan-Mahmood, M.T.C. McCrudden, B.M. Torrisi, *et al.* (2013). Microneedles for intradermal and transdermal drug delivery. *Eur. J. Pharm. Sci.* 50: 623–637.
66 K. Berg (2001). Basic principles of 5-aminolevulinic acid-based photodynamic therapy. In: *Photodynamic Therapy and Fluorescence Diagnosis in Dermatology* (115–143). Elsevier Science B.V.
67 A.K. Jain, C.H. Lee and H.S. Gill (2016). 5-Aminolevulinic acid coated microneedles for photodynamic therapy of skin tumors, *J. Control. Release* 239: 72–81.
68 L.-C. Su and M.-C. Chen (2017). Efficient delivery of nanoparticles to deep skin layers using dissolvable microneedles with an extended-length design. *J. Mater. Chem. B* 5: 3355–3363.

69 R.F. Donnelly, T.R.R. Singh, M.J. Garland, *et al.* (2012). Hydrogel-forming microneedle arrays for enhanced transdermal drug delivery. *Adv. Funct. Mater.* 22: 4879–4890.

70 X. Huang, P.K. Jain, I.H. El-Sayed and M.A. El-Sayed (2008). Plasmonic photothermal therapy (PPTT) using gold nanoparticles. *Lasers Med. Sci.* 23: 217–228.

71 M.C. Chen, Z.W. Lin and M.H. Ling (2016). Near-infrared light-activatable microneedle system for treating superficial tumors by combination of chemotherapy and photothermal therapy. *ACS Nano* 10: 93–101.

72 M.-C. Chen, H.-A. Chan, M.-H. Ling and L.-C. Su (2017). Implantable polymeric microneedles with phototriggerable properties as a patient-controlled transdermal analgesia system. *J. Mater. Chem. B* 5: 496–503.

73 Y. Hao, M. Dong, T. Zhang, *et al.* (2017). Novel approach of using near-infrared responsive PEGylated gold nanorod Coated poly(L-lactide) microneedles to enhance the antitumor efficiency of docetaxel-loaded MPEG-PDLLA micelles for treating an A431 tumor. *Appl. Mater. Interfaces* 9: 15317–15327.

9

Microneedles in Improving Skin Appearance and Enhanced Delivery of Cosmeceuticals

Emma McAlister, Maelíosa T.C. McCrudden and Ryan F. Donnelly

School of Pharmacy, Queen's University Belfast, Belfast BT9 7BL, UK

9.1 Introduction

Since first postulated as a drug delivery mechanism in 1976, microneedle (MN) technology has garnered tremendous interest internationally. In the recent past, MNs have been exploited in the cosmeceutical industry. Cosmetic MN devices have been designed and developed with the distinct function of disrupting the barrier properties of the *stratum corneum (SC)*, thus enabling skin rejuvenation and improved skin appearance *via* induced collagen synthesis and deposition. In addition, they are being used in combination with topical agents or light sources, and also as vehicles to deliver cosmeceutical molecules across the skin.

This chapter focuses on describing the evolution of skin microneedling technologies leading to the emergence and use of MN devices in this field. Commercially available MN devices used in cosmetic applications are described and areas of debate surrounding the safety profiles of MNs, including patient and healthcare providers acceptability of the same, are addressed. Finally, ongoing developments and new advancements in the MN-mediated delivery of cosmeceuticals are highlighted.

9.2 The Skin

Skin health is influenced by numerous factors such as lifestyle, environment (chronic sun exposure and ultraviolet (UV) radiation), genetics, hormones and nutrition [1]. Aging of the skin is a multifactorial biological phenomenon. For example, extrinsic aging, so called "photoaging," results from the cumulative effects of chronic exposure to the elements, primarily UV radiation, and intrinsic aging involves the degradation of elastin fibres and marked collagen reduction causing the development of wrinkles [2]. Facial scarring, arising for any number of reasons such as depigmentation, acne or burn related scars and the formation of large pores, is also a distressing phenomenon encountered by many people [3, 4]. In addition to all of these factors, innovative means of combating stretchmark development continues to undergo review [5]. One of the most common skin diseases that results in scaring is acne vulgaris, a potentially psychologically distressing condition, the pathogenesis of which is underpinned by

Microneedles for Drug and Vaccine Delivery and Patient Monitoring, First Edition.
Edited by Ryan F. Donnelly, Thakur Raghu Raj Singh, Eneko Larrañeta, and Maelíosa T.C. McCrudden.

active inflammation leading to damage to the elastic support structures beneath the skin surface [4, 6]. A recently published paper concisely details variability in acne scarring and documents different treatment approaches that can be used, with choice of treatment dependent on the type of acne scar presented [7]. Ablative methods such as laser resurfacing or dermabrasion are currently available for the treatment of scarred and aging skin but are invariably associated with significant post-operative changes in the skin and require lengthy healing times. When considering this, it was highlighted by Aust *et al.* [8] that there was a need for less invasive treatment options to combat these conditions. A thorough review in 2014 charts the wide range of skin resurfacing therapies currently available to clinicians, in addition to innovative therapies under development [9]. Microneedling and subsequently MN devices are examples of such innovative therapies.

9.3 Microneedling Technologies: An Evolutionary Step Towards MN Usage

The concept of skin microneedling technology in dermatology was pioneered in 1995 when Orentreich and Orentreich successfully reported the use of needles in the treatment of acne scars by a methodology termed subcutaneous incisionless surgery or subcision [10]. This approach involves pricking or puncturing the skin and then scarifying the dermis with the needle to build up connective tissue beneath the scars. This technique could not be used over large surface areas however, due to bleeding and unacceptable bruising. Building upon his work, Camirand and Doucet in 1997 used a tattoo pistol (without ink) to "...needle abrade..." scars [11]. Although this work could have been used on larger areas, this technique was slow and laborious [12]. Based on these principles, an innovator, Fernandes, designed a drum-shaped device with multiple fine protruding needles, initiating the new technological approach of microneedling [13]. Also referred to as MN therapy, percutaneous collagen induction (PCI) [12, 13], collagen induction therapy, dry tattooing (no pigment) and intra-dermabrasion, this innovative microneedling technology has shown promising results as a simple means of minimising skin imperfections such as acne scars, surgical scars, fine lines, wrinkles, stretch marks and cellulite, in addition to improving skin texture, firmness and hydration [14, 15].

The principle of this technique is to break collagen strands, which tether the scar to the base of the dermis, *via* minute controlled injuries, caused by multiple microscopic needles piercing the skin. This technique induces angiogenesis and collagenesis *via* induction of the natural post-traumatic inflammatory cascade. The basis of microneedling is the controlled mechanical stimulation of the wound healing response (Figure 9.1), which is divided into three phases: (I) initiation/inflammatory, (II) proliferation and (III) remodelling. Briefly, in phase I, after MN penetration, platelets, neutrophils and macrophages recruited to the injury site stimulate the release of growth factors, which in turn initiate the release of cytokines [16]. Growth factors including epidermal growth factor (EGF), transforming growth factor-alpha (TGF-α), TGF-beta (-β) and platelet derived growth factor (PDGF), in turn, activate fibroblasts and fibroblast proliferation, which all facilitate the production and propagation of intercellular matrix proteins [12, 16]. At this stage, neutrophils are the dominant leucocytes but are gradually

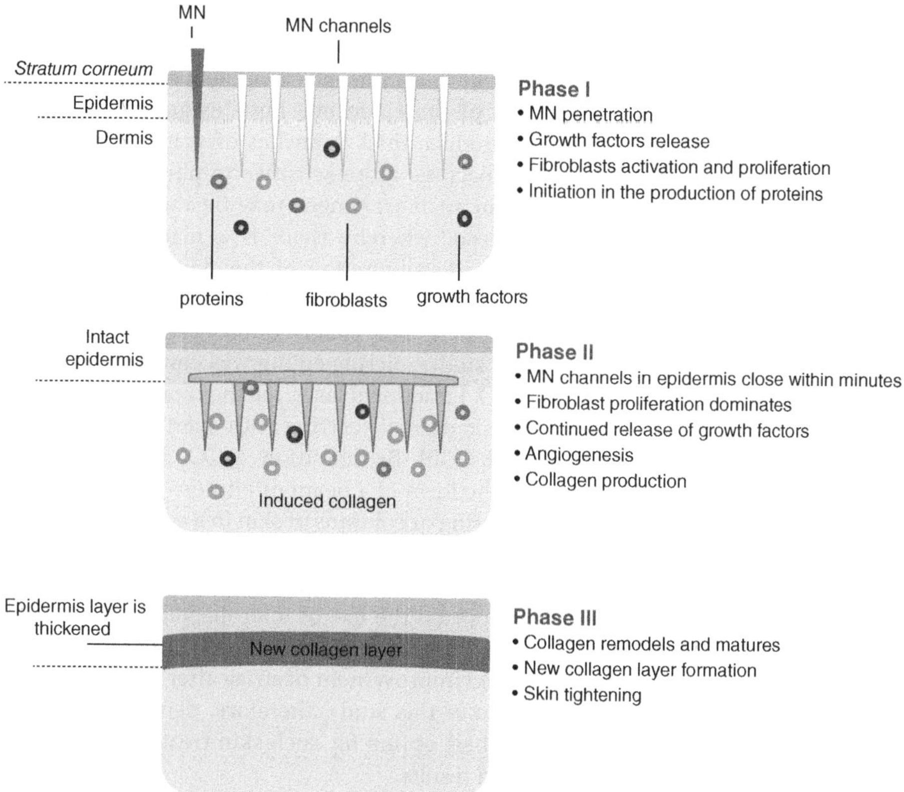

Figure 9.1 Schematic representation of the mode of action of MN devices. Phase I is characterised by growth factor release, following the initial microneedling injury. Phase II is dominated by fibroblast proliferation. Keratinocytes stimulate growth of the epidermis and release growth factors to promote collagen production by the fibroblasts. New blood vessels are created and there is a surge of collagen/matrix deposition. In the final remodelling phase, phase III, collagen III is converted into collagen I and the skin becomes tighter. Blood supply is normalised and the skin also becomes smoother.

replaced by monocytes, the dominant leucocytes in phase II. In phase II, fibroblast proliferation dominates and monocytes, keratinocytes and fibroblasts continue to release growth factors [12]. As a result, angiogenesis occurs whereby a fibronectin matrix is formed and fibroblasts ultimately deposit collagen, along with other matrix proteins including elastin, proteoglycans and glycosaminoglycans (GAGs) at the wound site [16]. In phase III, the conversion of collagen III, through tissue remodelling and vascular maturation, into collagen I ensues, resulting in skin tightening [12]. The final remodelling phase of healing after microneedling can take up to several months.

9.4 Benefits of Microneedling

One of the most important advantages of microneedling is in overcoming the use of traditional ablative methodologies. Ablative methods including subcision, chemical peels, collagen injections, cortisone-like injections, cryosurgery, dermabrasion and laser resurfacing are, by their nature, extremely physically disruptive to the targeted epidermis and superficial dermis [17]. In ablative methodologies such as these, undesired postoperative changes in the skin can result in significantly lengthened healing times. The destruction, rather than disruption, of the epidermis initiates an inflammatory response that stimulates fibroblasts to produce thick branches of scar collagen [18]. As a result, the skin has been shown to become more sensitive to photodamage and one study has stated that a side-effect of such treatment may be the development of dyschromias, the common skin disorder whereby there is a marked alteration in normal skin pigmentation, resulting in discolouration of the skin, hair and nails [19]. In contrast, microneedling (a non-ablative method), with needles of appropriate heights, would negate the risks and negative side-effects often seen with invasive ablative approaches. For example, in one study, microneedling was compared with an ablative methodology, carbon dioxide (CO_2) fractional laser, for the treatment of striae distensae (stretchmarks, a form of atrophic dermal scarring with overlying epidermal atrophy) in the abdomen and lower limbs [20]. By definition, a CO_2 fractional laser involves the emission of light energy, in the form of a beam of photons from CO_2 gas. This fractional approach results in the lasering of columns of skin in a grid-like manner, leaving the skin surrounding each column intact, thus aiding healing [21] (https://www.advdermatology.com/c02re--co2-fractional-laser--pages-235.php (accessed 7 June 2017). In this study, the results supported the use of microneedling over a CO_2 fractional laser, as 90% of microneedling treated patients showed improvement of striae, whereas only 50% of patients showed improvement of striae after treatment with a CO_2 fractional laser [20]. The outcome of this study, therefore, demonstrates that ablative methods are not necessarily the best option for such skin treatments, both in terms of significant recovery time and end results.

The suitability of non-ablative approaches for use on skin areas where laser treatments and chemical peels cannot be performed, for example, areas very close to the eyes, is another benefit of microneedling compared with the traditional ablative methods. Furthermore, in one comparative study, greater collagen deposition was compared between two non-ablative methods, intense pulsed light and microneedling to combat scarring [14]. A study was conducted using 54 imprinting control region mice and four weeks after the last treatment, skin thickness measurements using calipers, microscopic examination, western blot analysis for type I collagen and enzyme-linked immunosorbent assay for total collagen content were performed. Microneedling was determined in this study to be more efficacious than intense pulsed light treatment, based on the fact that increased collagen deposition was observed from the biopsied specimens, as well as increased expression level of type I collagen and total collagen content, thus leading to improved smoothing of scars [14]. Therefore, this study highlights the successful outcomes of microneedling in comparison with an alternative non-ablative method.

9.5 Commercially Available MN Devices

9.5.1 Dermaroller®

The original concept design, developed by Fernandes, that is used for microneedling is known commercially as a Dermaroller®. Manufactured by Dermaroller® Deutschland GmbH and registered with the Food and Drug Administration (FDA) as a class I

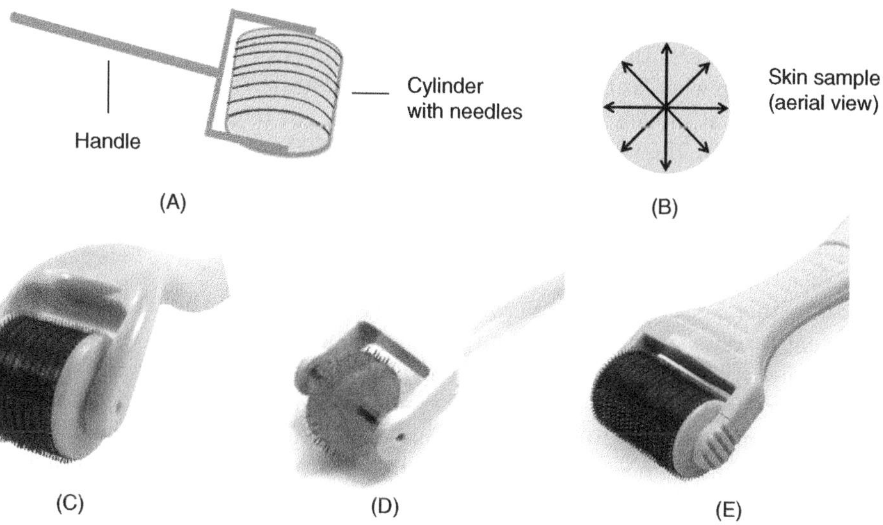

Figure 9.2 (A) Schematic representation of the Dermaroller® based upon the principles of Fernandes initial PCI innovation. It consists of 24 circular arrays of 8 needles each located on the roller (total 192 needles). (B) Schematic representation of an exemplar skin sample. The application of a roller MN device is applied across the skin, vertically, horizontally and diagonally, as depicted. (C, D and E) Commercially available Dermarollers®. (C) Royal Derma Roller with 540 needles. Reproduced with permission from [26] Royal Derma Roller (2015). Derma Rollers. https://www.royaldermaroller.com/collections/derma-rollers/products/derma-roller-1-0-acne-scars (accessed 25 May 2017). (D) Dermaroller® with narrow drum width, namely the DNS® Classic 3 Line Roller with 75 needles. Reproduced with permission from [27] DNS® (2006). DNS Classic 3 Line Roller. http://www.dnsroller.com/dns-classic-roller-dermaroller-micro-needling-derma-system-titanium-biogenesis-dns-london/dns-classic-3-line-roller-p10002#.WTgFf4WcHIU (accessed 25 May 2017). (E) Dermaroller® with large width of drum, namely Body Roller with 1080 needles. Reproduced with permission from [28] MTS Roller.com (2012). Derma Rollers, MTS Derma Roller, Body Roller. https://www.mtsroller.com/1080-titanium-needles-body-roller (accessed 7 June 2017).

medical device, a Dermaroller® (also known as a skin roller or MN roller) is described as a "...simple, hand-held device equipped with medical grade solid steel needles, projecting from a cylindrical roller..." (http://www.dermarollersystem.com/how-the-dermaroller-system-works.html (accessed 26 April 2017) [22]. Typically, a Dermaroller® consists of 24 circular arrays of 8 needles each located on the roller (total 192 needles) [23] (Figure 9.2A). The roller device is applied directly across the skin, vertically, horizontally and diagonally (Figure 9.2B). Dermarollers® can be used as homecare treatment MN devices or as a treatment conducted by a trained professional. Needle lengths are dependent on the nature of the treatment being employed [24]. For example, for treating acne and other scars, needle lengths of 1500–2000 μm are routinely employed. When microneedling is used as a procedure to treat aging skin and wrinkles, needle lengths of 500–1000 μm are usually recommended [25].

There are currently five types of Dermarollers® registered with the FDA, categorised according to their needle lengths. Home Dermarollers® consist of the C-8 and the C-8HE models. The cosmetic, C-8 and C-8HE models have needle lengths ranging between 130 and 200 μm, with a penetration diameter of 70 μm (http://www.dermarollersystem.com/how-the-derma-roller-system-works.html) (accessed 26 April 2017). The C-8HE was specifically designed for hair-bearing surfaces, such as the scalp whereas the C-8 model is described as the "...basic Dermaroller®..." [25]. The medical models, CIT-8, MF-8 and MS-4, are intended for use by trained professionals only. The CIT-8 has needle lengths of 500 μm and the MF-8 has needle lengths up to 1500 μm [23]. The MS-4 is the only Dermaroller® that has a small cylinder, 4 circular arrays of 24 needles (total 96 needles) that can have needle lengths up to 1500 μm. It is used on areas where better precision and deeper penetration is required, such as on facial acne scars [25].

As the therapeutic use of microneedling is now being extended beyond scar treatment, various modifications and advancements have evolved since the initial Dermaroller® was introduced. The current Dermaroller® market is ever-increasing with an assortment of Dermaroller® devices based on various needle lengths and drum sizes being introduced onto the market from a multitude of different manufacturing companies. Some such companies include Hansderma, White Lotus, Royal Derma Roller, bioGenesis London and MTS Roller. Hansderma have developed Genosys® rollers that feature needles 25% thinner than other brands along with more needles per device (450 needles per unit) [29]. They also provide Dermarollers® with detachable heads to facilitate re-use of the device handles. White Lotus have developed a hypoallergenic Dermaroller® (or Lotus Roller), which is metal-free and made from a biocompatabile polymer and therefore suitable for use by end-users who are allergic to metals [30]. Royal Derma Rollers manufacture Dermaroller® products that are made using titanium alloy needles [29] (Figure 9.2C) and bioGenesis London produce the DNS® Classic 3 (Figure 9.2D) and 8 Line Roller [31]. Lastly, MTS Roller manufacture a host of different Dermaroller® devices, namely the Dr. Roller, MT Roller, MRS Roller, Body Roller, Elimiscar Roller and the ZGTS Titanium Derma Roller [32]. These Dermarollers® differ according to handle and roller design. For example, the Body Roller, see Figure 9.2E, has a larger roller head (total 1080 titanium needles) to accommodate larger skin areas such as the chest, back, buttocks, thighs and arms [27].

Therefore, the microneedling field is growing constantly. The conventional Dermaroller® developed by Fernandes has evolved dramatically over the past 15 years through a variety of advancements. To this end, other MN devices will now be explored and described in more detail.

9.5.2 Beauty Mouse®

Listed as a medical device on the Australian Register of Therapeutic Goods (ARTG), the Beauty Mouse® is an approved MN device intended for home use, and again its usage is based on the same principles of the Dermaroller® technology (http://www.genuinedermarollerclinic.co.uk/beauty-mouse.html (accessed 5 June 2017)). It contains a total of 480 needles of length approximately 200 μm on three separate Dermaroller® heads strategically placed inside a computer mouse-shaped device (Figure 9.3A). Like the Body Roller, it has been developed to ensure coverage of larger skin surface areas, such as the arms, legs and buttocks, for the treatment of stomach or thigh stretch marks and cellulite [25, 33].

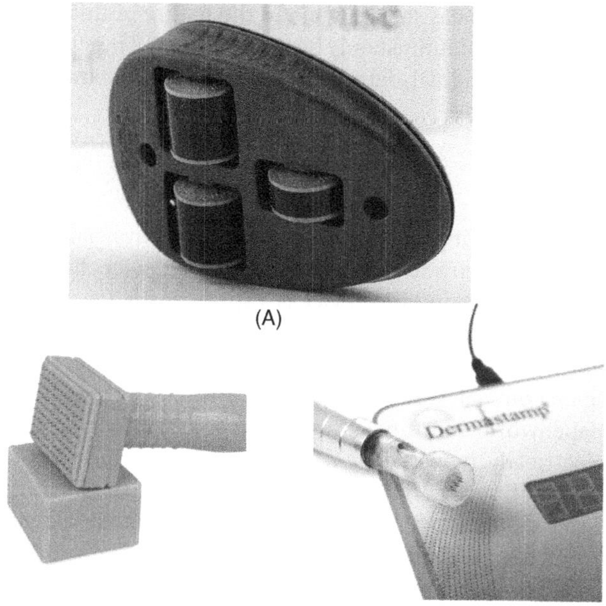

(B) (C)

Figure 9.3 Examples of various MN devices now available for use. (A) Image of Beauty Mouse®. Reproduced with permission from [34] Dermaroller®. Beauty Mouse® Micro-Needling Home Body. https://dermarollerus.com/color-box-product/891?width=600&height=600 (accessed 6 June 2017). (B) Image of White Lotus Dermastamp™. Reproduced with permission from [35] White Lotus Holistic Microneedling. Scar Dermastamp 1.0 mm. https://www.whitelotusantiaging.co.uk/scar-dermastamp-1-0mm (accessed 6 June 2017). (C) eDermastamp®. Reproduced with permission from [36] Consulting Room. eDermastamp™. http://www.consultingroom.com/Treatment/eDermastamp-eDS-Skin-Rejuvenation (accessed 6 June 2017).

9.5.3 Dermastamp™

The evolution of microneedling technologies is highlighted by the emergence onto the UK market of miniature versions of the Dermaroller®, termed Dermastamps™. These are sterile medical MN devices with a range of different needle heights (200–3000 μm) (Figure 9.3B) [25]. They are specifically designed to accommodate small, localised, confined areas where it is difficult for the conventional Dermaroller® to achieve optimal stimulation, for example, the upper lip. The device uses "…vertical penetration to create infusion channels…" in the skin and is considered ideal for use on isolated scars and wrinkles [37]. Another area where this device is being considered is in the stimulation of hair growth but no research studies have yet been published on this topic and MN device. Very recently, a more advanced device called the eDermastamp® has been developed, which is manufactured by the same makers of Dermaroller®, Dermaroller® Deutschland GmbH (Figure 9.3C) [38] (http://www.original-dermaroller.de/en/edermastamp.html (accessed 6 June 2017)). This device is an electronically powered stamping device consisting of six fine precision needles of maximum length 1500 μm, arranged in a circular design, the needles of which are made from medical grade stainless steel. Owing to the nature of this device, the manufacturers have recommended that it must only be used by trained practitioners and authorised clinics (http://www.original-dermaroller.de/en/edermastamp.html) (accessed 6 June 2017).

9.5.4 Dermapen®

Designed to overcome the issues of varying pressure application by physicians/users and the subsequent needle depth penetration achieved, an MN device has been launched termed the Dermapen®. This ergonomic device is a spring loaded, oscillating MN device, consisting of two parts: an electric hand piece and a sterile, individual, disposable needle cartridge, which carries out the function of "…fractional mechanical resurfacing…" (http://dermapen.com/safe-effective-dermapen-leads/ (accessed 26 April 2017), http://www.dermapenworld.com/dp-family/dermapen-3?tab=1 (accessed 26 April 2017)). It uses an electrically powered pen to deliver a vibrating, vertical, stamp-like motion to the skin, creating a series of micro-channels in it. Consisting of 12 stainless-steel needles, the penetration depth of the needles can be adjusted from 250 to 2500 μm by simply turning the adjustment ring on the MN device (http://www.dermapenworld.com/dp-family/dermapen-3?tab=1) (accessed 26 April 2017). Treatment in acne scarring, burn scars and photoaging is being investigated by the manufacturers, although no available research studies focusing on this MN device have yet been published.

Owing to the substantial interest in these MN devices, more commercial products based on the same microneedling technology are currently being developed, of which four will be discussed. Within the Dermapen® family, AOVN™ technologies launched Dermapen 3™, the latest and most advanced model (Figure 9.4A) (http://www.dermapenworld.com/dp-family/dermapen-3?tab=1) (accessed 26 April 2017). Compared with the conventional Dermapen®, its key new feature is that it is able to produce over 1300 micro-holes in the skin per second, therefore allowing shorter treatment times. The manufacturers deem it to be more robust, more user-friendly and better suited for the rigors of clinical applications than previous models of the Dermapen®, with clinical reviews stating that it "…glides better because there is less rattling and vibration…" (http://www.dermapenworld.com/dp-family/dermapen-3?tab=1) (accessed 26 April 2017). There are currently two models of the Dermapen 3™ device available on the market, Dermapen 3MD™ and Dermapen 3PRO™. Not intended for home use, the Dermapen 3MD™ and Dermapen 3PRO™ are for clinical and professional use, respectively. Penetrating to depths up to 2500 μm, the Dermapen 3MD™ is suited for more aggressive skin remodelling treatments such as scars whereas the Dermapen 3PRO™ (penetration depth up to 1000 μm) is suitable for use in improving the appearance of fine lines, pigmentation and enlarged pores.

A new and improved version of the conventional Dermapen® that uses the patented microneedling tip technology, SurSpace™, is the MDerma™ FDS (http://dermapen.com/mderma-pen/ (accessed 27 April 2017)). Its patented 12 needle tip arrangement creates maximum pressure efficacy. Its disposable needle cartridge includes a scalloped edge and two opposing vents to eliminate suction and prevent device contamination (Figure 9.4B). The oscillating function of the treatment needle tip is also improved, using a new elastomeric spring with an updated motor for maximum power and

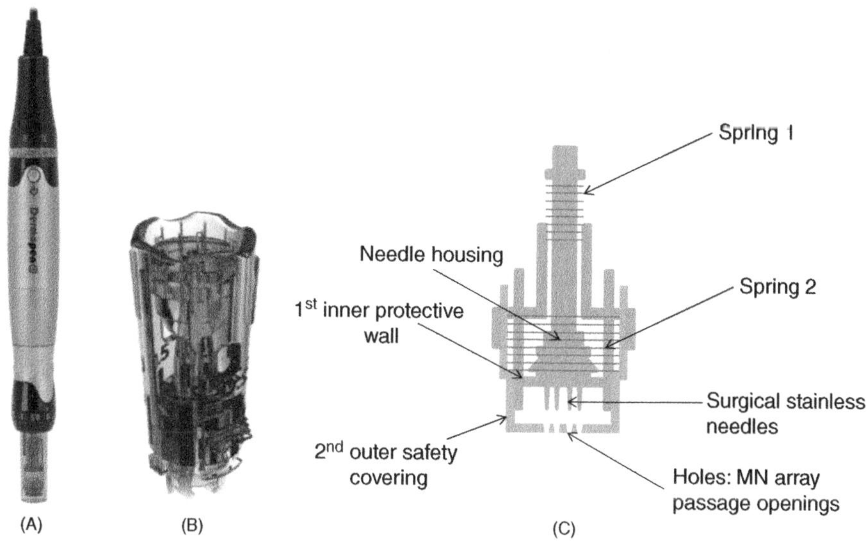

Figure 9.4 Commercially available Dermapens®. (A) Image of a Dermapen 3™. Reproduced with permission from [39] Advanced Skin Innovations (2017). Dermapen™. http://www.equipmed.com/LiteratureRetrieve.aspx?ID=150517 (accessed 7 June 2017). (B) Image of a MDerma™ FDS disposable needle cartridge. Reproduced with permission from MDerma™ marketing PDF – The most advanced, effective and safest micro-needling tip. https://www.mdermaaustralia.com.au/wp-content/uploads/2015/12/MDermaFeatures.pdf (accessed 7 June 2017). (C) Schematic diagram of the patented disposable needle cartridge INNOTip™, highlighting the double-protective, sealed tip and dual-spring system.

performance. From the patented INNO™ Technology, another Dermapen®-like MN device that has been combined with the patented disposable needle cartridge, termed the INNOTip™ (engineered by Clinical Resolution Lab, Inc.), is the INNOPen™ (https://www.uniqueskin.co.uk/products/innopen-microneedling-professional-use-only (accessed 27 April 2017)). In contrast to competitor MN devices that use single-walled open tip cartridges, the INNOTip™ provides a unique, double-protective, sealed tip system, see Figure 9.4C, which consists of an inner protective wall and an outer safety covering made of sterile, medical grade polycarbonate resin (https://www.uniqueskin.co.uk/products/innopen-microneedling-professional-use-only and http://www.clinicalresolution.com/main/pdf/INNOTip.pdf (accessed 6 June 2017)). Consisting of 13 surgical stainless-steel needles, the first inner protective wall stabilises the INNOTip™ at all speeds, which serves to prevent the needle housing from shaking and becoming off-centred due to vibrations caused by the device's motor. Furthermore, it ensures safety *via* uniformity in the vertical lining of the needles therefore preventing unnecessary needle trauma and injury from slanted needle insertion. The second outer safety covering provides complete enclosure of the system with only the 13 passage openings. As a result, all 13 needles are required to pass through these tiny holes before perforating the skin. This outer safety covering detects any deformity and/or misalignment of the needles therefore ensuring accurate needle penetration and patient safety (https://www.uniqueskin.co.uk/products/innopen-microneedling-professional-use-only and http://www.clinicalresolution.com/main/pdf/INNOTip.pdf) (accessed 6 June 2017). The dual-spring system is another unique feature of the INNOTip™

that allows the user to have full control over the needle motion. The INNOPen™ is available in two models, INNOPen MD™ and INNOPen PRO™. The INNOPen MD is for medical professionals whereas the INNOPen PRO™ is for clinical use only. Like the Dermapen 3™ models, the INNOPen MD™ penetrates up to 2500 μm and therefore is used for deep wrinkles and atrophic scars. Similarly, the INNOPen PRO™ has a maximum needle length of 1000 μm, so is suitable for treating skin conditions such as anti-aging and hyperpigmentation.

As the microneedling market continues to grow and develop, new innovations will emerge which will expand upon the unique engineering features of the aforementioned MN devices. In line with this, the combinatorial use of MN devices with other novel cosmeceutical treatments will undoubtedly be explored. On this theme, MN devices have undergone exploration in the field of photodynamic therapy (PDT) for enhanced topical treatments.

9.5.5 Light Emitting MN Devices

Microneedling has previously been used in combination with PDT to enhance topical delivery of aminolevulinic acid (ALA), in the treatment of actinic keratosis, the dry scaly patches of skin caused by long term sun exposure [40]. In this study, skin was pre-treated with an MN device, Roll-CIT™, (MN width = 108 μm, MN length = 300 μm) and ALA was then applied to the skin for a defined period. Following this, the use of red light and broadband pulsed light allowed for deeper activation of ALA, resulting in statistically significant improvements in photoaging scores [40]. For a more detailed review of the use of MNs in conjunction with PDT, please refer to Kearney *et al.* [41] and Chapter 8 of the present book. Other research teams have also developed microscale optical diffusers, or fibre optic MNs, for the enhancement of clinical laser procedures and homogeneous light emission, while minimising photothermal damage in non-target tissues [42]. In one such study, results indicated that MNs with needles of smaller diameter (base width = 33–48 μm) were favourable, in terms of their light delivery abilities [42]. The authors agreed, however, that further research is required to mechanically strengthen these MNs due to the risk of damage/breakage whilst being inserted into the skin. Evolving from these innovations, new CE approved MN devices, termed light emitting diode (LED) MicroNeedling Rollers (MN length of 1000 μm) have been launched. These incorporate titanium needles and LED light to combat wrinkles and scarring and are used in a fashion akin to that described for the Dermaroller® device (http://www.dhgate.com/product/new-led-540-needle-roller-acne-scars-cellulite/133208643.html#sb1-8-1b;hot|4293359023 (accessed 30 April 2017)). Worth noting, however, is that despite the fact that these MN devices are readily available for purchase online, there is very limited technical information available regarding their development and use.

9.6 Patient Factors Relating to MN Devices

Despite the promising therapeutic benefits of MN devices and their increasing availability and usage, issues surrounding patient acceptability of such devices, in addition to queries surrounding potential erythema, irritation and patient safety in usage and sterilisation considerations have been raised [43–45].

9.6.1 Acceptability of MN Devices by Patients and Healthcare Providers

The future commercial and clinical success of MN devices will undoubtedly depend not only upon their ability to fulfil a designated function, but also on their acceptability by both patient and healthcare professionals [46]. To this end, initial user perspectives of MNs were documented in an informative study carried out by the Birchall group [47] and this data has been built upon by recent pilot studies conducted by the Donnelly research team [48–50]. With reference to one of those studies, in all instances, study participants acknowledged the potential benefits of MN delivery systems with 80% of the participants having a "strongly positive" perception of the MNs [48]. Potentially detrimental to the impact of such studies however are reports, such as that published in the *Daily Mail* newspaper in 2009, which stated that the Dermaroller®, specifically, resembled a "…miniature medieval instrument of torture…" [43]. Despite this poor description of the MN device, the article concluded that the use of the Dermaroller® resulted in signs of improved skin appearance [43]. Media reports, such as this, reinforce the importance of clear message dissemination to the general public about these MN devices and their uses so that their reputation may not be harmed unnecessarily.

9.6.2 Potential Irritation and Erythema

With reference to the potential for local irritation or erythema of the skin following the use of an MN device (a fundamental concern based on the potent immune-stimulatory nature of the skin) [23], research has proven that prompt recovery of skin barrier function is achieved within a matter of hours of MN device usage [3, 51]. These studies also address the issue that the time frame for skin recovery is dependent on age, skin elasticity, skin application site and application pressure (Dermaroller®) [3, 51]. In relation to the reddening of the skin, one study proved that erythema normalised within 24–48 h in all groups under investigation with no significant difference observed in the erythema index ratios between the different needle lengths employed in the study [3].

9.6.3 Patient Safety

The potential long-term effects of MN application, or indeed repeated MN application, on skin have not yet been fully evaluated with only one study having addressed this question. This study described the local and systemic effects of repeat applications of polymeric MNs on hairless Crl: SKH1-Hr^{hr} mice *in vivo* [52]. Two control groups of mice had no MNs inserted into their skin, while the remaining groups had one of two polymeric MN formulations (dissolving MNs or hydrogel-forming (swelling) MNs) inserted into their skin repeatedly. Dissolving MNs were inserted once per week for five consecutive weeks whereas hydrogel-forming MNs were inserted twice weekly for three successive weeks. Skin appearance and skin barrier function, as illustrated by measurement of transepidermal water loss (TEWL), were not measurably altered during the entire study period. Biomarkers of infection (C-reactive protein (CRP)), immunity (immunoglobulin G (IgG)) and inflammation/irritation (tumour necrosis factor-alpha (TNF-α) and interleukin 1-β (IL-1β)) were also statistically unchanged, regardless of the MN formulation, needle density or number of applications of the MN [52]. Therefore, this preliminary study suggests that repeated use of the specific polymeric MN investigated does not cause undesirable local or systemic side-effects in the animal model employed. It is clear that similar studies must be carried out for all MN delivery platforms under development to ensure safety profiles in advance before proceeding to MN usage with patients.

Moving beyond this, it is without question that inappropriate use of any of these novel technologies could cause problems. With specific reference to cosmetic MN devices, three cases of allergic granulomatous and systemic hypersensitivity in female patients, following application of a non-sterile topical product and subsequent MN treatment, have been reported [44]. Worth noting however is that, in each case, medical supervision was absent. In all three cases, the deleterious side-effects witnessed were determined to have been due to the inappropriate intradermal tattooing of the skin with antigenic topical products, rather than usage of the MN device alone [44]. The combined use of these topical products with MN application is unlicenced and so the artificially enhanced delivery of these products to the dermis in this report resulted in the hypersensitivity reactions observed. To explain this further, as the name suggests, these topical products were licenced for use topically rather than designed specifically to penetrate across the *SC*. In all instances, full or partial recovery of the skin was achieved following corticosteroid or tetracycline treatments [44]. These cases highlight however the need for caution with regards to inappropriate use of topical medicines in combination with enhanced transdermal delivery methods. As a result, MN application should only be used in conjunction with fully licenced and tested therapeutics or cosmetics, designed for this intended purpose. Furthermore, despite the fact that MN devices are often categorised into home and medical use in the product literature, they are widely available for purchase by any individual online from a multitude of commercial websites (https://www.amazon.co.uk/Derma-Roller-Anti-Ageing-Stretch-Cellulite/dp/B00L63C468/ref=sr_1_7_a_it?ie=UTF8&qid=1496847802&sr=8-7&keywords=derma+rollers (accessed 7 June 2017)). With no restriction on purchasing of these products, it is clearly evident that there is potential for abuse and misuse of these MN devices.

9.6.4 Sterilisation Considerations

The question of whether MN devices, for use in cosmetic applications, should be terminally sterilised has also been brought to the fore. This question must be addressed in line with the recommendations of regulatory bodies, as inappropriate use of MN devices could potentially lead to the early, unwarranted rejection of these devices. An example of one such damaging report, which involved the inappropriate use of non-adequately sterilised MN devices between patients, was published in the *Daily Mail* in August 2011 [45]. Regardless of how MN devices are classified by the pharmaceutical and cosmeceutical industries, for example as drug delivery systems, consumer products or medical devices, they are not equivalent to conventional transdermal patches, in that they do not simply adhere to the skin surface [46]. Considered to be more akin to a conventional hypodermic injection, this may mean that they will be required to be terminally sterilised. The Dermaroller®, Beauty Mouse®, Dermastamp™ and Dermapen® are supplied as sterile products and as they are all fabricated from metal, they can be easily re-sterilised. In particular, the Dermapen® explicitly highlights this issue by describing it as a "...patent pending MN tip..." (http://dermapen.com/safe-effective-dermapen-leads/ (accessed 26 April 2017)) that is supplied sterilised in individual packages which are easily replaced and loaded into the spring automated device. It has been demonstrated in published work that the holes created in the skin by the individual needles on the MN, elicit minimal microbial penetration [2, 53, 54]. In order to ensure this is the case, particularly in a home environment, we, the authors, feel that appropriate sterilisation of such MN devices prior to and following use is essential. With this in mind, the UK Health Centre does provide guidance for at-home sterilisation of Dermaroller® devices, whereby it is suggested that they be sterilised using those agents routinely used to sterilise babies' bottles (http://www.healthcentre.org.uk/cosmetic-treatments/derma-roller-sterilising.html (accessed 9 May 2017)).

9.7 Delivery of Cosmeceutical Compounds

9.7.1 A Role for Hyaluronic Acid in MN Delivery Systems

Hyaluronic acid (HA), also termed hyaluronan, is a ubiquitous component of the extracellular matrix, similar to collagen and elastin, which maintains the suppleness of the skin. HA is a naturally occurring linear, polyanionic, polysaccharide with repeating disaccharide units composed of β-glucuronic acid and *N*-acetyl glucosamine [55], see Figure 9.5. HA is lost from the skin upon aging but has been deemed to play a pivotal role in tissue rejuvenation [56]. It has previously been shown, when applied topically, to be absorbed from the surface of the skin through the epidermis, thus restoring moisture and elasticity [57]. It was, at that time, suggested by the authors that HA could be a suitable candidate to act as a unique vehicle for the transport of other drugs into the deeper layers of the dermis. Prior to this, a HA-based gel was implanted intradermally in augmentation therapy studies of facial soft tissues, once again highlighting the potential of this compound in skin rejuvenation [58]. Based on these previous studies and the unique, versatile properties of HA, it has been studied as a suitable candidate for MN formulation and subsequent delivery of incorporated active pharmaceutical ingredients (APIs) [56, 59, 60]. One such HA-based MN product is MicroHyala®. Based on the dissolving MN design, as previously addressed in earlier studies [61], the salt form of HA, namely, sodium hyaluronate, acts as the base material for these MNs [56]. Despite abstract reference to the use of this product in cosmetic applications [56], there are, to date, no publications on the use of this, or similar HA-based patches, in cosmetic applications specifically. The mechanical properties of MicroHyala®, incorporating model drugs, has been evaluated however and the authors have conceded that the product, although promising, does require further development [56]. With the emergence of HA as a novel cosmetic and MN scaffold material, it was agreed recently that a fuller understanding of its inherent biochemical functions and interactions is warranted, in order to ensure HA homeostasis [62]. Although it is evident that further research must be performed on HA-based MN products, the viscoelastic properties of this compound, together with the excellent biocompatibility

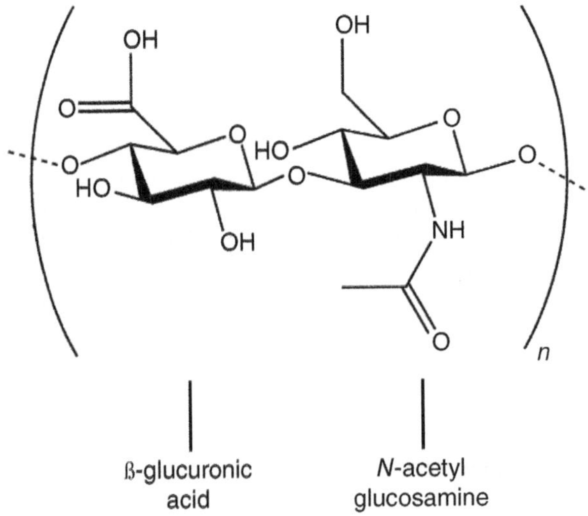

Figure 9.5 Schematic representation of the structure of hyaluronic acid (HA), highlighting the β-glucuronic acid and *N*-acetyl glucosamine repeating subunit pattern.

and non-immunogenicity of HA, suggest it to be an ideal candidate for investigation in MN-based cosmetic, medical and pharmaceutical products.

9.7.2 MN-mediated Peptide Delivery

Protein-based compounds do not readily permeate across the skin, due to their large molecular weights and hydrophilic nature [77]. The delivery of protein- and peptide-based therapeutics across the skin continues to undergo extensive investigation, with current research focusing on the delivery of insulin and vaccines across the skin [24, 64, 65]. In relation to the cosmetic industry, the intradermal delivery of peptide cosmeceuticals *via* MNs is now undergoing investigation. Three such peptides are melanostatin, rigin and pal-KTTKS. Melanostatin is a novel melanin synthesis inhibitor and rigin can reduce inflammation. Pal-KTTKS is a peptide that, when delivered into the dermis, has been proposed to stimulate collagen production [66]. A recent study used confocal scanning laser microscopy and fluorescently tagged melanostatin, rigin and pal-KTTKS to assess the influence of MN usage on the penetration and distribution of these peptides in the skin, compared with passive diffusion of the same peptides [66]. Pre-treatment of the skin with solid, stainless-steel MNs, followed by peptide application, resulted in increased penetration and distribution of the lowest molecular mass peptide, namely melanostatin. This trend was not maintained with the peptides of higher molecular masses however. The authors, in interpreting these findings, perhaps could have commented on the MN modality enlisted in this work. It would have been interesting to compare and contrast delivery of the same peptides using various MN approaches, such as dissolving and hydrogel-forming MNs. These alternative approaches may, in time, prove more suitable for the delivery of protein-based therapeutics and cosmeceuticals when compared with solid MNs.

9.7.3 The Delivery of Other Cosmeceutical Agents

The exploitation of MN technologies in facilitating efficient delivery of cosmeceuticals across the skin is constantly advancing with many interesting studies having been published in the recent past. A study carried out in mice detailed the *in vivo* efficacy of eflornithine cream, a topical product used to reduce facial hirsutism, following pre-treatment of mouse skin with MNs [67]. The positive results achieved in this preliminary study promoted the continuation of work in this area. In this respect, MN devices have also been used in combination with topical products, such as minoxidil, to stimulate hair regrowth. Following on from pilot studies carried out in mice [68, 69], a 2013 study evaluated the influence of microneedling and minoxidil topical treatment, on human subjects suffering from androgenetic alopecia (AGA) [70]. A Dermaroller®, with needle heights of 1500 μm, was rolled over the shaven scalp in longitudinal, vertical and diagonal directions and minoxidil was applied 24 h post-procedure. Using the three primary efficacy measures of hair count and patient/investigator assessment, the Dermaroller® and minoxidil treated group were determined to exhibit statistically superior hair growth when compared with those treated with minoxidil only [70].

With a view to improving skin appearance, an interesting study was conducted *in vitro* on the delivery of epigallocatechin-3-gallate (EGCG), a high molecular weight and lipid binding flavanol that has UV radiation protection, photoaging and collagen degradation prevention properties, using solid maltose MNs [71]. Permeation of EGCG from an aqueous solution, as well as a rheologically optimised hydrogel, through dermatomed porcine ear skin (untreated and MN treated) was evaluated. To quantitatively determine the amount of EGCG retained in the skin layers, after 24 h, the *SC* was separated from the underlying epidermis and dermis. MN treated skin showed significant enhancement in the delivery of EGCG to the viable epidermis and dermis from the aqueous solution (38.67 ± 2.96 μg/cm^2) as well as the hydrogel (24.60 ± 2.62 μg/cm^2) in comparison with the untreated skin (24.16 ± 2.11 and 15.62 ± 0.24 μg/cm^2 for hydrogel and aqueous solution, respectively) [71]. To explain this, due to EGCG aqueous solubility, EGCG was able to diffuse through the hydrophilic micro-channels, but at the same time, due to the lipophilicity rendered by the gallate group (in the molecule) and very high binding affinity to the skin tissues, particularly the collagen network, EGCG was found to have concentrated in the dermis. This is the ideal site of action for its antioxidant and photoprotective activities [71]. Therefore, this study provided evidence to indicate that solid maltose MNs facilitated the penetration of EGCG across the *SC* into the deeper skin layers.

In another study, skin depigmentation was investigated on the facial area of 45 women who had facial hyperpigmentation [72]. Subjects in this study were divided into two groups and dissolving HA-MNs containing the depigmentation agent, 4-*n*-butylresorcinol, were applied to the faces of the participants every four days for eight weeks (Group 1) and every three days for eight weeks (Group 2). In both groups, a control MN patch (containing no 4-*n*-butylresorcinol) was applied to the other side of the face. For effective skin depigmentation, the skin depigmentation agent must be delivered to melanocytes, where melanin is synthesised. This study showed that this novel HA-MN patch was twice as effective in delivering 4-*n*-butylresorcinol than the control patch in both the four- (Group 1) and three-day (Group 2) interval tests, with no allergic reactions and only mild skin irritation observed throughout the study [72]. Comparing the four- and three-day interval tests, the three-day interval was more effective, therefore indicating that more frequent patch application would lead to a better skin depigmentation effect [72]. Thus, the authors concluded that this novel dissolving HA-MN has the potential to be fabricated using other depigmentation and anti-aging agents because of its usability, safety and efficacy.

Two complementary examples of successful MN-mediated cosmeceutical delivery were published recently, evaluating the anti-wrinkle effect of an ascorbic acid-loaded dissolving HA-MN patch [73, 74]. In one of the studies, two anti-wrinkle compounds of different hydrophilicities, namely ascorbic acid and retinyl retinoate (Figure 9.6A and B, respectively) were incorporated into dissolving HA-MNs, separately, for evaluation in combating wrinkles, on 24 women for 12 weeks [73]. Patients were randomly allocated to either the ascorbic acid or the retinyl retinoate group. All subjects applied dissolving HA-MN patches to the "crow's feet" area of the face twice daily and these were left in place for 6 h. These HA-MN patches were deemed to have efficiently delivered their payload as both MN formulations displayed improved skin appearance in terms of reduction of roughness and diminished wrinkle appearance. In addition, the MN demonstrated sufficient mechanical robustness for insertion into the skin at drug loadings of 60% for ascorbic acid and 35% for retinyl retinoate [73]. However, in this study, skin problems including irritation and sensitisation, which are critical issues in development of the MN patch for cosmetic purposes, were not fully assessed. Therefore, leading on from this study, a second study, from the same research group, was conducted, evaluating the anti-wrinkle effect of the ascorbic acid-loaded dissolving MN patch *via* a double-blind, placebo-controlled clinical study [74]. The anti-wrinkle effect was similarly determined (as in the previous described study) using a Visiometer® [73] but additionally, skin irritation and sensitisation assessments were also performed using the modified Shelanski and Shelanski procedure [74]. There were no skin reactions, allergic contact dermatitis, irritant contact dermatitis or other side-effects observed as a result of the treatment in any of the 51 test subjects [74]. Both studies serve as promising indications of proof-of-concept for the incorporation of other anti-aging compounds of varying hydrophilicities into dissolving HA-MNs. In addition, this work supports the possibility of using ascorbic acid-loaded dissolving MN devices in future cosmetic treatments.

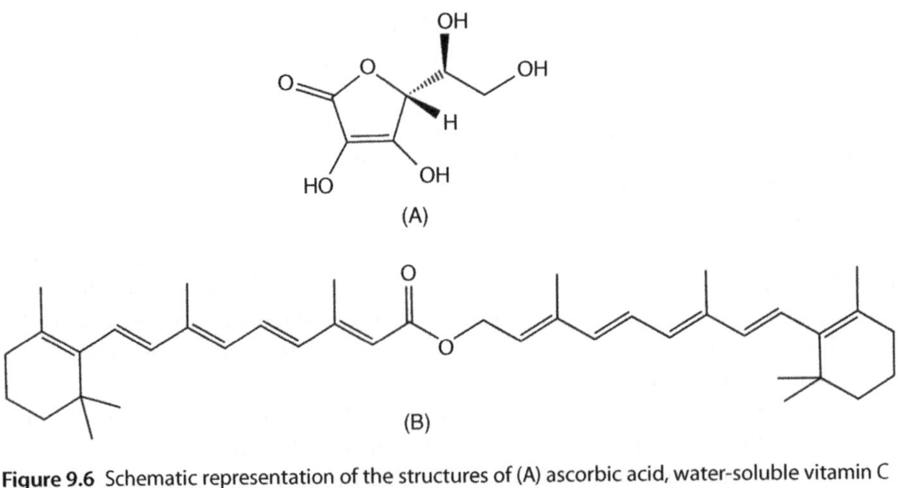

Figure 9.6 Schematic representation of the structures of (A) ascorbic acid, water-soluble vitamin C and (B) retinyl retinoate, a practically water-insoluble derivative of vitamin A.

9.8 Recent Developments

9.8.1 Human Stem Cells

The combinatorial effects of microneedling and human stem cell delivery as an innovative anti-aging treatment, is a new and exciting area of research. This combinatorial approach gained momentum when it was demonstrated that endothelial precursor cells (EPCs), differentiated from human embryonic stem cells (hESCs), improved blood perfusion in damaged tissues as a result of the secretion of high levels of growth factors and cytokines [75]. Conditioned medium (CM) of hESC-derived EPCs (which are comprised of several growth factors and cytokines), significantly enhanced the proliferation and migration of dermal fibroblasts and epidermal keratinocytes, as well as increasing collagen synthesis by fibroblasts [75]. In this respect, growth factors may promote the reduction of the signs of skin aging, with the beneficial role of growth factors in skin rejuvenation having been studied and discussed recently [76, 77]. With this in mind, a study conducted on 25 women investigated the effects of the secretory factors of hESC-EPC CM (compared with saline, a control medium) in the treatment of wrinkles and depigmentation [78]. The left and right sides of the face of each participant were randomly assigned to treatment with microneedling plus saline (control) or microneedling plus hESC-EPC CM. DTS Dermarollers® (MN length of 250 μm) were used to enhance the penetration of hESC-EPC CM into the skin layers [78]. In terms of wrinkle and pigmentation improvement, treatment with microneedling plus hESC-EPC CM, measured by a Visiometer® and a Mexameter®, respectively, resulted in a significantly greater decrease in maximum roughness, average roughness and melanin index (MI), two weeks after the final session. In terms of MI, a decrease in this value provides confirmation of skin lightening. Of note, no serious adverse events were encountered during this study protocol. This is the first, and only, *in vivo* experiment known to the authors that demonstrates the efficacy of microneedling plus hESC CM for improving the signs of skin aging with regards to reducing wrinkles and pigmentation. It signals an innovative move forwards in the use of stem cells and microneedling.

9.8.2 Fractional Radiofrequency

In 2009, non-ablative fractional radiofrequency (RF) microneedling was introduced as a new approach in facilitating facial rejuvenation, and it is continuing to receive much recognition for its unique "...deep dermal heating with epidermal sparing..." feature [79, 80]. RF is non-ionising electromagnetic radiation in the frequency range of 3 kHz to 300 GHz [81]. Fractional RF microneedling uses insulated MNs at a pre-set depth to penetrate the skin and release radiofrequency currents only from the needle tips, producing thermal zones in the dermal structural components and accessory glands [25, 82]. The production of thermal energy at fixed spacing induces long-term dermal remodelling, neoelastogenesis and neocollagenesis, allowing cells to promote healing, resulting in skin tightening [83]. Since RF microneedling allows the delivery of energy at a very specific point, tissue damage is minimised, thus sparing the epidermis and skin adnexal structures that contribute to rapid healing, such as growth factors and fibroblasts. Moreover, as fractional RF microneedling can control the depth and subsequently the induction of RF thermal zones, this allows for deeper penetration into dermal layers but lowers the risk of pain and adverse events such as blistering, burns and post-inflammatory hyperpigmentation associated with epidermal injury [81]. With these advantages in mind, this technology has been very successfully used in the treatment of acne vulgaris and acne scars, with the treatment of hyperhidrosis also being considered [84, 85].

Expanding upon this, wrinkles as a result of photoaging are currently being explored using a similar approach. Specifically, a study carried out in 20 patients examined the *in vivo* efficacy of this system for the treatment of periorbital wrinkles (wrinkles around the eyes) [80]. The patients were treated three times at four-week intervals with the fractional RF MN system and evaluated using a five-point Wrinkle Assessment Scale. Digital imaging confirmed dramatic improvement of periorbital wrinkles, with only mild hyperpigmentation documented [80]. In an extension of this work, a more thorough study was then undertaken evaluating the *in vivo* efficacy of MN fractional RF, in comparison with the synergistic effects of the same technique used in combination with stem cell CM [81]. In a split-face comparative study, one side of each of 15 subjects' faces were treated with fractional RF alone, and the other sides of the faces were treated with fractional RF, in addition to stem cell CM. Patients received three sessions of treatment at four-week intervals. Skin roughness was measured using a Visiometer® and histologic evaluations were obtained *via* quantitative assessments of procollagen-I using biopsy specimens. Among these measurements, the combined treatment using fractional RF microneedling and stem cell CM provided a significant improvement in skin roughness, dermal thickness and dermal collagen content [81]. It was also very promising to note that there were no serious adverse events documented in this study. Therefore, these two studies demonstrate the success of combining microneedling and RF energy.

Building upon these preliminary studies and highlighting a new dimension in microneedling technology, a 2-in-1 MN device, combining RF and microneedling, resembling the Dermapen® has been launched onto the market, named the INTRAcel™ (http://www.intraceluk.com/intracel-fractional-rf-microneedling-videos.htm (accessed 27 April 2017)). Each needle on this device is insulated, with only 300 μm of the apex emitting a high-tensioned RF pulse onto the target area of the skin. Therefore, this innovative device provides a new treatment option in skin rejuvenation.

9.9 Conclusion

Research on the use of MN devices in the cosmeceutical industry has intensified in the recent past and continues to evolve and develop, with the emergence of superior materials, fabrication methods and designs. One area of MN research which is undergoing sustained development is the facilitated delivery of cosmeceutical agents. In this chapter, the integration and use of MNs in cosmetic procedures has been discussed. A wide range of MN devices have been described, including the Dermaroller®, Beauty Mouse®, Dermastamp™ and Dermapen®, with further commercial products currently in development. Additionally, the role of HA in MN delivery systems, the use of MN platforms in the delivery of cosmetic peptides and emerging fractional RF/MN combinatorial treatments were also discussed. As this field progresses towards further commercialisation of MN devices, however, it is clear from the safety and public perception studies outlined herein, that interaction and engagement with the relevant regulatory agencies is essential in order to address issues regarding MN sterile manufacture and to avoid misuse of these devices.

References

1 S.K. Schagen, V.A. Zampeli, E. Makrantonaki and C.C. Zouboulis (2012). Discovering the link between nutrition and skin aging. *Dermato-Endocrinology* 4 (3): 298–307.

2 M.E.I. Domyati and W. Medhat (2013). Minimally invasive facial rejuvenation: current concepts and future expectations. *Expert Review of Dermatology* 8 (5): 565–580.

3 T.Y. Han, K.Y. Park, J.Y. Ahn, *et al.* (2012). Facial skin barrier function recovery after microneedle transdermal delivery treatment. *Dermatologic Surgery* 38 (11): 1816–1822.

4 I. Majid (2009). Microneedling therapy in atrophic facial scars: an objective assessment. *Journal of Cutaneous and Aesthetic Surgery* 2 (1): 26–30.

5 L. Liu, H. Ma and Y. Li (2014). Interventions for the treatment of stretch marks: a systematic review. *Cutis* 94 (2): 66–72.

6 A. Harvey and T.T. Huynh (2014). Inflammation and acne: putting the pieces together. *Journal of Drugs in Dermatology* 13 (4): 459–463.

7 M. Sanchez-Viera (2015). Management of acne scars: fulfilling our duty of care for patients. *British Journal of Dermatology* 172 (suppl. 1): 47–51.

8 M.C. Aust, D. Fernandes, P. Kolokythas, *et al.* (2008). Percutaneous collagen induction therapy: an alternative treatment for scars, wrinkles, and skin laxity. *Plastic and Reconstructive Surgery* 121 (4): 1421–1429.

9 M.M. Loesch, A.K. Somani, M.M. Kingsley, *et al.* (2014). Skin resurfacing procedures: new and emerging options. *Clinical, Cosmetic and Investigational Dermatology* 7: 231–241.

10 D.S. Orentreich and N. Orentreich (1995). Subcutaneous incisionless (subcision) surgery for the correction of depressed scars and wrinkles. *Dermatologic Surgery* 21 (6): 543–549.

11 A. Camirand and J. Doucet (1997). Needle dermabrasion. *Aesthetic Plastic Surgery* 21 (1): 48–51.

12 D. Fernandes and M. Signorini (2008). Combating photoaging with percutaneous collagen induction. *Clinics in Dermatology* 26: 192–199.

13 D. Fernandes (2002). Percutaneous collagen induction: An alternative to laser resurfacing. *Anaesthetic Surgery Journal* 22 (3): 307–309.

14 S.E. Kim, J.H. Lee, H.B. Kwon, *et al.* (2011). Greater collagen deposition with the microneedle therapy system than with intense pulsed light. *Dermatologic Surgery* 37 (3): 336–341.

15 Dermaroller Planet (2009). What a dermaroller does. http//:www.dermaroller-planet.com/new-to-dermarollers/what-a-dermaroller-does (accessed 3 May 2017).

16 M.C. Aust, K. Reimers, H.M. Kaplan, *et al.* (2011). Percutaneous collagen induction - regeneration in place of cicatrisation. *Journal of Plastic, Reconstructive and Aesthetic Surgery* 64 (1): 97–107.

17 K.M. Hassan and A.V. Benedetto (2013). Facial skin rejuvenation: ablative laser resurfacing, chemical peels, or photodynamic therapy? Facts and controversies. *Clinics in Dermatology* 31 (6): 737–740.
18 J.M. Rawlins (2006). Quantifying collagen type in mature burn scars: a novel approach using histology and digital image analysis. *Journal of Burn Care and Research* 27 (1): 60–65.
19 M.H. Gold and J.A. Biron (2012). Treatment of acne scars by fractional bipolar radiofrequency energy. *Journal of Cosmetic and Laser Therapy* 14 (4): 172–178.
20 M.H. Khater, F.M. Khattab and M.R. Abdelhaleem (2016). Treatment of striae distensae with needling therapy versus CO_2 fractional laser. *Journal of Cosmetic and Laser Therapy* 18 (2): 75–79.
21 T. Omi and K. Numano (2014). The role of the CO_2 laser and fractional CO_2 laser in dermatology. *Laser Therapy* 23 (1): 49–60.
22 Dermaroller® (2013). Dermaroller® Germany. https://dermarollerus.com/about-dermaroller-germany (accessed 28 May 2017).
23 M.M. Badran, J. Kuntsche and A. Fahr (2009). Skin penetration enhancement by a microneedle device (Dermaroller®) *in vitro*: Dependency on needle size and applied formulation. *European Journal of Pharmaceutical Sciences* 36 (4-5): 511–523.
24 Y.C. Kim, J.H. Park and M.R. Prausnitz (2012). Microneedles for drug and vaccine delivery. *Advanced Drug Delivery Reviews* 64 (14): 1547–1568.
25 A. Singh and S. Yadav (2016). Microneedling: Advances and widening horizons. *Indian Dermatology Online Journal* 7 (4): 244–254.
26 Royal Derma Roller (2015). Derma Rollers. https://www.royaldermaroller.com/collections/derma-rollers/products/derma-roller-1-0-acne-scars (accessed 25 May 2017).
27 DNS® (2006). DNS Classic 3 Line Roller. http://www.dnsroller.com/dns-classic-roller-dermaroller-micro-needling-derma-system-titanium-biogenesis-dns-london/dns-classic-3-line-roller-p10002#.WTgFf4WcHIU (accessed 25 May 2017).
28 MTS Roller.com (2012). Derma Rollers, MTS Derma Roller, Body Roller. https://www.mtsroller.com/1080-titanium-needles-body-roller (accessed 7 June 2017).
29 Hansderma (2013). Genosys devices. http://www.hansderma.biz (accessed 27 April 2017).
30 White Lotus Holistic Microneedling (2016). Dermaroller. https://www.whitelotusantiaging.co.uk/dermaroller/ (accessed 27 April 2017).
31 DNS® (2006). DNS Classic Series. http://www.dnsroller.com/dns-classic-roller-dermaroller-micro-needling-derma-system-titanium-biogenesis-dns-london#.WThDfoWcHIU (accessed 7 June 2017).
32 MTS Roller (2012). Derma Rollers, MTS Derma Roller. https://www.mtsroller.com/ (accessed 25 May 2017).
33 Dermastamp® (2010). Beauty Mouse. http://dermastamp.com.au/beauty-mouse/ (accessed 26 April 2017).
34 Dermaroller® (2013). Beauty Mouse® Micro-Needling Home Body. https://dermarollerus.com/color-box-product/891?width=600&height=600 (accessed 6 June 2017).
35 White Lotus Holistic Microneedling (2016). Scar Dermastamp 1.0 mm. https://www.whitelotusantiaging.co.uk/scar-dermastamp-1-0mm (accessed 6 June 2017).

36 Consulting Room (2013). eDermastamp™. http://www.consultingroom.com/Treatment/eDermastamp-eDS-Skin-Rejuvenation (accessed 6 June 2017).

37 Dermique (2014). Micro Dermastamp Treatment. http://www.dermique.com.my/micdermtreatment.php (accessed 5 June 2017).

38 Aestheticare® (2013). EDS ®. http://www.aestheticare.co.uk/devices-treatments/eds/ (accessed 5 June 2017).

39 Advanced Skin Innovations (2017). *Dermapen™*. http://www.equipmed.com/LiteratureRetrieve.aspx?ID=150517 (accessed 7 June 2017).

40 M.T. Clemmentoni, M.B. Roscher and G.S. Munavalli (2010). Photodynamic photorejuvenation of the face with a combination of microneedling, red light and broadband pulsed light. *Lasers in Surgery and Medicine* 42: 150–159.

41 M.C. Kearney, S. Brown, M.T.C. McCrudden, *et al.* (2014). Potential of microneedles in enhancing delivery of photosensitising agents for photodynamic therapy. *Photodiagnosis and Photodynamic Therapy* 11(4): 459–466.

42 M.A. Kosoglu, R.L. Hood, J.H. Rossmeisl, *et al.* (2011). Fiberoptic microneedles: novel optical diffusers for interstitial delivery of therapeutic light. *Lasers in Surgery and Medicine* 43 (9): 914–920.

43 Mail Online (2009). It costs £250 and looks like an instrument of medieval torture - but can the Dermaroller make you look younger? http://www.dailymail.co.uk/femail/beauty/article-1226193/Can-Dermaroller-make-look-younger.html (accessed 21 October 2014).

44 R. Soltani-Arabshahi, J.W. Wong, K.L. Duffy and D.L. Powell (2014). Facial allergic granulomatous reaction and systemic hypersensitivity associated with microneedle therapy for skin rejuvenation. *JAMA Dermatology* 150 (1): 68–72.

45 Mail Online (2011). Chinese women warned over 'potentially lethal' microneedle roller after beauty treatment takes Hong Kong by storm. http://www.dailymail.co.uk/femail/article-2026700/Microneedle-Therapy-System-potentially-lethal-Chinese-women-warned.html (accessed 21 October 2014).

46 R.F. Donnelly and A.D. Woolfson (2014). Patient safety and beyond: What should we expect from microneedle arrays in the transdermal delivery arena? *Therapeutic Delivery* 5 (6): 653–662.

47 J.C. Birchall, R. Clemo, A. Anstey and D.N. John (2011). Microneedles in clinical practice--an exploratory study into the opinions of healthcare professionals and the public. *Pharmaceutical Research* 28 (1): 95–106.

48 R.F. Donnelly, K. Moffatt, A.Z. Alkilani, *et al.* (2014). Hydrogel-forming microneedle arrays can be effectively inserted in skin by self application: A pilot study centred on pharmacist intervention and a patient information leaflet. *Pharmaceutical Research* 31 (8): 1989–1999.

49 E.M. Vicente-Pérez, H.L. Quinn, E. McAlister, *et al.* (2016). The use of a pressure-indicating sensor film to provide feedback upon hydrogel-forming microneedle array self-application *in vivo*. *Pharmaceutical Research* 33 (12): 3072–3080.

50 A. Ripolin, J. Quinn, E. Larrañeta, *et al.* (2017). Successful application of large microneedle patches by human volunteers. *International Journal of Pharmaceutics* 521: 92–101.

51 H. Kalluri, C.S. Kolli and A.K. Banga (2011). Characterization of microchannels created by metal microneedles: formation and closure. *AAPS Journal* 13 (3): 473–481.

52 E.M. Vicente-Pérez, E. Larrañeta, M.T.C. McCrudden, *et al.* (2017). Repeat application of microneedles does not alter skin appearance or barrier function and causes no measurable disturbance of serum biomarkers of infection, inflammation or immunity in mice *in vivo*. *European Journal of Pharmaceutics and Biopharmaceutics* 117: 400–407.

53 R.F. Donnelly, T.R.R. Singh, A.Z. Alkilani, *et al.* (2013). Hydrogel-forming microneedle arrays exhibit antimicrobial properties: Potential for enhanced patient safety. *International Journal of Pharmaceutics* 451 (1-2): 76–91.

54 R.F. Donnelly, T.R.R. Singh, M.M. Tunney, *et al.* (2009). Microneedle arrays allow lower microbial penetration than hypodermic needle *in vitro*. *Pharmaceutical Research* 26 (11): 2513–2522.

55 M. Farwick, P. Lersch and G. Strutz (2008). Low molecular weight hyaluronic acid: its effect on epidermal gene expression and skin ageing. *SOFW Journal* 134: 2–6.

56 Y. Hiraishi, T. Nakagawa, Y.S. Quan, *et al.* (2013). Performance and characteristics evaluation of a sodium hyaluronate-based microneedle patch for a transcutaneous drug delivery system. *International Journal of Pharmaceutics* 441 (1-2): 570–579.

57 T.J. Brown, D. Alcorn and J.R. Fraser (1999). Absorption of hyaluronan applied to the surface of intact skin. *Journal of Investigative Dermatology* 113 (5): 740–746.

58 F. Duranti, G. Salti, B. Bovani, *et al.* (1998). Injectable hyaluronic acid gel for soft tissue augmentation-a clinical and histological study. *Dermatologic Surgery* 24 (12): 1317–1325.

59 S.G. Lee, J.H. Jeong, K.M. Lee, *et al.* (2014). Nanostructured lipid carrier-loaded hyaluronic acid microneedles for controlled dermal delivery of a lipophilic molecule. *International Journal of Nanomedicine* 9: 289–299.

60 S. Liu, M.N. Jin, Y.S. Quan, *et al.* (2014). Transdermal delivery of relatively high molecular weight drugs using novel self-dissolving microneedle arrays fabricated from hyaluronic acid and their characteristics and safety after application to the skin. *European Journal of Pharmaceutics and Biopharmaceutics* 86 (2): 267–276.

61 M.T.C. McCrudden, A.Z. Alkilani, C.M. McCrudden, *et al.* (2014). Design and physicochemical characterisation of novel dissolving polymeric microneedle arrays for transdermal delivery of high dose, low molecular weight drugs. *Journal of Controlled Release* 180: 71–80.

62 U. Anderegg, J.C. Simon and M. Averbeck (2014). More than just a filler - the role of hyaluronan for skin homeostasis. *Experimental Dermatology* 23 (5): 295–303.

63 M.T.C. McCrudden, T.R.R. Singh, K. Migalska and R.F. Donnelly (2013). Strategies for enhanced peptide and protein delivery. *Therapeutic Delivery* 4 (5): 593–614.

64 M. Hultström, N. Roxhed and L. Nordquist (2014). Intradermal insulin delivery. *Journal of Diabetes Science and Technology* 8 (3): 453–457.

65 H. Suh, J. Shin and Y.C. Kim (2014). Microneedle patches for vaccine delivery. *Clinical and Experimental Vaccine Research* 3 (1): 42–49.

66 Y.H. Mohammed, M. Yammadda, L.L. Lin, *et al.* (2014). Microneedle enhanced delivery of cosmeceutically relevant peptides in human skin. *PLoS One* 9 (7): e101956.

67 A. Kumar, Y.W. Naguib, Y. Shi and Z. Cui (2016). A method to improve the efficacy of topical eflornithine hydrochloride cream. *Drug Delivery* 23 (5): 1495–1501.

68 K. Jeong, Y.J. Lee, J.E. Kim, *et al.* (2012). Repeated microneedle stimulation induce the enhanced expression of hair-growth-related genes. *International Journal of Trichology* 4: 117.

69 B.J. Kim, Y.Y. Lim, H.M. Kim, *et al.* (2012). Hair follicle regeneration in mice after wounding by microneedle roller. *International Journal of Trichology* 4: 117.

70 R. Dhurat, M.S. Sukesh, G. Avhad, *et al.* (2013). A randomized evaluator blinded study of effect of microneedling in androgenetic alopecia: A pilot study. *International Journal of Trichology* 5 (1): 6–11.

71 A. Puri, H.X. Nguyen and A.K. Banga (2016). Microneedle-mediated intradermal delivery of epigallocatechin-3-gallate. *International Journal of Cosmetic Science* 38 (5): 512–523.

72 S. Kim, H. Yang, M. Kim, *et al.* (2016). 4-*N*-Butylresorcinol dissolving microneedle patch for skin depigmentation: a randomized, double-blind, placebo-controlled trial. *Journal of Cosmetic Dermatology* 15 (1): 16–23.

73 M. Kim, H. Yang, H. Kim, *et al.* (2014). Novel cosmetic patches for wrinkle improvement: retinyl retinoate- and ascorbic acid-loaded dissolving microneedles. *International Journal of Cosmetic Science* 36 (3): 207–212.

74 C. Lee, H. Yang, S. Kim, *et al.* (2016). Evaluation of the anti-wrinkle effect of an ascorbic acid-loaded dissolving microneedle patch via a double-blind, placebo-controlled clinical study. *International Journal of Cosmetic Science* 38 (4): 375–381.

75 M.J. Lee, J. Kim, K.I. Lee, *et al.* (2011). Enhancement of wound healing by secretory factors of endothelial precursor cells derived from human embryonic stem cells. *Cytotherapy* 13 (2): 165–178.

76 W.S. Kim, B.S. Park, S.H. Park, *et al.* (2009). Antiwrinkle effect of adipose-derived stem cell: activation of dermal fibroblast by secretory factors. *Journal of Dermatologic Science* 53 (2): 96–102.

77 B.S. Park, K.A. Jang, J.H. Sung, *et al.* (2008). Adipose-derived stem cells and their secretory factors as a promising therapy for skin aging. *Dermatologic Surgery* 34 (10): 1323–1326.

78 H.J. Lee, E.G. Lee, S. Kang, *et al.* (2014). Efficacy of microneedling plus human stem cell conditioned medium for skin rejuvenation: a randomized, controlled, blinded split-face study. *Annals of Dermatology* 26 (5): 584–591.

79 B.M. Hantash, B. Renton, R.L. Berkowitz, *et al.* (2009). Pilot clinical study of a novel minimally invasive bipolar microneedle radiofrequency device. *Lasers in Surgery and Medicine* 41 (2): 87–95.

80 S.J. Lee, J. Kim, Y.J. Yang, *et al.* (2015). Treatment of periorbital wrinkles with a novel fractional radiofrequency microneedle system in dark-skinned patients. *Dermatologic Surgery* 41 (5): 615–622.

81 K.Y. Seo, D.H. Kim, S.E. Lee, *et al.* (2013). Skin rejuvenation by microneedle fractional radiofrequency and a human stem cell conditioned medium in Asian skin: a randomized controlled investigator blinded split-face study. *Journal of Cosmetic and Laser Therapy* 15 (1): 25–33.

82 B.J. Simmons, R.D. Griffith, L.A. Falto-Aizpurua and K. Nouri (2014). Use of radiofrequency in cosmetic dermatology: focus on non-ablative treatment of acne scars. *Clinical, Cosmetic and Investigational Dermatology* 7: 335–339.
83 G. Hruza, A.F. Taub, S.L. Collier and S.R. Mulholland (2009). Skin rejuvenation and wrinkle reduction using a fractional radiofrequency system. *Journal of Drugs in Dermatology* 8 (3): 259–265.
84 B.A. Chandrashekar, R. Sriram, R. Mysore, *et al.* (2014). Evaluation of microneedling fractional radiofrequency device for treatment for acne scars. *Journal of Cutaneous and Aesthetic Surgery* 7 (2): 93–97.
85 S.T. Kim, K.H. Lee, H.J. Sim, *et al.* (2014). Treatment of acne vulgaris with fractional radiofrequency microneedling. *Journal of Dermatology* 41 (7): 586–591.

10

Microneedles for Ocular Drug Delivery and Targeting: Challenges and Opportunities

Ismaiel A. Tekko[1,2] and Thakur Raghu Raj Singh[1]

[1] *School of Pharmacy, Queens University Belfast, Belfast BT9 7BL, UK*
[2] *Department of Pharmaceutics and Pharmaceutical Technology, Faculty of Pharmacy, Aleppo University, Aleppo, Syria*

10.1 Introduction

Ocular drug delivery presents an important and interesting area of research for formulation scientists and pharmacologists. Its importance comes from the fact that visual impairment and blindness are potentially the most devastating health problems worldwide. The World Health Organization (WHO) estimates that globally about 285 million people are visually impaired of which 39 million are blind and 246 million have low vision [1]. Diseases affecting the anterior segment of the eye, such as corneal neovascularisation (CNZ), glaucoma, bacterial/fungal keratitis, uveitis, herpes simplex keratitis, blepharitis and dry eye syndrome, can cause serious vision impairment and discomfort. On the other hand, diseases of the posterior segment of the eye, such as age-related macular degeneration (AMD), diabetic retinopathy, diabetic macular edema, cytomegalovirus retinitis and other chorioretinal diseases, could lead to a permanent loss of vision [2].

Interestingly, there are a reasonable number of current and emerging drugs and pharmacological therapies that offer to reduce the progression of principal sight-threatening eye diseases. However, delivery of these drugs and therapies to the target ocular tissues in an efficient and minimally invasive manner is still a challenging task [3, 4]. The challenge in drug delivery comes from the extremely delicate nature of the ocular tissues concerned, their relative inaccessibility and the barrier properties of ocular tissues that hinder effective drug diffusion to the target tissues [5, 6]. For example, the posterior segment of the eye, which includes the retina, choroid and vitreous body, is difficult to access due to the recessed location within the orbital cavity.

Multiple approaches have been used to deliver drugs to the eye such as systemic, topical, periocular (or transscleral) and intravitreal (IVT) routes. Topical (e.g. eye drops) and systemic (e.g. oral tablets) routes result in low or sub-therapeutic drug levels due to multiple ocular barriers, requiring administration of unnecessarily high concentrations of drug that cause drug-related toxicity and produce low treatment efficacy [7, 8]. To overcome the barrier function of the eye and to enhance localisation of the drug close to the target tissues, injections are given either directly into the eye (IVT), around the outer surface of the eye (periocular or transscleral route) or within the tissues (intracorneal

Microneedles for Drug and Vaccine Delivery and Patient Monitoring, First Edition.
Edited by Ryan F. Donnelly, Thakur Raghu Raj Singh, Eneko Larrañeta, and Maelíosa T.C. McCrudden.

and intrascleral). These injections are given using conventional hypodermic needles. Although the periocular route is considered less invasive than the IVT, transient diffusion of a drug across the sclera is limited. Drug diffusion across the scleral membrane is dependent upon the drug's solubility, molecular weight/molecular radius, charge and polarity [9]. However, this method has shown low intraocular bioavailability due to a delay in diffusion through the sclera, systemic clearance and loss of drug before reaching the target tissues (e.g. retina) [4]. One of the standard treatments to overcome limitations of periocular injections is by either IVT, for posterior segment diseases, or intracorneal injections, for anterior segment diseases.

Injection into the eye, for both anterior and posterior segments, using conventional hypodermic needles causes considerable discomfort and pain and requires a specialised set of skills. Notably, hypodermic needle-based injections given on a frequent basis and over the long term may increase the chance of severe ocular complications and poor patient compliance. Therefore, there is great demand for less invasive technologies that not only enhance patient compliance but also allow localised and targeted drug delivery to the ocular tissues with appropriate precision and accuracy.

In this regard, microneedles (MNs) have emerged as an outstanding technology that has revolutionised transdermal drug and vaccine delivery. MNs have also been investigated for ocular drug delivery application as a minimally invasive and efficient drug delivery system. MNs can target specific ocular tissues for localised delivery of drug-loaded formulations with greater precision and accuracy than the hypodermic needles. However, to date, limited work has been done in this area, unlike transdermal MN studies.

This chapter seeks to provide an overview of mainstream challenges that are often faced in order to achieve adequate ocular drug levels within targeted tissue(s) of the eye, the problems encountered with the use of conventional hypodermic needles and the importance of MNs to overcome current challenges in treating sight-threatening eye diseases. It also provides an overview of the limitations and difficulties of application of MNs in ocular drug delivery and its prospects.

10.2 Anatomy of the Eye and Barriers to Drug Delivery

The human eye is an anatomically and physiologically unique sensory organ [5], which can be divided into two segments for the purpose of drug delivery (see Figure 10.1):

- The anterior segment: consists of cornea, iris, ciliary body, aqueous humour and crystalline lens.
- The posterior segment: consists of sclera, choroid, retina, macular, optic nerve and vitreous humour.

10.2.1 The Anterior Segment and its Barrier Function

The anterior segment occupies one-third of the eye covered by a thin layer called the conjunctiva. The conjunctiva is a thin transparent mucous membrane that covers the cornea, the inner surface of the eyelids and the anterior surface of the sclera and extends to the end of corneal limbus (Figure 10.1). It is a highly vascularised layer,

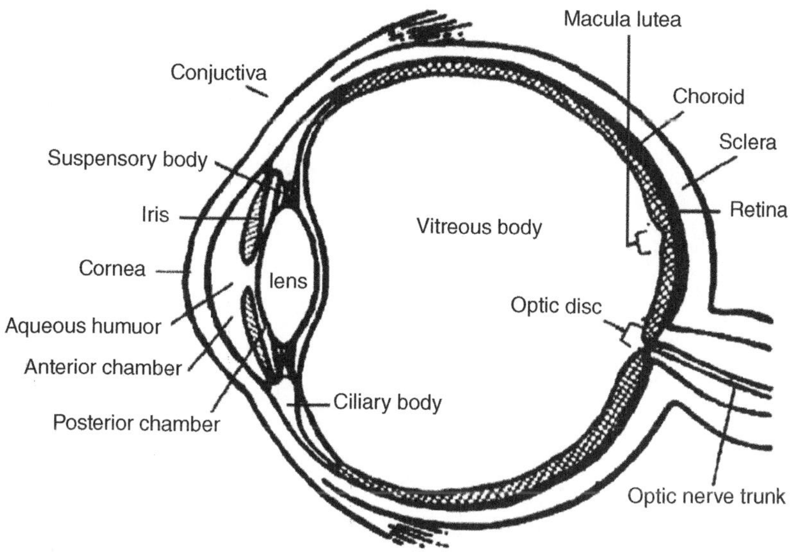

Figure 10.1 Schematic diagram representing the anatomy of the human eye. Adapted with permission from [5] F. Idrees, D. Vaideanu, S.G. Fraser, *et al.* (2006). A review of anterior segment dysgeneses. *Surv. Ophthalmol.* 51: 213–231.

richly supplied with lymphatics and is involved in the production and maintenance of tear film. It facilitates lubrication in the eye by generating the mucous and helps with tear film adhesion. This membrane provides the first physical barrier to anterior drug delivery, but not as much as the cornea. The vast presence of localised blood capillaries and a rich lymphatic system within the conjunctiva results in a rapid clearance of drug molecules. This significant drug loss into the systemic circulation raises the issue of not only lowering the ocular bioavailability but it can lead to unwanted systemic exposure of the drug [10].

The cornea is a clear outer layer of the eyeball, which is composed of five distinct layers: from anterior to posterior they are the epithelium, Bowman's layer, stroma, Descemet's membrane and endothelium (Figure 10.2).

The epithelium is the outermost layer of the cornea, which is approximately 50 μm in thickness, and consists of 5–7 layers of superficial, wing and basal epithelial cells. The stroma, around 500 μm thick, is the second layer and accounts for >90% of the cornea thickness. It mainly consists of an extracellular matrix, stromal cells and approximately 4% glycosaminoglycans [2]. Water-soluble molecules readily traverse this layer and even high molecular weight drugs diffuse with ease [5]. However, the stroma restricts the movement of lipophilic drugs and macromolecules with a molecular weight above 50 000 Da [12]. The endothelium layer consists of a monolayer of cuboidal cells with an approximate thickness of 5 μm [6]. Both the epithelium and the endothelium are hydrophobic, providing a barrier to the movement of hydrophilic molecules across the cornea; however, the endothelium is approximately 2.7 times more permeable than the epithelium [10].

The cornea has a dual action, which is limiting the entry of exogenous substances into the eye and protecting the ocular tissue. The cornea has average dimensions of

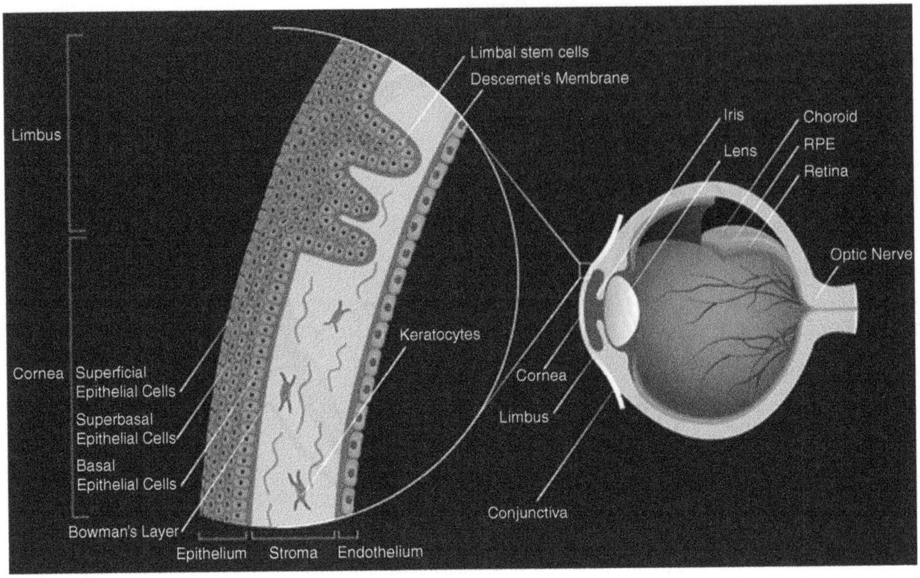

Figure 10.2 Schematic diagram showing the anatomy and structure of the adult human cornea. Reproduced with permission from LifeMap Sciences, [11] R. Edgar, Y. Mazor, A. Rinon, *et al.* (2013). LifeMap Discovery™: the embryonic development, stem cells, and regenerative medicine research portal. *PloS One* 8 (7): e66629.

11.5 mm horizontally, 10.5 mm vertically, with a mean surface area of 1.3 cm^2, representing around 7% of the total surface area of eyeball. The thickness in the central region is around 0.52 mm and increases towards the periphery [13]. The cornea forms a significant barrier to ophthalmic topical formulation, especially for hydrophilic and macromolecular drugs, due to its lipoidal nature and the tight junctions between the cells, especially in the superficial epithelium cells [14]. Drug molecules require a partition coefficient of greater than 1 to permeate the epithelium adequately [14]. The molecular weight of hydrophilic molecules also plays a major factor in their permeation through the corneal epithelium [15], with those larger than 60–100 Da being unable to pass [15, 16].

It is worth noting that in the anterior segment of the eye, the first challenge for the delivery of a topically applied drug is the precorneal lacrimal fluid. Lacrimal fluid turnover and clearance is approximately 1 μl/min [17] via the nasolacrimal duct, meaning formulations instilled to the eye can be cleared from the ocular surface in a matter of minutes [18]. Additionally, the lacrimal fluid is rich in peptides and proteins, which are capable of binding drug molecules and inhibiting their release or permeation [19].

10.2.2 The Posterior Segment and its Barriers Function

The posterior segment forms the remaining two-thirds of the eye globe; it consists of the sclera, choroid, retina, macula, vitreous humour and optic nerve [20]. In general, the posterior segment has an array of consecutive barriers that hinder drug permeation, resulting in numerous challenges to drug delivery. These barriers include the sclera, choroid and the blood retinal barrier (BRB).

The sclera forms the outer layer of the posterior segment (Figure 10.1), which offers mechanical support and strength to the eye. The sclera covers approximately 80% of the eyeball surface and forms a relatively large surface area, 16.3 cm^2 [21]. It is an elastic, tough, vascular, opaque white–yellow, microporous tissue composed of collagen and elastin fibres entwined with proteoglycans [22]. The scleral thickness varies throughout its circumference. In humans, the mean scleral thickness is reported to be 0.53 mm, with the thickest portion being approximately 1 mm at the posterior, near the optic nerve, and the thinnest portion being 0.39 mm at the equator [21]. It consists of four layers, from the outer side to the inner side these are: Tenon's capsule, the episclera, the stroma and the lamina fusca [13]. Apart from not having epithelium and endothelium layers, the scleral tissue differs from the corneal tissue primarily in the uniformity of the arrangement of the collagen fibres and the degree of hydration [23].

Unlike the cornea, the sclera has irregular collagen fibres and a fourfold lower concentration of proteoglycans, resulting in a lower water content, namely, 68% in comparison with 78% in the corneal stroma [23]. Also, the sclera is perforated by blood vessels and has an extensive nerve supply [23, 24]. Owing to the sclera's large aqueous content, hydrophilic molecules can diffuse through this layer more readily than hydrophobic molecules. The sclera is permeable to high molecular weight compounds and even proteins of 150 kDa [25]. However, the permeability declines exponentially with an increase in molecular radius of the compound [26]. The charge of the drug molecule also presents a challenge to penetration through the sclera, as positively charged molecules are at risk of interacting with the negatively charged proteoglycans within the sclera [22].

The choroid is one of the most highly vascularised regions of the body, and its primary function is to supply blood, rich in oxygen and nutrients, to the retina [27]. Bruch's membrane is a thin layer located between the choroid and the retinal pigment epithelium (RPE), and functions as a barrier to the movement of vessels from the choroid into the RPE and retina. The choroid has been shown to thin on aging [28, 29]. In contrast, Bruch's membrane thickens with increasing age, causing a disruption to its barrier activity, giving rise to a number of ocular diseases [30]. Changes in thickness within the choroid and Bruch's membrane can affect successful drug permeation and penetration from the subconjunctiva and sclera, resulting in decreased drug delivery to the retina [27]. The BRB acts to restrict entry of unwanted molecules from the choroid into the retina. It is the most significant barrier to systemic drug delivery. Following systemic administration, drug molecules can easily enter the highly vascularised choroid tissue, but in the main they are unable to pass the BRB. The BRB is extremely effective at performing this restricting function due to its unique composition. The outer portion of the BRB is formed by the RPE, and the inner portion is formed from tight junctions of retinal capillary endothelial cells [18, 31].

The retina is the inner layer, which is the intended site of action for most drugs delivered to the posterior segment of the eye. Although it does not have its own barrier function, the inner limiting membrane that separates the retina and the vitreous humour is composed of ten distinct extracellular matrix proteins, and is thought to prevent the penetration of some drug molecules into the retina [27]. However, penetration of anti-vascular endothelial growth factors (VEGF), such as bevacizumab

(Avastin®, Genentech Inc.), with a molecular weight of 149 kDa, into the RPE via an IVT route has been seen [32].

10.3 Ocular Diseases and Treatments

There are a series of chronic and acute diseases that affect the eye. Diseases that affect the anterior segment of the eye, such as CNZ, conjunctivitis, glaucoma, bacterial/fungal keratitis, uveitis, blepharitis and dry eye syndrome, could cause serious vision impairment or discomfort. The diseases that originate in the posterior segment of the eye, such as AMD, diabetic retinopathy, diabetic macular edema (DME) and other chorioretinal diseases, could lead to permanent loss of vision and account for most cases of blindness [1, 33, 34]. Table 10.1 summarises some of the most common ocular diseases, classifications, signs and symptoms and available treatment options.

10.4 Current Ocular Drug Delivery Systems and Administration Routes

Effective treatment of eye disease requires delivery of the drug molecules close to the target tissue at sufficient concentrations and over a suitable time scale. Three commonly employed administration routes include topical (e.g. eye drops, suspension, ointments and gels), oral/systemic administration (e.g. oral tablets and capsules) and direct injection in the eye (e.g. intravitreal and subconjunctival injections). The choice of route of administration or type of delivery system is dependent on the target tissue and potential barriers that need to be overcome. However, each route and method of administration has its advantages and disadvantages, as discussed next.

10.4.1 Topical Route

The topical route is widely employed due to its ease of administration in treatment of anterior segment diseases. A range of topical formulations that are commonly employed include eye drops, gels, suspensions and emulsions, to name just a few. However, the most commonly formulated preparation is topical eye drops, which have the advantage of being easily self-administered with good patient compliance. However, lacrimation, tear production, blood flow and corneal/scleral barriers lead to low (<5%) bioavailability [35]. Therefore, frequent application of eye drops becomes necessary in treating diseases such as glaucoma, CNZ and fungal/bacterial keratitis. On the other hand, topical administration is inefficient to treat posterior segment eye diseases due to the barrier properties of the eye such as the cornea and sclera, as discussed earlier.

10.4.2 Oral/Systemic Administration Route

While systemic delivery in the form of a patient taking an oral formulation would be highly convenient and acceptable to patients, it is not an automatic choice in ocular drug delivery. Only a small proportion of systemically delivered drug gets to the posterior segment of the eye, largely due to the barrier properties of the BRB [36] and low

Table 10.1 Major ocular diseases, signs and symptoms and current treatment. Adapted from [34] J.N. Shah, H.J. Shah, A. Groshev, *et al*. (2014). Nanoparticulate transscleral ocular drug delivery. *J. Biomol. Res. Ther*. 3: 116

Disease	Classification	Signs and symptoms	Treatment
AMD	Dry AMD (non-exudative)	Breakdown of photoreceptors, retinal pigment epithelium (RPE) and choriocapillaries	Specific high dose formulation containing antioxidants, zinc and vitamin supplements
	Wet AMD (exudative)	Growth of abnormal blood vessels behind the retina, macula, disruption of Bruch's membrane and degeneration of RPE leading to complete loss of vision	Intravitreal injection of anti-VEGF agents such as Ranibizumab, Pegaptanib sodium and Bevacizumab
DME	Focal or non-cystoid DME	Small aberrations in retinal blood vessels followed by intra-retinal leakage	Focal or grid lasers and steroids
	Diffuse or cystoids DME	Formation of microcrysts and dilation of retinal capillaries	—
Proliferative vitreoretinopathy (PVR)	Inflammation of the retinal tissue can be classified into focal, diffuse, subretinal, circumferential and anterior displacement. On the other hand, inflammation of the scar tissue can be classified into anterior and posterior	Simple scar formation and proliferation of cells in vitreous and retina	Surgery and adjunctive treatment after surgery, to avoid relapses (5-Fluorouracil and low molecular weight heparin)
Uveitis	Anterior uveitis, intermediate uveitis, posterior uveitis, panuveitic uveitis	Inflammation occurs in the middle layer of eye (uvea)	Corticosteroids and immunosuppressive agents
Cytomegalovirus retinitis (CMV)	—	Inflammation of the retina, retinal detachment and complete blindness	Cidofovir, Ganciclovir (GCV) and Foscarnet
Glaucoma	Primary open angle glaucoma (POAG), angle closure glaucoma (ACG)	Obstruction to the outflow of aqueous humour from the anterior segment	Prostaglandin analogues, beta-adrenergic receptor antagonists, alpha-2 adrenergic agonists, parasympathomimetics, carbonic anhydrase inhibitors
Conjunctivitis/red eye	Based on the structures: blepharoconjucntivitis, keratoconjucntivitis inflammation and episcleritis	Infection of conjunctiva, redness, irritation, grittiness and watering of the eyes	Allergic infections are treated using anti-histamine and non-steroid anti inflammatory agents and bacterial conjunctivitis is treated using antibiotics and corticosteroids

cardiac output to the retina. Thus, large systemic doses are often required to achieve a therapeutically effective drug concentration, which, in turn, can lead to significant side-effects [36]. Indeed, one study showed the intravitreal drug levels of poorly lipid soluble antibiotics, such as penicillins and cephalosporins, are at maximum 10% of serum levels, resulting in frequent administration being necessary and, as a result, systemic side-effects [37]. Furthermore, with many drugs now being proteinaceous in nature, utilisation of the systemic route would, consequently, lead to issues such as denaturation and a low therapeutic effect.

10.4.3 Ocular Injections

Injections into the ocular tissues are clinically employed to gain direct access to the target tissues thereby overcoming the barrier function of the eye in treating both anterior and posterior segment eye diseases. Accordingly, ocular injections can be classified into anterior injections and posterior injections.

10.4.3.1 Anterior Segment Injections

Anterior injections include subconjunctival, intrastromal, intracameral and intracorneal injections (Figure 10.3). These routes of drug delivery facilitate high drug concentrations within the selected tissue of the anterior segment. Injections are found to be particularly beneficial in the emergency management of acute conditions, such as in CNZ and fungal keratitis.

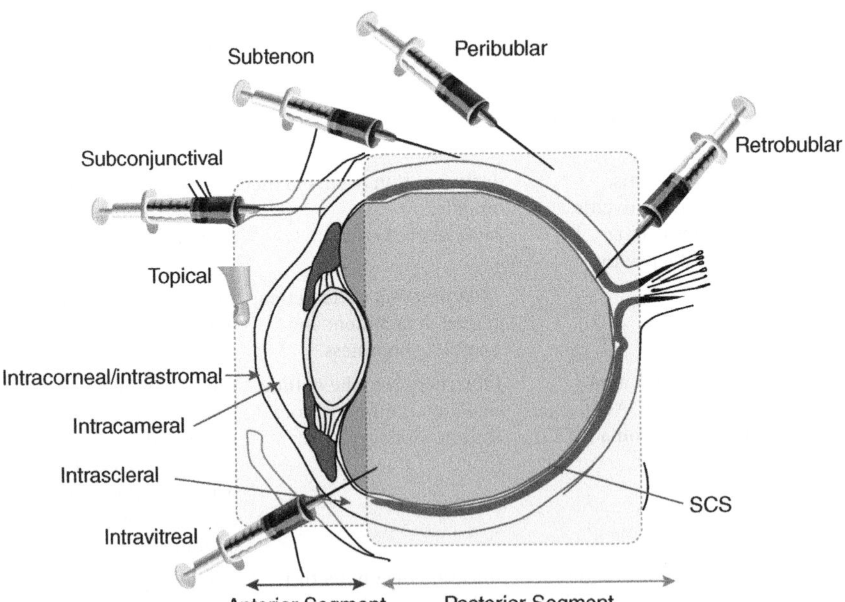

Figure 10.3 Schematic representation showing various routes of ocular drug delivery. Reproduced from [3] Thakur Raghu Raj Singh, I. Tekko, K. McAvoy, *et al.* (2017). Minimally invasive microneedles for ocular drug delivery. *Expert Opin. Drug Deliv.* 14: 525–537.

Subconjunctival Injections As the name indicates, the medication is delivered directly into the subconjunctival space (Figure 10.3). It is considered to be more patient friendly than any of the other types of ocular injections. Most commonly, hypodermic needles with sizes ranging from 21–30G are used for this purpose, with injections volumes of up to 0.1 ml. For example, in treating CNZ, bevacizumab (Avastin®) was administered by a topical route, with a 10 μl (2.5 mg/ml) eye drop given five times per day [38]. Frequent administration is required due to poor penetration of the bevacizumab, which is a hydrophilic and high molecular weight (149 kDa) drug. Alternatively, lower drug concentrations and reduced frequency of administration were achieved by subconjunctival injection [39–41], which demonstrates the advantages of injections over topical delivery. Although subconjunctival injections guarantee better delivery than topical eye drops, local side-effects, such as haemorrhages, have been reported [42]. Additionally, rapid drug elimination following subconjunctival administration is also well documented, which results in drainage of formulation into the systemic circulation thereby lowering ocular bioavailability [35]. The short residence time limits the effective permeation of drug molecules through the multiple ocular barriers before they reach their intended site of action at either the back or front of the eye.

Intrastromal, Intracameral and Intracorneal Injections These are other forms of injections that can be used to allow direct administration of the medication to the target tissues within the cornea. For example, intrastromal injections are widely used as a means of effective drug delivery in the management of CNZ [42, 43] and fungal keratitis [44–46]. A 31G needle was used for intrastromal injection (10 μl) of bevacizumab in human eyes. In certain cases, multiple intrastromal injections were given within the same eye [42], to accommodate higher amounts of drug per eye. Although intrastromal injection using a hypodermic needle has shown promising results [43–45], it is unpleasant for patients. This is due to a series of ocular complications and side-effects, including being painful and highly invasive, the possibility of imposing bacterial infections, inflammation and tissue damage, and it requires expertise in clinical administration [44, 47–49]. More importantly, delivering precise volumes (<25 μl) of drug solutions/suspensions within the thin corneal tissue is technically challenging and it is virtually impossible to produce reproducible results in each patient. Thus, varying dosages will lead to different levels of therapeutic efficacy among the patient group.

10.4.3.2 Posterior Segment Injections

The most commonly employed injections for treating posterior segment eye diseases such as AMD are by intravitreal injections (IVTs). IVTs were first utilised in 1911 to introduce air into the eye to repair retinal detachment [50]. Since then their use has evolved as a method of improving ocular ailments and delivering a range of therapeutics for the treatment of numerous ocular conditions, especially those of the posterior segment. IVT has become the "gold standard" to allow localised delivery of drugs to the back of the eye, with millions of injections given each year worldwide for patients suffering from a range of eye diseases.

Over recent decades, the use of IVTs has risen considerably; with these injections being one of the most frequently performed medical procedures in the United States [51]. It is also estimated that in the United Kingdom, in a department with around 500 000 patients in their care, 50–100 of these injections are performed weekly [52].

IVTs allow the local delivery of therapeutics and therefore reduce the systemic adverse effects [53]. According to the Royal College of Ophthalmologists guidelines on IVTs, the needles used should be 30G needles with non-colloidal clear solutions and 27G for particulate preparations. The injection needle length should be 12–15 mm, that is from 1/2 to 5/8 inch, with a maximum injection volume of 100 µl [54].

Although IVTs are not overly patient-friendly, they are able to overcome the multiple barriers discussed earlier, and deliver adequate drug concentrations locally [55]. Nevertheless, IVT, being an invasive method, is associated with multiple adverse effects and complications such as raised intraocular pressure (IOP), discomfort or pain (despite the use of anaesthesia), intraocular inflammation, retinal detachment, haemorrhage, endophthalmitis, cataract, lens damage and potentially blindness [6, 56–58]. These issues require supplementary medication. In treating chronic ocular diseases such as AMD, repeated injections are required every 4–6 weeks, indefinitely. Frequent injections will significantly increase the burden on patients and physicians. Furthermore, intravitreal delivery with conventional hypodermic needles should strictly adhere to numerous safeguards to avoid mechanical injury to the lens and retina [56]. These risks are dependent upon the needle type, where lower gauge needles cause more pain and higher damage to the eye. Therefore, smaller 27–30G needles are preferable.

Drug formulations can also be injected on the outer surface of the eyeball, through periocular injections (transscleral delivery) such as sub-tenon, retrobulbar, peribulbar and posterior juxtascleral (Figure 10.3), which are considered to be less invasive than IVT. Transscleral delivery via periocular administration is thought to be one of the safest means of achieving stable drug concentrations within the vitreous and retina, although there have been reports of anterior segment complications after periocular injection, such as raised IOP, cataract and strabismus. Other challenges to drug delivery via the transscleral are dependent on the nature of the drug molecule. Interestingly the sclera is highly permeable to large drug molecules; however, the RPE is a major barrier to diffusion for both large molecules and hydrophilic drug molecules, and this may be the rate-limiting feature in the delivery of such molecules to the retina via the transsceleal route [6].

Long-acting ocular drug delivery systems, such as micro/nanoparticles, liposomes, *in situ* implant forming gels and preformed solid implants, are attracting tremendous interest due to their ability to maintain constant drug levels following a single administration [58]. However, administration of these formulations using either standard hypodermic needles or surgical implantation would still hamper patient compliance. For example, some studies have previously developed and evaluated the administration of sustained release preformed intrascleral implants [60–62]. Although these intrascleral implants showed sustained drug release, they necessitate surgical administration within the thin scleral tissue, which presents concerns greater than those seen with IVT injections. Furthermore, any surgical procedure will only impose additional costs and technical challenges with the treatment modality.

In conclusion, conventional hypodermic injections are capable of delivering drug formulations to the target site but are associated with numerous adverse effects and risks. Although there is tremendous ongoing research interest in developing novel sustained release formulations, technologies that enable safe delivery of the existing or new sustained release formulations are still limited. Therefore, in an attempt to overcome invasive ocular injections using conventional hypodermic needles, and for the safer delivery of medication, minimally invasive MN-based devices were found to be of significant interest.

10.5 Microneedles in Ocular Drug Delivery

As described in earlier chapters, MNs were originally developed as a painless, minimally invasive and effective technology for transdermal drug delivery [63–66]. MNs are typically around 25–2000 μm in height and are fabricated from a wide range of materials, including silicon, stainless steel, glass and polymers, and in different shapes, either solid or hollow types. MN application to the eye could offer several advantages over invasive intraocular injections that utilise long, conventional, hypodermic needles. The dimensions of MNs are long enough to overcome the ocular barriers of both the anterior and posterior segments of the eye, allowing localised delivery within the tissues such as the sclera, stroma and suprachoroidal space [3]. Unlike hypodermic needles, MNs reduce the chances of pain, tissue damage and the risk of infection. Using MNs in ocular drug delivery is a relatively new concept since very little research has been carried out in this field. Ocular drug delivery, using MNs, can be achieved by employing any of the four MN application strategies described in previous chapters, that is, solid MNs with "poke and patch," solid MNs with "coat and poke," dissolving MNs and hollow MNs [67].

10.5.1 Hollow MNs

This strategy involves a two-step process. In the first step, solid MN arrays are applied to the ocular tissue and subsequently removed to create micro-channels, thereby enhancing drug permeation. Permeation of the active molecules from this formulation occurs via passive diffusion through the created micro-channels. In 2014 Thakur *et al.* [68] investigated this approach using hollow MN devices 400, 500 and 600 μm in height fabricated from hypodermic needles (i.e. 27, 29 and 30G). In an *in vitro* setup, these hollow MNs were used to inject a thermoresponsive poloxamer-based hydrogel containing a model fluorescent drug (i.e. sodium fluorescein) into the scleral tissue of a rabbit to form an *in situ* implant within the micro-channels (Figure 10.4), to provide a sustained release of fluorescein sodium over 24 h. This method of implant formation, without the need for surgical intervention, would enhance patient acceptability; furthermore, it could provide sustained drug levels thereby reducing the frequency of eye drops, for example.

10.5.2 Solid MNs with "Coat and Poke" Strategy

This approach relies on coating a drug formulation on the MNs and subsequent insertion into the ocular tissue to enable dissolution of the coated formulation leading to localisation within the tissue. Prausnitz and coworkers were the first to demonstrate the application of coated MNs to the eye [48]. This study provided the first reported drug delivery into the anterior segment of the eye using coated stainless-steel MNs measuring 500–750 μm in length and 200×50 μm in width, and with a tip angle of 55° (Figure 10.5a), which were tested for anterior and posterior drug delivery via either intrascleral or intracorneal routes, respectively. The MNs were coated (Figure 10.5b) with the model drug sodium fluorescein (approximately 280 ng) and inserted halfway into the cornea of a

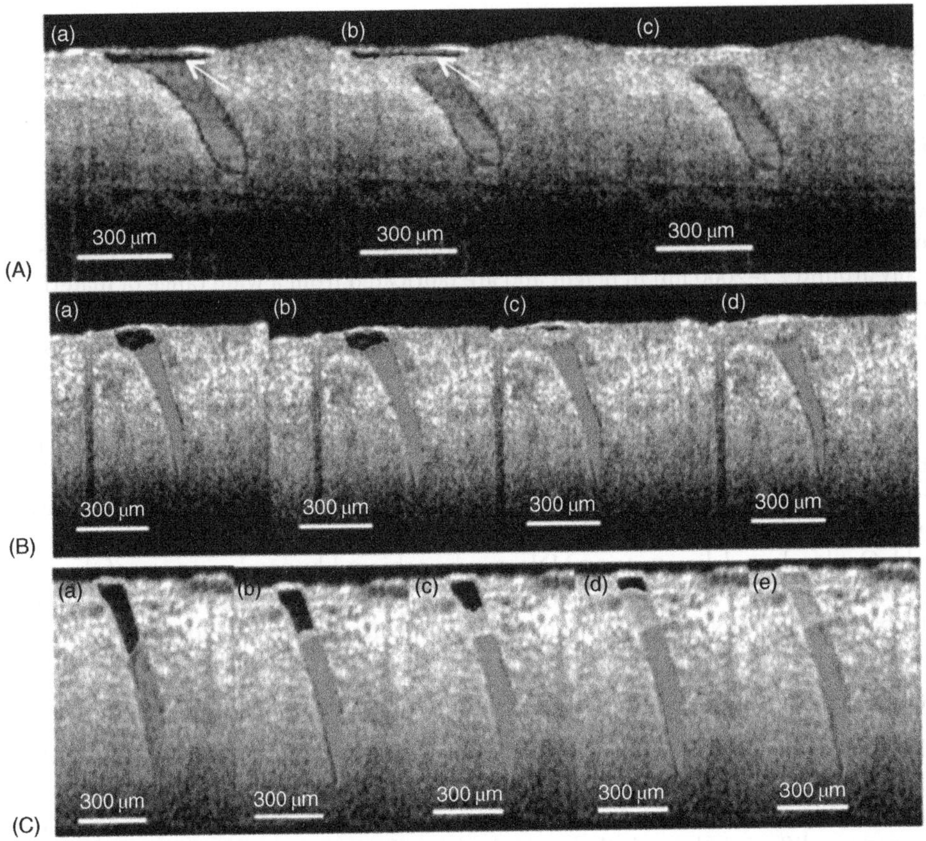

Figure 10.4 Optical coherence tomography images showing 30G hollow microneedle injection of 50 μl F6 formulation (coloured in red) injected into equatorial sclera to a depth of: (A) 400 μm at (a) 0, (b) 1 and (c) 2 h; (B) 500 μm at (a) 0, (b) 1, (c) 2 and (d) 2.5 h; and (C) 600 μm at (a) 0, (b) 1, (c) 2, (d) 2.5 and (e) 3 h. The arrow indicates empty space in the sclera created following hollow microneedle application and its subsequent closure over time. Reproduced from [68] Thakur Raghu Raj Singh, S.J. Fallows, H.L. McMillan, *et al.* (2014). Microneedle-mediated intrascleral delivery of in situ forming thermoresponsive implants for sustained ocular drug delivery. *J. Pharm. Pharmacol.* 66: 584–595.

rabbit eye and left in place for 2 min and then retrieved back out. Following MN insertion, after 1 min there was a sharp increase in intraocular fluorescein concentration and then gradually further increases peaked at 3 h and then gradually decreased to background within 24 h. The study showed that the drug depot was formed within the cornea (Figure 10.5c), which steadily released fluorescein into the anterior segment for hours. Although a small abrasion was noted at the site of MN insertion, it disappeared after 3 h. The study showed that MNs were able to achieve a 60-fold increase in fluorescein in comparison with topical application.

In this study, another pilocarpine-coated MN produced nearly 45-fold increase in its bioavailability relative to topical administration. Interestingly, these MNs were able to deliver other molecules such sulforhodamine, protein and DNA in an *in vitro* study targeted for posterior segment delivery. The study revealed that MNs penetrated the

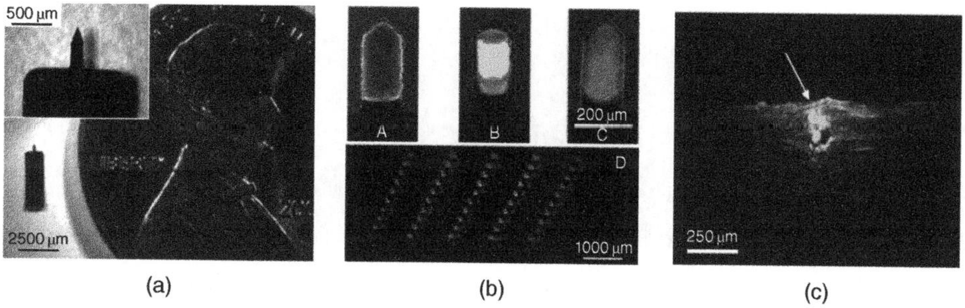

Figure 10.5 (a) Brightfield microscopy image of a single solid stainless-steel MN used for intrascleral and intracorneal drug delivery shown next to a US penny for size comparison. Inset: magnified view of the needle, which is 500 μm in length and with a 45° tip angle. (b) MNs coated with model drugs, including (A) sodium fluorescein, (B) fluorescein-labelled bovine albumin and (C) fluorescein-labelled plasmid DNA, using a dip-coating method; (D) a 50-needle array of microneedles coated with sodium fluorescein, which exhibited good needle-to-needle uniformity and no contamination of the base substrate with excess coating material. (c) Histologic sections of human cadaveric sclera pierced with a single microneedle (750 μm in length and 55° tip angle) coated with fluorescein-labelled bovine serum albumin, inserted to deposit their model drug coatings within the sclera. Arrow: site of microneedle insertion through the scleral surface. Adapted with permission from [48] J. Jiang, H.S. Gill, D. Ghate, *et al.* (2007). Coated microneedles for drug delivery to the eye. *Invest. Ophthalmol. Vis. Sci.* 48: 4038–4043.

human cadaver sclera to a depth of 300 μm. The drug coating rapidly dissolved off the needles within the scleral tissue within 20 s of insertion.

10.5.3 Dissolving MNs

Dissolving MNs are made of a soluble/biodegradable matrix that includes the active substance. Conventionally, micromoulding techniques are used to produce this type of MNs [3, 63]. Following insertion, the needle matrices dissolve/biodegrade in the tissue to release the payload.

This approach was first investigated by Thakur *et al.* in 2016 [69]. In this study, dissolving 3×3 MN arrays with conical shape measuring about 800 μm in height (Figure 10.6a) were fabricated from poly(vinylpyrrolidone) (PVP) polymer of various molecular weights (MWs) to enhance ocular drug delivery of small and large molecules, namely fluorescein sodium and fluorescein isothiocyanate–dextrans (with a MW of 70kDa and 150 kDa) incorporated into the MN shafts. The experiments were performed *in vitro* employing porcine cornea and sclera tissues. The study revealed that PVP MNs are strong enough to be inserted into both scleral (Figure 10.6b) and corneal (Figure 10.6c) tissues and then rapidly dissolve within less than 3 min from application, which was dependent upon the PVP's MW. Interestingly, the dissolving MNs enhanced the model drugs permeation across both the corneal and scleral tissues, in comparison with topically applied aqueous solutions. Confocal images (Figure 10.7A and B) showed that the macromolecules formed depots within the tissues, which led to sustained delivery.

This can be considered of great importance for delivery of high MW anti-VEGF drugs in a minimally invasive manner. Because MNs dissolve in the ocular tissue fluid, there is no sharps waste, which makes dissolving MNs impossible to reuse and

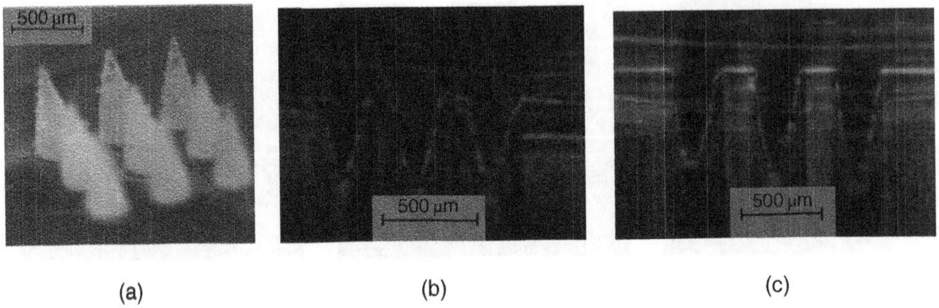

Figure 10.6 (a) A light microscopy image of a PVP-based dissolving MN array loaded with sodium fluorescein (as a model drug), the MNs measuring 800 μm in height and 300 μm in width at the baseplate. (b) OCT image of MN array after insertion into a pig's scleral tissue (*in vitro*). (c) OCT image of MN array after insertion into a pig's corneal tissue (*in vitro*). Adapted from [69] Thakur Raghu Raj Singh, I.A. Tekko, F. Al-Shammari, *et al.* (2016). Rapidly dissolving polymeric microneedles for minimally invasive intraocular drug delivery. *Drug Delivery Transl. Res.* 6: 800–815.

thereby eliminates the risks of biohazard sharps, unlike intravitreal injections that use hypodermic needles.

10.5.4 Hollow MN Strategy

Hollow MNs, as the name suggests, consist of hollow needles that, following insertion, facilitate injection of a fluid medication (as a drug solution or its nano-/microparticle formulations) into the ocular tissue for localised and/or sustained drug release. Jiang *et al.*, in 2009, first demonstrated this strategy for intrascleral delivery using a hollow glass MN for a model drug (sulforhodamine) and micro/nanoparticle formulations [70]. The MN was fabricated from a borosilicate cylindrical glass micropipette tube with 1.5 mm outer diameter and 0.86 mm internal diameter (Figure 10.8a). The MN was initially inserted into the tissue at a depth of 700–1080 μm, and retracted out of the tissue in increments of 60 μm during the solution injection. Sulforhodamine solution was infused at a pressure of 15 psi. Essentially no solution was delivered into the tissue after the initial insertion. Upon further retraction from 200 to 300 μm, the delivery was achieved at volumes of from 10 to 35 μl of fluid containing either the soluble drug molecule sulforhodamine B (Figure 10.8b) or nanoparticles suspensions (Figure 10.8c) from an individual MN. However, microparticles were only delivered in the presence of hyaluronidase and collagenase spreading enzymes. The enzymes in this case were used to breakdown the tissue components to accommodate the microparticles.

Hollow MNs were also used to deliver drug into the superchoroidal space (SCS). The SCS is a potential space between the sclera and choroid that extends circumferentially around the eye. Being immediately adjacent to the choroid and retina, delivery into the SCS can offer targeted drug delivery to these tissues.

In 2011 Patel *et al.* investigated posterior drug delivery into the SCS using hollow MNs [71]. A single glass hollow MN measuring 800–1000 μm in length was used to infuse nanoparticle and microparticle suspensions into the SCS in *ex vivo* rabbit, pig and human eyeballs. MNs were also shown to deliver sulforhodamine B as well as nanoparticle and microparticle suspensions into the SCS of rabbit, pig and human eyes. Volumes up to 35 μl were administered consistently. The study suggested that particles of 20 and

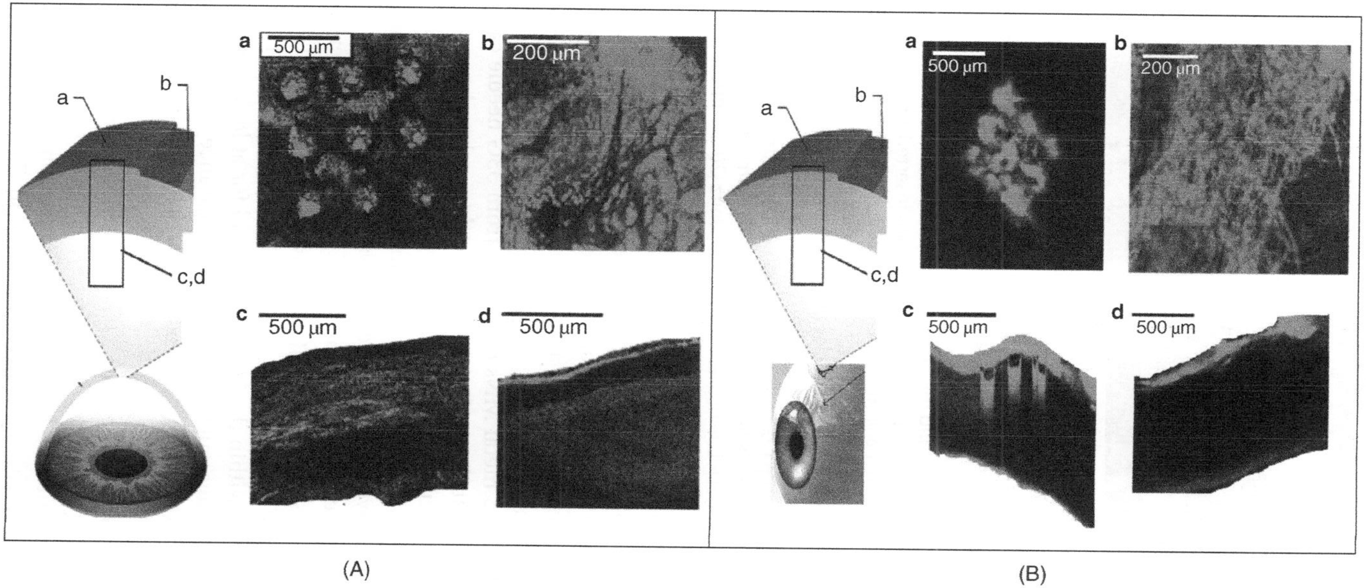

Figure 10.7 (A) Schematic diagram on the left-hand side represents the collection and processing of confocal images of the corneal tissue following application of FD70-loaded PVP K29/32 MN arrays, where confocal images represent (a) after 5 min of MN application, (b) topical image at a depth of 80 μm from surface of the tissue after 1 h following MN array insertion, (c) cross-section image of tissue after 1 h following MN array insertion and (d) cross-section image of tissue 1 h after applying an aqueous solution of FD 70 (at 2 mg/ml). (B) Schematic diagram on the left-hand side represents the collection and processing of confocal images of scleral tissues following application of FD70-loaded PVP K29/32 MN arrays: (a) topical image of tissue after 5 min following insertion of MN array, (b) cross-section image of tissue after 5 min following MN array insertion, (c) topical image at a depth of 80 μm from surface of the tissue after 1 h following MN insertion and (d) cross-section image of tissue 1 h after applying an aqueous solution of FD 70 (at 2 mg/ml). Reproduced from [69] Thakur Raghu Raj Singh, I.A. Tekko, F. Al-Shammari, *et al*. (2016). Rapidly dissolving polymeric microneedles for minimally invasive intraocular drug delivery. *Drug Delivery Transl. Res.* 6: 800–815.

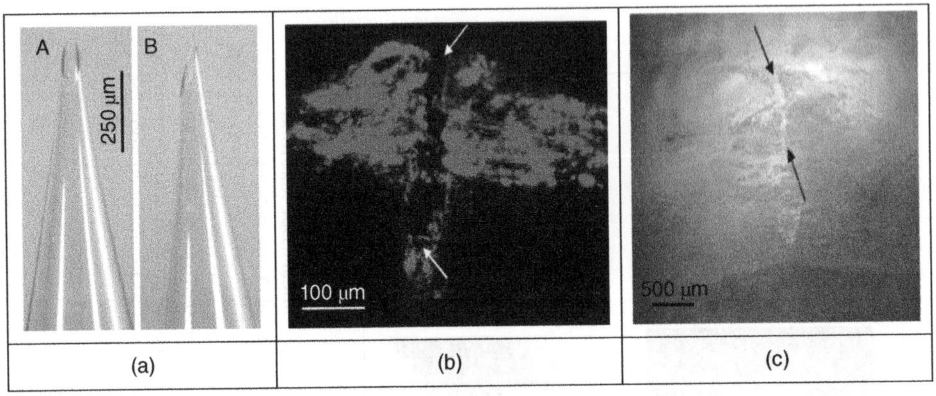

Figure 10.8 (a) Representative hollow microneedle: (A) front view and (B) side view. The microneedle shown has an approximately elliptical tip opening with a long axis of ~100 μm and a short axis of ~40 μm with a bevel tip angle of 25°. (b) Histological section using fluorescence microscopy showing the site of microneedle insertion (indicated by arrows) and the distribution of injected sulforhodamine preferentially localised to the upper portion of the tissue. A microneedle was inserted 720 μm into the sclera and then retracted 60 μm every 3 min to a maximum retraction of 240 μm. Sulforhodamine solution was infused into the tissue at a pressure of 15 psi [70]. (c) Representative histological image of human cadaver sclera after infusion of fluorescent nanoparticles (280 μm diameter). The microneedle was inserted 720 μm into the anterior region of the sclera and then retracted 60 μm every 3 min to a maximum retraction of 240 μm. A 1.0 wt% nanoparticles suspension was infused into the tissue at a pressure of 15 psi over 15 min. The site of microneedle insertion is indicated by arrows. Reproduced with permission from [70] J. Jiang, J.S. Moore, H.F. Edelhauser and M.R. Prausnitz (2009). Intrascleral drug delivery to the eye using hollow microneedles. *Pharm. Res.* 26: 395–403.

100 nm can spread within the sclera as well as the SCS, whereas particles of 500 and 1000 nm were localised exclusively in the suprachoroidal space. The study also suggested that to deliver 500–1000 nm particles in the SCS, a minimum MN length of 1000 μm and a pressure of 250–300 kPa were necessary.

Similarly, Patel *et al.* [72] then used metal MNs fabricated from 33G needle cannulas, with 750 μm length and a bevel at the orifice, to evaluate the ocular pharmacokinetics of molecules (sodium fluorescein, fluorescein isothiocyanate dextrans of 40 kDa and 250 kDa, bevacizumab tagged with Alexa-Fluor 488) and particles (FluoSpheres) injected into the SCS of the rabbit eye. Here, the metal MNs were attached to a 1-ml syringe. In general, the molecules were cleared from the SCS within 1 day, therefore, particles were injected into the SCS so that the drug could be localised and remain for months. Particles of from 20 nm to 10 μm diameter were injected into the SCS of rabbit eyes, *in vivo*. The particles remained within the SCS and choroid for at least 2 months. It was noted that the capillary drainage might play a role in clearance from the SCS. Nevertheless, this study demonstrated the ability to localise particles within the SCS for sustained drug release.

In an *in vivo* study, Gilger *et al.* [73] used the same 33G hollow MNs, 850 μm in height, to deliver triamcinolone acetonide (TA) to the SCS. The study has demonstrated that 0.2 mg of SCS-delivered TA was as effective in reducing inflammation as 2.0 mg of TA delivered by IVT in a model of acute posterior uveitis inflammation. Furthermore, there was no evidence of any adverse effects, increase in IOP, drug toxicity or haemorrhage following MN application. Likewise, Chiang *et al.* 2016 [74] have recently investigated circumferential distribution of particles in the SCS of rabbit and human cadaver eyes. The same hollow MNs were used as reported by Patel *et al.* [71], with a 33G needle of 750 µm height. Injection of red-fluorescent microspheres of 200 nm diameter with volumes ranging from 50 to 200 µl were performed in the SCS. In rabbit eyes, particles, when injected into the superior or inferior hemispheres, did not significantly cross into the other hemisphere due to a barrier formed by the long posterior ciliary artery. In human eyes, the short posterior ciliary arteries prevented circumferential spread towards the macula and optic nerve. Therefore, considering that the anatomical barriers could hinder the even spreading of the administered drug or formulation within the SCS, judicious selection of the region for injection is essential. Kim *et al.* [75], from the Prausnitz group, investigated using single solid stainless-steel MNs measuring 400 µm in length coated with bevacizumab to treat CNZ. Results revealed that the drug was delivered intrastromally and needed a lower dose compared with subconjunctival and topical eye drops, that is, just 4.4 µg of the drug are needed to produce the same effect with as much as 2500 µg via subconjunctival injection and 52 500 µg when delivered via eye drops.

10.5.5 Other Strategies

Unlike the single HMNs or coated MNs, Palakurthi *et al.* [76] investigated MNs that were fabricated into an array of 3×3 biodegradable MN arrays loaded with methotrexate. The MNs measuring 2 mm in length, 2 mm in width and 2.3 mm in height were surgically placed in the deep lamellar scleral pocket in a rabbit eye, *in vivo*. These were found to be effective in sustaining and controlling drug delivery and to be safe. The central advantage of using MNs is the opportunity for minimally invasive application. However, in this case we are of the opinion that the term *microneedle* perhaps needs reconsidering, importantly due to the growing interest in MNs and to a large extent because the MNs are fabricated with heights of less than 1000 µm and widths of less than 500 µm.

10.6 MN Application Devices

The reliability and reproducibility of insertion of MNs into the ocular tissues in an *in vivo* setting could present a major challenge for progressing MNs as a platform in ocular drug delivery. Few studies have addressed this issue and come up with innovative designs for MN applicators.

Song *et al.* [77] designed an MN-based pen type device (Figure 10.9) to enhance the reliability of MN insertion, allowing easy insertion into a small target region of ocular tissue. A solid SU-8 resin based MN was fabricated and attached to a macro-scale applicator to create the MN pen. The resulting MN had a base area of 200×200 µm^2 with a height of 140 µm. Rhodamine B, Evans blue or Sunitinib malate were used, along with polymer carrier, as model drugs to dip coat the MN. It was shown that the MN pen enabled precise localisation of drug within the stromal membrane of the cornea, which is otherwise difficult to achieve when given topically due to the corneal epithelium.

Matthaei *et al.* [78], to improve the reproducibility of injection methods using hand-held syringes, compared different types of hollow MNs and syringes and quantified the intrastromal distribution of Indian ink in mouse cornea by injections

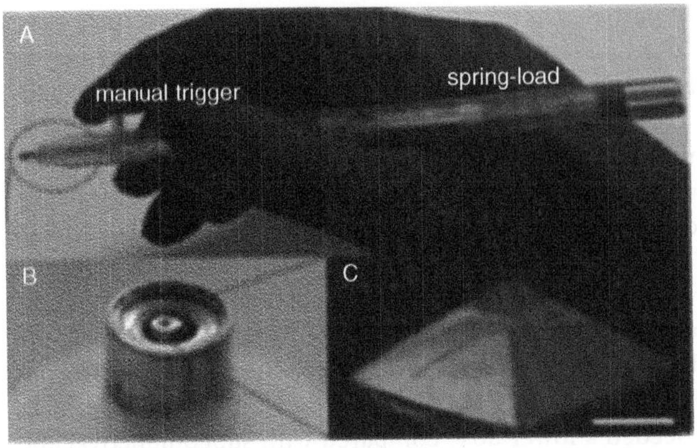

Figure 10.9 Images show a photograph of: (a) spring-loaded MN pen; (b) MN guiding structure at the end of MN pen; and (c) transfer moulded MN structure on the tip end of MN pen. Scale bar is 100 μm. Adapted with permission from [77] H.B. Song, K.J. Lee, I.H. Seo, *et al.* (2015). Impact insertion of transfer-molded microneedle for localized and minimally invasive ocular drug delivery. *J. Control. Release* 209: 272–279.

of different volumes (1 and 2 μl). Needle types and syringes tested were namely: 33G (attached to a 2.5 μl syringe), 35G needles (attached to a 10 μl syringe) and glass MNs bevelled to 25° and an inner tip diameter of approximately 50 μm (attached to a 2.5 μl syringe), respectively. Injections of 1 and 2 μl resulted in an overall mean of 49 and 73%, respectively, of the total corneal area involved. The use of 33G metal needles provided the most reliable and effective outcomes, whereas the glass MN tips broke within the stroma in 25% of cases, which is undesirable and creates potential safety concerns. Irrespective of needle type, a small amount of leakage was noted in all cases.

10.7 MN Safety Concerns

In the literature, not too many studies have been reported regarding the safety of using MNs for ocular applications. Kim *et al.* evaluated the safety of using MNs in an *in vivo* setting [75]. They analysed rabbit corneas with and without MN treatment and with and without suture placement to assess the safety of MN insertion by both magnified inspection of the corneal surface *in vivo* and histological examinations. A small puncture in the corneal epithelium is evident with a size in the order of 200 μm. After 24 h, it was not possible to locate the insertion site due to apparent repair of the epithelium. Similarly, at later times the corneal surface continued to look intact and normal. They also examined eyes treated with bevacizumab-coated MNs and again saw only a microscopic puncture in the corneal epithelium that disappeared within 1 day, which was not associated with any complications. They examined these injection sites on a daily basis throughout the 18 days of the experiments, but they never saw evidence of corneal opacity in any of the 22 eyes treated with MNs in this study.

In addition to examining the corneal surface, animals were killed at different time points to look for changes in the corneal micro-anatomical structure. Histological analyses of treated and untreated eyes were evaluated. Eyes treated by insertion of non-coated MNs exhibited no significant changes in the cornea, with no evidence of corneal puncture being found. No significant presence of macrophages or vascularisation was observed. They also compared histological sections from eyes that had only been sutured and then treated 4 days later with 1.1 μg of bevacizumab using an MN. In the sutured eyes, there were large numbers of inflammatory cells present, but no notable differences were seen between sutured eyes with and without MN treatment. Altogether, these data indicate that insertion of MNs into the cornea is minimally invasive and thereby has few or no adverse effects on the eye. Although this analysis suggests that MN insertion into the eye may be well tolerated, additional studies are needed to assess safety more fully.

Thakur *et al.* investigated the safety of using PVP-based dissolving MNs by testing the biocompatibility of the polymer when exposed to the retinal cells [76]. The viability of the cells was found to be >83 % at all PVP concentrations below 2 mg/ml. It was concluded that the PVP K29/32 MN is non-toxic in *in vitro* testing of the retinal cells, and therefore, it was deemed as biocompatible.

10.8 Conclusion

Drug delivery to the eye remains a significant challenge due to its recessed location and barrier function. Although delivery of conventional formulations is relatively easy for treatment of anterior segment eye diseases, low bioavailability remains a significant obstacle. In contrast, treating sight threatening eye diseases that originate in the posterior segment of the eye is even more demanding due to the inaccessibility of tissues such as the retina, where direct injection of medication is currently practiced to overcome the barrier issues. Conventional hypodermic needle based injections create significant tissue damage and increase the risk of infections. Other side-effects of using hypodermic needles include poor localisation of drug/formulation, increase in IOP, damage to adjacent tissues and poor patient compliance. Additionally, to maintain constant drug levels, frequent injections are necessary, which only further complicate this mode of drug delivery.

Minimally invasive MNs could offer numerous advantages to overcome the current issues surrounding hypodermic injections, as demonstrated by a number of studies presented in this chapter. Unlike hypodermic needles, MNs can aid in the localisation of drug delivery formulations within the selected ocular tissue, which will overcome the barrier issue and allow therapeutic drug levels to be maintained within the eye for prolonged time periods. MNs can address treatment of both anterior and posterior disorders. However, the dimensions of the needles should be small enough to avoid unintended damage to adjacent tissues, whilst being efficient in overcoming the barrier function. Significant research has been done in developing applicators for MN administration to the skin; however, more research is needed in developing appropriate devices for the eye. Safety of MN applications to the eye also requires further research, as, to date, limited work has been done in this area. In particular, safety related to MN application forces, tissue type, depth of penetration, extent of application, formulation types and suitable animal model need further investigation. Overall, MNs hold the potential to revolutionise the way drug formulations are administered in the eye.

Nevertheless, current limitations and challenges of MN applications warrant further research to enable their widespread clinical use.

References

1 J. Olver, L. Cassidy, G. Jutley and L. Crawley (2014). *Ophthalmology at a Glance*, 2_{nd} edn, 123. Hoboken, New Jersey: Wiley Blackwell.
2 V.K. Yellepeddi, R. Sheshala, H. McMillan, *et al.* (2015). Punctal plug: a medical device to treat dry eye syndrome and for sustained drug delivery to the eye. *Drug Discov. Today* 20: 884–889.
3 Thakur Raghu Raj Singh, I. Tekko, K. McAvoy, *et al.* (2017). Minimally invasive microneedles for ocular drug delivery. *Expert Opin. Drug Deliv.* 14: 525–537.
4 D. Ghate and H.F. Edelhauser (2006). Ocular drug delivery. *Expert Opin. Drug Deliv.* 3: 275–287.
5 F. Idrees, D. Vaideanu, S.G. Fraser, *et al.* (2006). A review of anterior segment dysgeneses. *Surv. Ophthalmol.* 51: 213–231.
6 D.H. Geroski and H.F. Edelhauser (2000). Drug delivery for posterior segment eye disease. *Invest. Ophthalmol. Vis. Sci.* 41: 961–964.
7 S.S. Lee, P. Hughes, A.D. Ross and M.R. Robinson (2010). Biodegradable implants for sustained drug release in the eye. *Pharm. Res.* 27: 2043–2053.
8 J.P. Shin, Y.C. Park, J.H. Oh, *et al.* (2009). Biodegradable intrascleral implant of triamcinolone acetonide in experimental uveitis. *J. Ocul. Pharmacol. Ther.* 25: 201–208.
9 V. Ranta and A. Urtti (2006). Transscleral drug delivery to the posterior eye: Prospects of pharmacokinetic modeling. *Adv. Drug Deliv. Rev.* 58: 1164–1181.
10 U.B. Kompella, R.S. Kadam and V.H. Lee (2010). Recent advances in ophthalmic drug delivery. *Ther. Delivery* 1: 435–456.
11 R. Edgar, Y. Mazor, A. Rinon, *et al.* (2013). LifeMap Discovery™: the embryonic development, stem cells, and regenerative medicine research portal. *PloS One* 8 (7): e66629.
12 L. Rabinovich-Guilatt, P. Couvreur, G. Lambert and C. Dubernet (2004). Cationic vectors in ocular drug delivery. *J. Drug Target.* 12: 623–633.
13 J. Fischbarg (2006). *The Biology of the Eye*. Burlington: Elsevier.
14 R.D. Schoenwald and R.L. Ward (1978). Relationship between steroid permeability across excised rabbit cornea and octanol-water partition coefficients. *J. Pharm. Sci.* 67: 786–788.
15 H. Sasaki, M. Ichikawa, K. Yamamura, *et al.* (1997). Ocular membrane permeability of hydrophilic drugs for ocular peptide delivery. *J. Pharm. Pharmacol.* 49: 135–139.
16 W.G. Bachman and G. Wilson (1985). Essential ions for maintenance of the corneal epithelial surface. *Invest. Ophthalmol. Vis. Sci.* 26: 1484–1488.
17 A. Urtti and L. Salminen (1993). Minimizing systemic absorption of topically administered ophthalmic drugs. *Surv. Ophthalmol.* 37: 435–456.
18 A. Urtti (2006). Challenges and obstacles of ocular pharmacokinetics and drug delivery. *Adv. Drug Deliv. Rev.* 58: 1131–1135.
19 T.J. Mikkelson, S.S. Chrai and J.R. Robinson (1973). Competitive inhibition of drug–protein interaction in eye fluids and tissues. *J. Pharm. Sci.* 62: 1942–1945.

20 K. Cholkar, A. Patel, A. Dutt Vadlapudi and A. K Mitra (2012). Novel nanomicellar formulation approaches for anterior and posterior segment ocular drug delivery. *Recent Pat. Nanomed.* 2: 82–95.
21 T.W. Olsen, S.Y. Aaberg, D.H. Geroski and H.F. Edelhauser (1998). Human sclera: thickness and surface area. *Am. J. Ophthalmol.* 125: 237–241.
22 J.R. Dunlevy and J.A.S. Rada (2004). Interaction of lumican with aggrecan in the aging human sclera. *Invest. Ophthalmol. Vis. Sci.* 45: 3849–3856.
23 P.G. Watson and R.D. Young (2004). Scleral structure, organisation and disease. A review. *Exp. Eye Res.* 78: 609–623.
24 A. Lens, T. Langley, S.C. Nemeth and C. Shea (1999). *Ocular Anatomy and Physiology.* Slack Inc.
25 J. Ambati, C.S. Canakis, J.W. Miller, *et al.* (2000). Diffusion of high molecular weight compounds through sclera. *Invest. Ophthalmol. Vis. Sci.* 41: 1181–1185.
26 A. Edwards and M.R. Prausnitz (1998). Fiber matrix model of sclera and corneal stroma for drug delivery to the eye. *AIChE J.* 44: 214.
27 N. Kuno and S. Fujii (2011). Recent advances in ocular drug delivery systems. *Polymers* 3: 193–221.
28 Y. Ikuno, K. Kawaguchi, T. Nouchi and Y. Yasuno (2010). Choroidal thickness in healthy Japanese subjects. *Invest. Ophthalmol. Vis. Sci.* 51: 2173–2176.
29 R. Margolis and R.F. Spaide (2009). A pilot study of enhanced depth imaging optical coherence tomography of the choroid in normal eyes. *Am. J. Ophthalmol.* 147: 811–815.
30 N.V. Chong, J. Keonin, P.J. Luthert, *et al.* (2005). Decreased thickness and integrity of the macular elastic layer of Bruch's membrane correspond to the distribution of lesions associated with age-related macular degeneration. *Am. J. Pathol.* 166: 241–251.
31 M. Hornof, E. Toropainen and A. Urtti (2005). Cell culture models of the ocular barriers. *Eur. J. Pharm. Biopharm.* 60: 207–225.
32 P. Heiduschka, H. Fietz, S. Hofmeister, *et al.* (2007). Penetration of bevacizumab through the retina after intravitreal injection in the monkey. *Invest. Ophthalmol. Vis. Sci.* 48: 2814–2823.
33 N.R. Galloway, W.M.K. Amoaku, P.H. Galloway and A.C. Browning (2006). *Common Eye Diseases and their Management*, 3rd edn. London: Springer.
34 J.N. Shah, H.J. Shah, A. Groshev, *et al.* (2014). Nanoparticulate transscleral ocular drug delivery. *J. Biomol. Res. Ther.* 3: 116.
35 R. Gaudana, H.K. Ananthula, A. Parenky and A.K. Mitra (2010). Ocular drug delivery. *AAPS J.* 12: 348–360.
36 P.M. Hughes, O. Olejnik, J. Chang-Lin and C.G. Wilson (2005). Topical and systemic drug delivery to the posterior segments. *Adv. Drug Deliv. Rev.* 57: 2010–2032.
37 M. Barza (1978). Factors affecting the intraocular penetration of antibiotics. The influence of route, inflammation, animal species and tissue pigmentation. *Scand. J. Infect. Dis. Suppl.* (14): 151–159.
38 E. Yoeruek, F. Ziemssen, S. HenkeFahle, *et al.* (2008) Safety, penetration and efficacy of topically applied bevacizumab: evaluation of eyedrops in corneal neovascularization after chemical burn. *Acta Ophthalmol.* 86: 322–328.

39 C. Kim, Y. Ahn, P. Wilder-Smith, *et al.* (2010). Efficient and facile delivery of gold nanoparticles in vivo using dissolvable microneedles for contrast-enhanced optical coherence tomography. *Biomed. Opt. Express* 1 (Jan 1): 106–113.

40 J. Chang, N.K. Garg, E. Lunde, *et al.* (2012). Corneal neovascularization: an anti-VEGF therapy review. *Surv. Ophthalmol.* 57: 415–429.

41 M.H. Dastjerdi, K.M. Al-Arfaj, N. Nallasamy, *et al.* (2009). Topical bevacizumab in the treatment of corneal neovascularization: results of a prospective, open-label, noncomparative study. *Arch. Ophthalmol.* 127: 381–389.

42 A.C. Vieira, A.L. Höfling-Lima, J.Á. Gomes, *et al.* (2012). Intrastromal injection of bevacizumab in patients with corneal neovascularization. *PubMed – NCBI, Arq. Bras. Oftalmol.* 75 (4): 277–279.

43 M.N. Hashemian, M.A. Zare, F. Rahimi and M. Mohammadpour (2011). Deep intrastromal bevacizumab injection for management of corneal stromal vascularization after deep anterior lamellar keratoplasty, a novel technique. *Cornea* 30: 215–218.

44 X. You, J. Li, S. Li and W. Shi (2015). Effects of lamellar keratectomy and intrastromal injection of 0.2% fluconazole on fungal keratitis. *J. Ophthalmol.* 2015: 656027.

45 N. Sharma, P. Agarwal, R. Sinha, *et al.* (2011) Evaluation of intrastromal voriconazole injection in recalcitrant deep fungal keratitis: case series. *Br. J. Ophthalmol.* 95: 1735–1737.

46 G. Kalaiselvi, S. Narayana, T. Krishnan and S. Sengupta (2015). Intrastromal voriconazole for deep recalcitrant fungal keratitis: a case series. *Br. J. Ophthalmol.* 99: 195–198.

47 D. Maurice (2001). Review: practical issues in intravitreal drug delivery. *J. Ocul. Pharmacol. Ther.* 17: 393–401.

48 J. Jiang, H.S. Gill, D. Ghate, *et al.* (2007). Coated microneedles for drug delivery to the eye. *Invest. Ophthalmol. Vis. Sci.* 48: 4038–4043.

49 D.M. Maurice (2002). Drug delivery to the posterior segment from drops. *Surv. Ophthalmol.* 47 (suppl. 1): S41–S52.

50 J. Ohm (1911). Über die Behandlung der Netzhautablösung durch operative Entleerung der subretinalen Flüssigkeit und Einspritzung von Luft in den Glaskörper. *Graefe's Arch. Clin. Exp. Ophthalmol.* 79: 442–450.

51 P.Y. Ramulu, D.V. Do, K.J. Corcoran, *et al.* (2010). Use of retinal procedures in medicare beneficiaries from 1997 to 2007. *Arch. Ophthalmol.* 128: 1335–1340.

52 R. Tailor, R. Beasley, Y. Yang, Y. and N. Narendran (2011). Evaluation of patients' experiences at different stages of the intravitreal injection procedure–what can be improved? *Clin. Ophthalmol.* (Auckland, NZ) 5: 1499.

53 M. Englander, T.C. Chen, E.I. Paschalis, *et al.* (2013). Intravitreal injections at the Massachusetts Eye and Ear Infirmary: analysis of treatment indications and postinjection endophthalmitis rates. *Br. J. Ophthalmol.* 97: 460–465.

54 The Royal College of Ophthalmologists. (2009). Guidelines for Intravitreal Injections Procedure. Available at: https://www.rcophth.ac.uk/wpcontent/uploads/2015/01/2009-SCI 012_Guidelines_for_Intravitreal_Injections_Procedure_1.pdf (accessed 26 January 2018).

55 E.M. del Amo and A. Urtti (2008). Current and future ophthalmic drug delivery systems: A shift to the posterior segment. *Drug Discov. Today* 13: 135–143.

56 G.A. Peyman, E.M. Lad and D.M. Moshfeghi (2009). Intravitreal injection of therapeutic agents. *Retina* 29: 875–912.
57 D. Kurz and T.A. Ciulla (2002). Novel approaches for retinal drug delivery. *Ophthalmol. Clin. North Am.* 15: 405–410.
58 K.G. Falavarjani and Q.D. Nguyen (2013). Adverse events and complications associated with intravitreal injection of anti-VEGF agents: a review of literature. *Eye.* 27: 787–794.
59 J. Wang, A. Jiang, M. Joshi and J. Christoforidis (2013). Drug delivery implants in the treatment of vitreous inflammation. *Mediators Inflamm.* 2013: 780634.
60 E. Sakurai, M. Nozaki, K. Okabe, *et al.* (2003). Scleral plug of biodegradable polymers containing tacrolimus (FK506) for experimental uveitis. *Invest. Ophthalmol. Vis. Sci.* 44: 4845–4852.
61 J. Okabe, H. Kimura, N. Kunou, *et al.* (2003). Biodegradable intrascleral implant for sustained intraocular delivery of betamethasone phosphate. *Invest. Ophthalmol. Vis. Sci.* 44: 740–744.
62 Y. Kim, J. Lim, H. Kim, *et al.* (2008). A novel design of one-side coated biodegradable intrascleral implant for the sustained release of triamcinolone acetonide. *Eur. J. Pharm.Biopharm.* 70: 179–186.
63 H.L. Quinn, L. Bonham, C.M. Hughes and R.F. Donnelly (2015). Design of a dissolving microneedle platform for transdermal delivery of a fixed-dose combination of cardiovascular drugs. *J. Pharm. Sci.* 104: 3490–3500.
64 S. Henry, D.V. McAllister, M.G. Allen and M.R. Prausnitz (1998). Microfabricated microneedles: a novel approach to transdermal drug delivery. *J. Pharm. Sci.* 87: 922–925.
65 N. Roxhed, B. Samel, L. Nordquist, *et al.* (2008). Painless drug delivery through microneedle-based transdermal patches featuring active infusion. *IEEE Trans. Biomed. Eng.* 55: 1063–1071.
66 S.A. Coulman, A. Anstey, C. Gateley, *et al.* (2009). Microneedle mediated delivery of nanoparticles into human skin. *Int. J. Pharm.* 366 (1): 190–200.
67 H.L. Quinn, M. Kearney, A.J. Courtenay, *et al.* (2014). The role of microneedles for drug and vaccine delivery. *Expert Opin. Drug Deliv.* 11: 1769–1780.
68 Thakur Raghu Raj Singh, S.J. Fallows, H.L. McMillan, *et al.* (2014). Microneedle-mediated intrascleral delivery of in situ forming thermoresponsive implants for sustained ocular drug delivery. *J. Pharm. Pharmacol.* 66: 584–595.
69 Thakur Raghu Raj Singh, I.A. Tekko, F. Al-Shammari, *et al.* (2016). Rapidly dissolving polymeric microneedles for minimally invasive intraocular drug delivery. *Drug Delivery Transl. Res.* 6: 800–815.
70 J. Jiang, J.S. Moore, H.F. Edelhauser and M.R. Prausnitz (2009). Intrascleral drug delivery to the eye using hollow microneedles. *Pharm. Res.* 26: 395–403.
71 S. Patel, A. Lin, H. Edelhauser and M. Prausnitz (2011). Suprachoroidal drug delivery to the back of the eye using hollow microneedles. *Pharm. Res..* 28 (Jan 1): 166–176.
72 S.R. Patel, D.E. Berezovsky, B.E. McCarey, *et al.* (2012). Targeted administration into the suprachoroidal space using a microneedle for drug delivery to the posterior segment of the eye. *Invest. Ophthalmol. Vis. Sci.* 53: 4433–4441.
73 B.C. Gilger, E.M. Abarca, J.H. Salmon and S. Patel (2013). Treatment of acute posterior uveitis in a porcine model by injection of triamcinolone acetonide into

the suprachoroidal space using microneedles. *Invest. Ophthalmol. Vis. Sci.* 54: 2483–2492.
74 B. Chiang, Y.C. Kim, H.F. Edelhauser and M.R. Prausnitz (2016). Circumferential flow of particles in the suprachoroidal space is impeded by the posterior ciliary arteries. *Exp. Eye Res.* 145: 424–431.
75 Y.C. Kim, H.E. Grossniklaus, H.F. Edelhauser and M.R. Prausnitz (2014). Intrastromal delivery of bevacizumab using microneedles to treat corneal neovascularization. *Invest. Ophthalmol. Vis. Sci.* 55: 7376–7386.
76 N.K. Palakurthi, Z.M. Correa, J.J. Augsburger and R.K. Banerjee (2011). Toxicity of a biodegradable microneedle implant loaded with methotrexate as a sustained release device in normal rabbit eye: a pilot study. *J. Ocul. Pharmacol. Ther.* 27: 151–156.
77 H.B. Song, K.J. Lee, I.H. Seo, *et al.* (2015). Impact insertion of transfer-molded microneedle for localized and minimally invasive ocular drug delivery. *J. Control. Release* 209: 272–279.
78 M. Matthaei, H. Meng, I. Bhutto, *et al.* (2012). Systematic assessment of microneedle injection into the mouse cornea. *Eur. J. Med. Res.* 17: 19.

11

Clinical Translation and Industrial Development of Microneedle-based Products

Ryan F. Donnelly

School of Pharmacy, Queen's University Belfast, Belfast BT9 7BL, UK

11.1 Introduction

Despite the limited number of drugs deliverable across the skin, due to the formidable barrier properties of the *stratum corneum*, the value of the worldwide transdermal product market is predicted to increase from $32 billion to >$80 billion by 2024 [1]. The unprecedented global shift in the adoption of unhealthy lifestyles is presumed to be responsible for the high prevalence of chronic diseases, such as cancer and diabetes, which is expected to drive the clinical urgency to incorporate transdermal drug delivery systems in future treatments. Moreover, the rising geriatric population base, which is highly susceptible to developing the aforementioned chronic diseases, are expected to propel the demand for highly efficacious pharmacological drugs. The market is driven by technological advancements in transdermal drug delivery devices, which is anticipated to serve the market with future growth opportunities. These advancements include advanced, third-generation, transdermal drug delivery technologies, including iontophoresis, ultrasound and microneedles or mechanical arrays. These technologies serve as effective transdermal drug diffusion alternatives that are capable of improving overall patient health and quality of life. For instance, the advent of matrix-controlled transdermal drug diffusion comprising a drug reservoir and a thermos-effector, enhances skin permeation and enables controlled drug delivery.

Transdermal delivery of biotechnology-derived drugs (e.g. proteins, peptides, antibodies) is an increasingly important area for development. The biopharmaceuticals market is projected to reach $291 billion by 2021 [2]. Biopharmaceuticals currently represent around 20% of the entire pharmaceutical market revenue, with further increases expected. Indeed, the revenue from sales of biopharmaceuticals represents a significant proportion of the total company revenue for several large pharma companies, such as Novartis, AbbVie and Pfizer. More than half of the current top 20 blockbuster drugs are biopharmaceuticals. Biopharmaceuticals can be considered as one of the most sophisticated and ground-breaking achievements in modern science. Unlike conventional pharmaceuticals that are synthesised by chemical processes, biopharmaceuticals are produced through biological synthesis in living systems. Owing to their biological origin, the biopharmaceuticals show much greater efficacy and efficiency when compared with conventional pharmaceuticals. Since biopharmaceuticals act

Microneedles for Drug and Vaccine Delivery and Patient Monitoring, First Edition.
Edited by Ryan F. Donnelly, Thakur Raghu Raj Singh, Eneko Larrañeta, and Maelíosa T.C. McCrudden.

by modulation of physiological processes at a cellular level, they can effectively treat the disorders that conventional pharmaceuticals could not treat. Biopharmaceuticals are the preferred choice of treatment for biomolecule deficiency disorders such as hormone deficiency or growth factor deficiency. They can also be used for stimulating and/or suppressing immune reactions, which makes them an ideal treatment option for autoimmune disorders. The most important contribution of biopharmaceuticals is in the field of targeted drug therapy, and the use of biopharmaceuticals is revolutionising this field. Since biopharmaceuticals can technically modulate any physiological pathway that has been fully understood, there is a huge growth potential for the biopharmaceuticals market. As a result of the huge advantages that biopharmaceuticals offer over conventional medicines, their use is continuously increasing. This, combined with the ability of biopharmaceuticals to treat diseases that were previously untreatable, has resulted in a huge market demand, which in turn has increased the profit margins for manufacturers. All these factors together are driving the biopharmaceuticals market, but despite these many growth factors the market faces severe challenges. These need to be overcome in order to achieve the full potential of this market. One such factor is the requirement of cold storage for almost all biopharmaceuticals. This limits their reach in developing countries. The necessity for hypodermic injection based delivery is also a major disadvantage

Owing to enzymatic breakdown and poor gastrointestinal (GI) absorption, biopharmaceuticals cannot be given orally and so must be injected. Microneedle (MN) based transdermal delivery systems have the potential to effectively overcome this problem. However, because of their hydrophilicity and large sizes, biopharmaceuticals cannot passively traverse the *stratum corneum*. As efficient transdermal transport will no longer be dependent on drug physicochemical properties, MN systems could significantly increase the size of the transdermal market. Conventional MNs have already successfully delivered a wide range of biomolecules, both *in vitro* and *in vivo*, as discussed in detail earlier in this book. Dry-state formulations may enhance thermal and chemical stability, thus obviating the cold chain. A 2012 report put the potential global market for MN-based drug delivery systems at just under $400 million [3]. However, given the need for enhanced, heat-stable delivery systems for biotechnology-derived drugs, the drive to extend patent lifetime in the face of competition from biosimilars and the potential for MN-based delivery systems to meet all of these challenges, the true market value for MN patches for administration of biopharmaceuticals could be substantially greater.

In order to achieve their undoubted potential, MN-based delivery systems must overcome current disadvantages related to the materials traditionally used to produce them, regulatory questions, scalable manufacture and the historical lack of major investment from the pharmaceutical industry.

11.2 Materials

The first two MN products successfully marketed, Micronject® and Soluvia®, are based on silicon and metal MNs, respectively. However, silicon is not biodegradable and implanted silicon is prone to biofouling [4–6]. Therefore, silicon MN left behind in skin due to brittle fracture of baseplates during insertion could cause skin problems.

Moreover, the possible problems resulting from inappropriate disposal of silicon or metal MNs, which both remain intact post-removal, have led to most researchers in this field focussing on MNs made from FDA-approved polymeric materials [7]. Initially, the hot polymer and carbohydrate melts used to make this second generation of MNs caused breakdown of biologics during processing [8]. Accordingly, the majority of recent research has focussed on dissolving MNs prepared from aqueous polymer blends [9–12].

To overcome the limited dosing capacity for macromolecules loaded into dissolving MNs, our group at Queen's University in Belfast developed MNs made from hydrogel-forming polymers. These MNs contain no active drug themselves [9, 13]. Instead, they are hard in the dry state, but rapidly take up skin interstitial fluid upon insertion to form discrete hydrogel pathways between an attached patch-type drug reservoir and the dermal microcirculation. In this way, the dose of drug is not limited to what can be loaded into, or coated on the surface of, the MNs themselves. These MNs possess inherent antimicrobial properties and deposit no polymer in the skin, yet they are sufficiently soft after insertion for just 1 min to prevent re-insertion, thus enhancing patient safety [9, 13].

While the idea of MN-based delivery systems was first proposed in the 1970s, the first practical demonstration was not until the late 1990s [14, 15]. Since then, the MN field has continued to develop, due to enhancements in methods of manufacture and the use of ever-more sophisticated designs. Researchers are gradually realising the importance of material biocompatibility and the challenges of keeping the bioburden low. The introduction of biocompatible polymeric MN devices may herald a new era in the development of MN technology, overcoming a number of disadvantages of previous MN designs. It is obvious that MN may have a major role to play in enhanced vaccine delivery strategies. By targeting the intradermal layer, which is replete with antigen-presenting cells, lower vaccine doses could potentially be used to achieve comparable levels of immune protection to conventional intramuscular and subcutaneous injections using needles and syringes [16–18]. Cold chain storage may be obviated by formulation of vaccines in the solid state in MN-based products. Needle-stick injuries and the need for reconstitution prior to administration will be eliminated and, if MNs can be shown to be inserted into skin without specialist training, it could mean significant savings for healthcare providers. Clearly, those in the developing world stand to benefit greatly from MN-based vaccination.

The compounds delivered by MNs to date have typically been of high potency, meaning only a low dose is required to achieve a therapeutic effect (e.g. insulin) [9, 16] or elicit the required immune response [17, 18]. Clearly, the majority of marketed drug substances, including many therapeutic antibodies, are not low dose, high potency molecules. Indeed, many drugs require doses of several hundred milligrams per day in order to achieve therapeutic plasma concentrations in humans. Up till now, such high doses could not be delivered transdermally from a patch of a reasonable size, even for molecules whose physicochemical properties are ideal for passive diffusion across the skin's *stratum corneum* barrier. Therefore, transdermal delivery has traditionally been limited to fairly lipophilic, low molecular weight, high potency drug substances. Since most drugs do not possess these properties, the transdermal delivery market has not expanded beyond around 20 drugs. Marketed MN-based patches are likely to increase this number of drugs in the coming years. However, this increase will only be maximised if high-dose molecules can also be delivered in therapeutic doses using MNs. We have shown that suitably formulated dissolving MN platforms can deliver therapeutic doses of a low potency, high dose drug substance [19]. However, deposition of a polymer in the skin from a dissolving MN system may be undesirable if the system is to be used on an ongoing basis. The dissolving MN system employed in our study would deposit approximately 5–10 mg of polymer per cm^2 in the skin [19]. If the patch size was 10 cm^2, then 50–100 mg of polymer would be deposited in the patient's skin every time the product is applied. While vaccines are used infrequently, most therapeutic agents need to be administered regularly. Accordingly, dissolving MN systems may be most appropriate for rapid delivery of low-dose vaccines, as we and others have described previously [17, 20]. We have now modified our novel hydrogel-forming MN system to facilitate delivery of clinically relevant doses of a low-potency, high-dose drug substance and rapid delivery of a model protein by increasing swelling capabilities and using a hygroscopic lyophilised drug reservoir [21].

11.3 Other Potential Applications

The utility of MN in bypassing formidable biological barriers, such as the skin's *stratum corneum*, opens up a range of additional applications, beyond transdermal and intradermal drug delivery. As discussed in detail earlier in this book, drug delivery into the eye and enhanced administration of active cosmeceutical ingredients are currently under intense investigation in both academia and industry, while we have recently reviewed the increasing investigations of the use of MNs for therapeutic drug monitoring purposes. If drug substances could be both monitored and delivered from the same, interconnected, device, then the possibility of an MN-based closed loop delivery system could become a reality.

As technological advances continue, MN arrays may well become one of the major pharmaceutical dosage forms and monitoring devices of the near future. However, in order for new pharmaceutical products and medical devices based upon MN arrays to realise their undoubted potential and provide benefits for patients and industry, a number of factors will need to be taken into account. These include reliable patient application, user acceptability, safety and cost-effective mass production.

11.4 Patient Application

If MN-based products are to be successfully developed and commercialised, then it will be important to know the dimensions of the micropores they create in patients' skin and how quickly normal skin barrier function recovers. Traditionally, coloured dyes have been used to stain the pores created and transepidermal water loss measurements used to quantify disturbances in skin barrier function following MN removal [22–24].

Although these techniques confirm that the skin's *stratum corneum* barrier has been compromised, they provide no information with respect to the true depth of MN penetration.

Optical coherence tomography (OCT), the optical analogue of ultrasound imaging, has been used to investigate MN-mediated skin puncture [22–24]. Since it is capable of imaging the skin down to depths of 2.5 mm, OCT has been used to study insertion depth and micropore width. This work indicates that MNs do not usually penetrate fully into skin. For example, approximately 80% of the shaft length of 600 μm MNs was shown to protrude beneath the *stratum corneum* in one study [24], with a micropore width of approximately 300 μm. Importantly, the technique allows the influence of different MN designs and application forces on the insertion depth to be studied and, if used in conjunction with transparent polymeric MNs, can be used to follow MN dissolution/swelling in real time *in vivo*. Micropore closure kinetics can also be studied in real time. At present, it appears that OCT studies on MN insertion, in-skin behaviour and skin recovery will be essential components of any regulatory submissions for MN-based products.

Indeed, in order to gain acceptance from healthcare professionals, patients and, importantly, regulatory authorities (e.g. the US Food and Drug Administration and the European Medicines Agency), it appears likely that some form of "dosing indicator" will need to be included within the overall MN "package." Whilst a wide variety of applicator designs have been disclosed within the patient literature, only a few, relatively crude, designs based upon high impact/velocity insertion, or rotary devices have been described [25]. Such device combinations are unlikely to enhance patient compliance and, currently, do not provide any feedback on successful skin insertion. Moreover, it is obvious that patients cannot "calibrate" their hands and will, therefore, apply MN with different forces unless properly instructed. With this in mind, we used OCT and transepidermal water loss measurements to illustrate that human volunteers can successfully insert our hydrogel-forming MN into their own skin by hand to consistent depths to yield consistent transient disturbances of skin barrier function when counselled by a pharmacist and having read a suitable Patient Information Leaflet [26]. A similar study by the Prausnitz Group also showed that consistent self-application is possible, once appropriate instructions are provided [27].

Moving towards commercialisation, it is likely that patients will need a level of assurance that the MN device has actually been inserted properly into their skin. This would be especially true in cases of global pandemics or bioterrorism incidents, where self-administration of MN-based vaccines becomes a necessity. Accordingly, a suitable means of confirming that skin puncture has taken place may need to be included within the MN product itself. To address this need, we used a pressure-indicating sensor film that changes colour from white to red if, and only if, a pressure greater than 18.6 N/cm^2 is applied [28]. Thus, a margin of safety exists, given that such MN can be reliably and reproducibly inserted into human skin with a pressure of approximately 10 N/cm^2. Volunteers were found to successfully insert hydrogel-forming MN into their forearms following suitable instruction and having read a Patient Information Leaflet. Of the volunteers, 75% stated that they preferred the MN patch with the indicating film. Given the fact that delivery of high doses of conventional drugs, as well as therapeutic antibodies, transdermally is likely to require a much larger MN patch than those required for delivery of potent vaccines, we used the same pressure-indicating sensor film in a human volunteer study where we investigated self-application of MN patches of similar size to conventional passive transdermal patches [29]. In this case, the sensor film was useful, since if one area of the larger patch did not have sufficient pressure applied to insert the underlying needles, then this would be obvious to the volunteer, who could then re-apply pressure over that area. We found that volunteers successfully inserted the needles of the larger patches into their own skin, with needles located at both the edges and the centre of the large patches inserted to statistically identical depths to those on a small 0.5 cm^2 patch. This is an important finding, becuase it supports the clinical viability of MN-based transdermal dosing of high-dose drugs.

11.5 Patient/Healthcare Provider Acceptability

Whether MN-based products are ultimately a commercial success will depend not only upon their ability to perform as designed, but also their acceptability to patients and healthcare professionals. A study conducted by Birchall's group [30] provided a range of opinions from healthcare professionals and members of the general public. The focus groups conducted showed that patient benefits, including reduced pain and needle stick injuries, increased acceptability by people with needle phobia and the potential for self-administration. However, concerns were raised about effectiveness and how a patient would know the device had been used properly. We also used focus groups to explore children's views on MN use as an alternative approach to blood sampling in monitoring applications [31]. A total of 86 children participated in 13 focus groups across seven schools in Northern Ireland. A widespread disapproval of conventional blood sampling using needles was evident, with pain, blood and traditional needle visualisation being particularly unpopular aspects. In general, MN had greater visual acceptability and caused less fear. A patch-based design enabled minimal patient awareness of the monitoring procedure, with personalised designs, (e.g. cartoon themes), favoured. Children's concerns included possible allergy and potential inaccuracies with this novel approach. However, many had confidence in the judgement of healthcare professionals if they deemed this technique appropriate. They considered paediatric patient education critical for acceptance of this new approach and called for an alternative name, without any reference to "needle." We concluded that a proactive response to these unique insights should enable MN array design to better meet the needs of this end-user group.

As potential prescribers for MN-based systems, paediatricians are also key potential end-users. In another study, we investigated the views of UK paediatricians on the use of MN technology within neonatal and paediatric care [32]. An online survey was developed and distributed among UK paediatricians to gain their opinions of MN technology and its use in the neonatal and paediatric care settings, particularly for MN-mediated monitoring. A total of 145 responses were obtained, with a completion response rate of 13.7%. Respondents believed an alternative monitoring technique to blood sampling in children was required. Furthermore, 83% of paediatricians believed there was a particular need in premature neonates. Overall, this potential end-user group approved of the MN technology and an MN-mediated monitoring approach. Minimal pain and the perceived ease of use were important elements in gaining favour. Concerns included the need for confirmation of correct application and the potential for skin irritation. The findings of this study provide an initial indication of MN acceptability among a key potential end-user group. Furthermore, the concerns identified present a challenge to those working within the MN field to provide solutions to further improve this technology.

Elderly patients receive the majority of prescription medicines worldwide. Accordingly, it is important to consider the future use of MNs by older people in order to ensure optimal therapeutic outcomes for this unique and increasing population group. We considered the use of MNs by those aged over 65 years, investigating insertion parameters in ageing skin, alongside the feasibility and acceptability of the technology [33]. Hydrogel-forming MN arrays were applied to seven subjects aged over 65 years, with breach of the *stratum corneum* confirmed using OCT. Insertion depths recorded in each case were similar to a comparative group, aged 20–30 years. Skin recovery was, however, demonstrated to occur at a slower rate in the older subjects, as measured using transepidermal water loss. Qualitative methods, including focus groups and semi-structured interviews, were employed to collect the views and opinions of older people and community pharmacists, respectively. The overall consensus was positive, with a number of benefits to MN-mediated drug delivery identified, such as reduced dosing frequency, improved adherence and an alternative delivery route where oral or injectable medication were precluded. Concerns centred on practical issues associated with age-related functional decline, including, for example, reduced dexterity and skin changes. This work provided the first convincing report of the importance of further translational research in this area to support future MN use in older people, ensuring an age-appropriate delivery platform.

Studies such as those discussed here, when appropriately planned to capture the necessary demographics, will undoubtedly aid industry in taking the necessary action to address concerns and develop informative labelling and patient counselling strategies to ensure safe and effective use of MN-based devices. Marketing strategies will, obviously, also be vitally important in achieving maximum market shares relative to existing and widely used conventional delivery and monitoring systems.

11.6 Patient Safety

Currently, little is known about any long-term effects that may occur due to repeatedly penetrating human skin with MNs. The skin is replete with antigen-presenting cells and, so, it is vital that delivery of biomolecules intradermally using MNs does not elicit immune responses to non-vaccine agents, such as insulin. It is likely that humanised versions of drug-like biomolecules will reduce this considerably, but non-immunogenicity will probably have to be shown on a drug-by-drug basis. Moreover, it will be important that no local or systemic reactions to the materials used to fabricate the MNs occur. It is statistically unlikely that MNs would ever be applied to exactly the same points on the skin's surface due to the very small size of the devices and so it is probable that MN-based systems will have very favourable safety profiles, especially over conventional hypodermic needles. While mild skin erythema post MN removal may initially be concerning to some patients, skin barrier function will recover within a matter of hours and any reddening of skin will be similarly transient, regardless of how long the MNs are in place [2, 4].

Polymer deposition from dissolving MN is of great interest currently. While the polymers used for MN production are typically approved pharmaceutical excipients, they have never before been used intradermally. Regulators may require information on the amounts of polymer left behind in skin after MN removal and information on clearance rates and routes. This may well be a non-issue for one-off vaccine administration, but could be important if a dissolving MN was to be used regularly for insulin delivery, for example.

Two isolated reports suggest that, when used inappropriately, MNs can indeed cause health problems, such as skin irritation and intradermal granulomas, as well as systemic hypersensitivity [34, 35]. While exact details are scarce (including the MN types used and precise skin sites affected), in both cases it is likely that MNs were used in combination with cosmetic products that were not intended for application to MN-punctured skin. Such products contain multiple excipients and, importantly, are not sterile. Medical supervision of MN use was not present in either case. MN devices are not equivalent to conventional transdermal patches, in that they are not simply applied to the skin surface. Rather, MNs function principally by breaching the skin's protective *stratum corneum* barrier and often penetrate into the viable epidermis and dermis [2, 4]. Such areas of the body are normally sterile. Accordingly, it is imperative that MNs, or products used with them, do not themselves contain microbial loads sufficient to cause skin or systemic infection. It is also important that bioburden be minimised to avoid immune stimulation, especially considering the rich immune cell population in the viable epidermis and dermis [36]. However, as we and others have shown [37, 38], microbial penetration through MN-induced holes is minimal. Indeed, there have never been any reports of MNs causing skin or systemic infections when they are used under medical supervision. This may be due to the action of the strong immune component of skin or the skin's non-immune enzyme-based defence mechanisms. As the micropores created by MN are aqueous in nature and very small in dimensions, there may not be a great tendency for microorganisms to traverse these openings. Accordingly, skin cleansing prior to MN insertion is unlikely to be necessary but should be investigated as part of product development studies. In an ideal situation, this would not be done. Accordingly, patients and healthcare professionals will not be unnecessarily inconvenienced and may be more reassured about the safety of the delivery system, making the use of the product in the domiciliary setting appear more akin to application of a conventional transdermal patch than a self-administered injection. In our own experience as pharmacists, patients often do not follow instructions provided verbally by healthcare professionals or in written form in product inserts, especially if they consider them unnecessary or inconvenient. Accordingly, skin cleansing will be practically unenforceable outside the hospital or GP surgery.

With these concerns in mind, we addressed, for the first time, the impact of skin insertion of polymeric MN arrays on multiple occasions in an animal model *in vivo* [39]. Dissolving MN arrays prepared from aqueous blends of 20% w/w Gantrez® S-97 BF and 40% w/w poly(vinylpyrrolidone) (58 kDa) and hydrogel-forming MN arrays prepared from aqueous blends of Gantrez S-97 BF and poly(ethylene glycol) (10 kDa) were repeatedly applied to the skin of immunocompetent hairless mice *in vivo*. Skin appearance and skin barrier function, as illustrated by measurement of transepidermal water loss, were not measurably altered during the entire study period. Biomarkers of infection, immunity and inflammation/irritation were also statistically unchanged, regardless of the MN formulation, needle density or number of applications. Mice remained healthy throughout and continued to gain weight during the study. For example, transepidermal water loss values were typically in the range

10–15 g/m^2/h immediately prior to MN insertion and 15–25 g/m^2/h immediately following MN removal, regardless of when they were measured during the study periods. Serum biomarker levels, measured immediately post-mortem were always in the range of 10–20 μg/ml for C-reactive protein, 0.5–1.5 mg/ml for Immunoglobulin G and 1000–2500 pg/ml for interleukin 1-β and were never statistically different from untreated controls. No measurable levels of tumour necrosis factor-α were found in any animals. These findings are encouraging for the formulations investigated, suggesting that their repeated use by patients will not cause undesirable side-effects.

Ultimately, regulatory authorities will need to decide, based on the weight of available evidence. MNs may be classed as drug delivery systems, consumer products or medical devices, depending upon the intended use stated by their manufacturers. If MNs are considered to be more akin to an injection than a transdermal patch, then they may need to be produced or rendered sterile. Any contained microorganisms would have to be identified and pyrogen content would need to be minimised. If sterile production is required, careful selection of the method to be used will be vital. Aseptic manufacture will be expensive and will present practical challenges if the MNs are to be made on a very large scale, for example, with vaccine delivery products. Terminal sterilisation using gamma irradiation, moist heat or microwave heating may damage the MNs or biomolecule cargoes, while ethylene oxide may permeate polymeric MN materials, thus contaminating the delivery system.

11.7 Manufacturing and Regulatory Considerations

Manufacturing MNs aseptically or employing terminal sterilisation procedures are likely to increase costs considerably. Scaling-up MN production will require considerable thought. This is especially true given the plethora of small-scale production methods described in the literature. Very often a number of steps are required, especially for coated MNs. Also, silicon MNs require cleanroom conditions. Overall, it is likely that any manufacturer wishing to develop MN products will need to make a substantial initial capital investment, given that equivalent manufacturing technologies are not currently available. Similarly, a range of new quality control tests will now also become necessary. It is likely that the regulatory requirements set for the first MN products to be approved for human use will set the standards for follow-on products. Packaging will be important in protecting the MNs from moisture and microbial ingress and suitable advice will need to be provided to avoid damage during patient handling and insertion.

Overall, from a regulatory perspective, it seems likely that MNs will be classed as a new dosage form, rather than an adjunct technology to existing transdermal patch drug delivery systems. In summary, the key regulatory questions that may need to be addressed are as follows.

- **Sterility of the MN dosage form** As an MN dosage form will penetrate the *stratum corneum* into the epidermis, rather than simply adhere to its surface (as in a conventional transdermal patch), sterility being a regulatory requirement, although a low bioburden, may be acceptable in cases where the system has inherent and demonstrable antimicrobial activity.

- **Uniformity of content (either from the system as a whole or, possibly, in respect of individual drug-loaded MNs within an array, depending on system design)** It is likely that this pharmacopoeial requirement, which is internationally harmonised, will be applied to MN systems, as it is for transdermal patch dosage forms.
- **Manufacturing aspects, including packaging** The normal aspects of quality, including security of packaging (which may also require a demonstration of adequate protection from, for example, water ingress) will apply.
- **Potential for re-use by patients or others** Many current MN systems, notably those made of silicon, can be removed intact from the skin and, therefore, could be re-used by the patient, or others. Thus, for reasons of safety, a self-disabling system ensuring single use only may be required.
- **Disposal** MN materials that are not dissolvable or biodegradable may be a hazard, therefore this environmental aspect of their use may be an issue.
- **Deposition of MN materials in the skin, particularly with respect to long-term use** Dissolvable, polymeric MNs will deposit in the skin the materials from which they were fabricated. This could lead to long- or short-term adverse skin effects, such as granuloma formation or local erythema, particularly where repeated use is a factor. This can be mitigated by varying the application site and may be less of an issue where the use is occasional, such as in vaccination.
- **Ease and reliability of application by patients** As with all dosage forms, patients must be able to use the product properly, without significant inconvenience.
- **Assurance of delivery (proper insertion)** Since there is no obvious sensation on applying an MN dosage form, some indication of correct application and delivery (particularly for vaccination applications) may be required.
- **Potential immunological effects** Repeated insult of the skin, an immunologically active site, by MNs may result in an immunological reaction, depending on the material involved. Some assurances as to immunological safety may be required by regulators.

11.8 Commercialisation of MN Technologies

Over the past decade, there has been a substantial increase in development of MN technologies. Indeed, the number of academic publications on the subject has more than tripled since 2007. While biological agents have been the main focus, water-soluble drugs not currently suitable for passive transdermal delivery are also of great interest. A number of companies are investing heavily in development of MN-based delivery systems, in particular. These include: 3M (www.3m.com (accessed October 2017)); Corium (www.coriumintl.com (accessed October 2017)); Zosano Pharma (www.zosanopharma.com (accessed October 2017)); Vaxxas (www.vaxxas.com (accessed October 2017)); Nemaura (www.nemaura.co.uk (accessed October 2017)); Becton-Dickinson (www.bdbiosciences.com (accessed October 2017)); LTS Lohmann and NanoPass Technologies (www.ltslohmann.de/home.html (accessed October 2017)).

Zosano Pharma develops transdermal delivery products based on the Macroflux® technology originally designed at Alza. Having apparently moved away from its initial focus on delivery of parathyroid hormone for management of post-menopausal osteoporosis, Zosano has recently announced successful results of a double-blind placebo controlled clinical trial focussed on delivery of zolmitriptan for treatment of migraine (www.zosanopharma.com).

Vaxxas is a venture capital-funded technology start-up company developing the coated MN Nanopatch™ technology that originated from Mark Kendall's research group at the Australian Institute of Bioengineering & Nanotechnology at The University of Queensland (www.vaxxas.com). In 2015, Vaxxas announced that it had secured equity funding of $20 million from new and existing investors. These funds represented the first closing of a Series B venture financing round, the proceeds from which were to be used to advance a series of clinical programmes and develop a pipeline of new vaccine products for major diseases using the Vaxxas Nanopatch platform. This round of financing brought the total capital raised by Vaxxas to $33 million.

NanoPass Technologies have shown their Micronjet™ device to be useful in the delivery of insulin, influenza vaccines and local anaesthetics in an extensive range of clinical studies, with evidence of dose-sparing compared with conventional routes of immunisation particularly notable (www.nanopass.com). However, it should be remembered that this device is more akin to a small array of very short silicon needles attached to the barrel of a conventional syringe, rather than a true microneedle array. Meanwhile, Beckton-Dickinson's Soluvia™ device, consisting of a single 1.5 mm 30-gauge stainless-steel needle on the end of a conventional syringe barrel, has been widely used for a number of years in Sanofi-Pasteur's market-approved intradermal influenza vaccine products Intanza® and Fluzone® (www.bdbiosciences.com).

3M's microstructured transdermal systems (MTS), based on either hollow or coated solid MNs, have been evaluated in a range of pre-clinical studies focussed on delivery of proteins, peptides and vaccines (www.3m.com). Nemaura Pharma, a specialist drug delivery firm based in the East Midlands in England, are developing MN systems for applications in cosmeceuticals, dermatology, analgesia, osteoporosis, immunology and oncology (www.nemaura.co.uk).

Whilst the these MN devices have been based upon solid or hollow MN systems, it is envisaged that devices based upon FDA-approved, biodegradable/dissolving polymeric MN formulations will, in the future, receive increased attention from pharma companies. This is due to the self-disabling nature of such systems. Once inserted into skin, these MNs will either dissolve or swell, thus making insertion into another patient post-removal virtually impossible. This will, therefore, reduce transmission of infection by preventing needlestick injuries associated with conventional needles. Disposal issues will also be bypassed, since there is no "sharp" remaining. Ultimately, the impact on healthcare in the developing world in particular could be significant. Encouragingly, Mark Prausnitz recently reported the first successful human clinical trial of a dissolving MN vaccine patch [40].

Corium have stated that they are exploring several applications of dissolving MNs with pharma partners, with a particular current interest in delivery of human parathyroid hormone (hPTH) and zolmitriptan, with the hPTH project using MNs manufactured at Corium now in Phase II clinical trials (www.coriumintl.com). Most notably, perhaps, LTS Lohmann (LTS), the world's largest transdermal patch manufacturer, have now entered into the MN field and are inviting partners to collaborate on development of new MN products based on such technology (www.ltslohmann.de/home.html). Given the manufacturing capabilities, expertise and customer base already possessed by LTS, it will be surprising if they do not claim a sizeable proportion of the developing MN market in the coming years. Indeed, LTS recently announced that they now hold Europe's first manufacturing licence for MN patches (www.ltslohmann.de/home.html). Fujifilm also appear to have considerable manufacturing capability for MN patches, but they have not made any significant announcements since 2012 [41].

It is notable that the firms publically engaged in MN development at the present time could all be reasonably categorised as "drug delivery" companies, or companies with a specialised drug delivery component. It is very unlikely that any of them will ever end up holding the product licence for an MN vaccine or drug product themselves. Their business models suggest that they would either sell their technology, or indeed the entire company, to a pharmaceutical company seeking to bring an MN product to market or, alternatively, as may be the case with Corium, Fujifilm and LTS, the drug delivery company would act as the product manufacturer, with the product taken through clinical trials, regulatory scrutiny and ultimate product registration by a sponsoring pharma firm. Rumours in the field abound about the involvement of big pharma companies in MN development. Indeed, GlaxoSmithKline hosted the 2016 Microneedles Conference in London. However, to date, involvement of major players has, perhaps understandably, been kept under wraps. It is possible that big pharma see MNs as simply a developing world vaccine product from which they are unlikely to make money. Alternatively, they may not want to make their interest public for fear of "tipping off" competitors, or they may be unsure of the technology, but want to keep an interest by sponsoring work with academic groups or specialist MN firms, so as to avoid "missing the boat" if MN products were to be marketed by a competitor. Whatever the reasons, the financial muscle of a large company would certainly expedite the first MN-based drug or vaccine product's route to market.

There may be an alternative, of course. The Bill & Melinda Gates Foundation have invested heavily in development of MN vaccines [42]. For example, \$6 million has recently been awarded to Vaxess to develop inactivated polio and live attenuated measles rubella vaccines (www.vaxess.com). PATH (formerly Program for Appropriate Technologies in Health) are now developing a Centre of Excellence in this area to focus on commercial development of MN delivery systems with applications in developing countries. The question remains here as to who will be the product licence holder, however. Whatever route is ultimately taken, once the first MN drug or vaccine product is finally approved for human use, one could reasonably expect numerous other products to quickly follow suit.

11.9 Conclusion

The future appears to be very bright for new delivery and, potentially, monitoring systems based upon MN technologies. The ever-increasing amount of fundamental knowledge appears to be feeding industrial development. MNs have many advantages over conventional needle-and-syringe based delivery systems for biological agents, in particular, in terms of reduced pain, infection risk and the ability to control administration. Skin barrier function disturbance is minimal and recovery rapid. Once regulatory hurdles are overcome and manufacturing processes developed, optimised and validated to current good manufacturing practice standards, the benefits for patients and, ultimately industry, will be considerable. It is interesting to note that the US FDA recently published draft guidance on "microneedling" for cosmetic applications, clearly illustrating the interest of regulators in the technology [43].

11.10 Future Perspectives

Given the inherent safety features of MN systems, it is easy to foresee a time within the next ten years when vaccination programmes in the developing world are based around MNs. The fact that most MN contain biomolecules, such as vaccine antigens, in the dry state, the cold chain will be circumvented. Needlestick injuries also will be obviated. Such an intervention could massively improve the quality of life, life expectancy and economic productivity of developing countries. Accordingly, the potential impact of MN research and ultimate commercialisation is vast. For translation to the clinic and, ultimately the market, methods of manufacture will need to be refined and scaled-up and the "microneedle" aspect of the name of the final patch systems removed. The term currently being used by the World Health Organisation for all MN-based systems is MicroArray Patches, or MAPs. Such an adjustment may seem minor, but for patient acceptance it could be important. Once vaccine products are accepted by regulators, healthcare providers and patients, other MN-based products for everyday patient and consumer use will become widely available, to the benefit of the pharmaceutical, medical devices and cosmetics industries and patients worldwide.

References

1 Grand View Research (2016). Transdermal drug delivery system market worth $81.4 billion by 2024. http://www.grandviewresearch.com/press-release/global-transdermal-drug-delivery-system-market (accessed October 2017).

2 Mordor Intelligence (2017). Biopharmaceuticals market – growth, trends & forecasts (2017–2022). http://www.mordorintelligence.com/industry-reports/global-biopharmaceuticals-market-industry (accessed October 2017).

3 Greystone Associates (2012). *Microneedles in Medicine: Technology, Devices, Markets and Prospects*. Amherst, NH: Greystone Associates.

4 Chow A.Y., Pardue, M.T., Chow, V.Y., *et al.* (2001). Implantation of silicon chip microphotodiode arrays into the cat subretinal space. *IEEE Trans. Neural Syst. Rehabil. Eng.* 9: 86–95.

5 Voskerician, G., Shive, M.S., Langer, R., *et al.* (2003). Biocompatibility and biofouling of MEMS drug delivery devices. *Biomaterials* 24: 1959–1967.

6 Schmidt, S., Horch, K. and Normann, R. (1993). Biocompatibility of silicon-based electrode arrays implanted in feline cortical tissue. *J. Biomed. Mater. Res.* 27: 1393–1399.

7 Pierre, M.B. and Rossetti, F.C. (2014). Microneedle-based drug delivery systems for transdermal route. *Curr. Drug Targets* 15: 281–291.

8 Donnelly, R.F., Morrow, D.I.J., Thakur, R.R.S., *et al.* (2009). Processing difficulties and instability of carbohydrate microneedle arrays. *Drug Dev. Ind. Pharm.* 35: 1242–1254.

9 Migalska, K., Morrow, D.I.J., Donnelly, R.F., *et al.* (2011). Laser-engineered dissolving microneedle arrays for transdermal macromolecular drug delivery. *Pharm. Res.* 28: 1919–1930.

10 Garland, M.J., Caffarel-Salvador, E., Donnelly, R.F., *et al.* (2012). Dissolving polymeric microneedle arrays for electrically assisted transdermal drug delivery. *J. Control. Release* 159: 52–59.

11 Demir, Y.K., Akan, Z. and Kerimoglu, O. (2013). Characterization of polymeric microneedle arrays for transdermal drug delivery. *PLoS One* 8: e77289.

12 Park, J.H. and Prausnitz, M.R. (2010). Analysis of mechanical failure of polymer microneedles by axial force. *J. Korean Phys. Soc.* 56: 1223–1227.

13 Donnelly, R.F., Thakur, R.R.S., Garland, M.J., *et al.* (2012). Hydrogel-forming microneedle arrays for enhanced transdermal drug delivery. *Adv. Funct. Mater.* 22: 4879–4890.

14 Gerstel, M.S. and Place, V.A. (1976). Drug delivery device. US Patent 3,964,482, filed 17 May 1971 and issued 22 June 1976.

15 Henry, S., McAllister, D.V., Prausnitz, M.R., *et al.* (1998). Microfabricated microneedles: a novel approach to transdermal drug delivery. *J. Pharm. Sci.* 87: 922–925.

16 McCrudden, M.T.C., Thakur, R.R.S., Donnelly, R.F., *et al.* (2013). Strategies for enhanced peptide and protein delivery. *Ther. Delivery* 4: 593–614.

17 Zaric, M., Donnelly, R.F., Kissenpfennig, A., *et al.* (2013). Targeting of skin dendritic cells *via* microneedle arrays laden with antigen encapsulated PLGA nanoparticles induces efficient anti-tumour and anti-viral immune responses. *ACS Nano* 7: 2042–2055.

18 Koutsonanos, D.G., Compans, R.W. and Skountzou, I. (2013). Targeting the skin for microneedle delivery of influenza vaccine. *Adv. Exp. Med. Biol.* 785: 121–132.

19 McCrudden, M.T.C., McCrudden, C., Donnelly, R.F., *et al.* (2014). Design and physicochemical characterisation of novel dissolving polymeric microneedle arrays for transdermal delivery of high dose, low molecular weight drugs. *J. Control. Release* 180: 71–80.

20 Raphael, A.P., Crichton ML, Kendall M.A., *et al.* (2010). Targeted, needle-free vaccinations in skin using multilayered, densely-packed dissolving microprojection arrays. *Small* 16: 1785–1793.

21 Donnelly, R.F., McCrudden, M.T.C., McAlister, E., *et al.* (2014). Hydrogel-forming microneedles prepared from "super swelling" polymers combined with lyophilised wafers for transdermal drug delivery. *PLoS One* 9 (10): e111547.

22 Enfield J., O'Mahony, C., Leahy, M., *et al.* (2010). *In vivo* dynamic characterization of microneedle skin penetration using optical coherence tomography. *J. Biomed. Opt.* 15: 046001.

23 Coulman S., Birchall, J.C. and Alex, A. (2011). *In vivo, in situ* imaging of microneedle insertion into the skin of human volunteers using optical coherence tomography *Pharm. Res.* 28: 66–81.

24 Donnelly, R.F., Garland, M.J., Morrow, D.I.J., *et al.* (2011). Optical coherence tomography is a valuable tool in the study of the effects of microneedle geometry on skin penetration characteristics and in-skin dissolution *J. Control. Release* 147: 333–341.

25 Thakur, R.R.S., Dunne, N.J., Donnelly, R.F., *et al.* (2011). Review of patents on microneedle applicators. *Recent Pat. Drug Delivery Formulation* 5: 11–23.

26 Donnelly, R.F., Moffatt, K., Zaid-Alkilani, A., *et al.* (2014). Hydrogel-forming microneedle arrays can be effectively inserted in skin by self-application: A pilot study centred on pharmacist intervention and a patient information leaflet. *Pharm. Res.* 31, 1989–1999.

27 Norman, J.J., Meltzer, M.I., Prausnitz, M.R., *et al.* (2014). Microneedle patches: Usability and acceptability for self-vaccination against influenza. *Vaccine* 32: 1856–1862.

28 Vicente-Pérez, E.M., Quinn, H.L., McAlister, E., *et al.* (2016). The use of a pressure-indicating sensor film to provide feedback upon hydrogel-forming microneedle array self-application *in vivo. Pharm. Rese.* 33: 3072–3080.

29 Ripolin, A., Quinn, J., Larrañeta, E., *et al.* (2017). Successful application of large microneedle patches by human volunteers. *Int. J. Pharm.* 521: 92–101.

30 Birchall J.C., Anstey, A., John, D., *et al.* (2011). Microneedles in clinical practice: An explanatory study into the views and opinions of healthcare professionals and the public *Pharm. Res.* 28: 95–106.

31 Mooney, K., McElnay, J.C. and Donnelly, R.F. (2014). Children's views on microneedle use as an alternative to blood sampling for patient monitoring. *Int. J. Pharm. Pract.* 22: 335–344.

32 Mooney, K., McElnay, J.C. and Donnelly, R.F. (2015). Paediatricians' opinions of microneedle-mediated monitoring: a key stage in the translation of microneedle technology from laboratory into clinical practice. *Drug Delivery Transl. Res.* 5: 346–359.

33 Quinn, H.L., Hughes, C.M. and Donnelly, R.F. (2017). *In vivo* and qualitative studies investigating the translational potential of microneedles for use in the older population. *Drug Delivery Transl. Res.* 2017 May 15. doi: 10.1007/s13346-017-0393-4.

34 Daily Mail (2011). Chinese women warned over 'potential lethal' microneedle roller after beauty treatment takes Hong Kong by storm. http://www.dailymail.co.uk/femail/article-2026700/Microneedle-Therapy-System-potentially-lethal-Chinese-women-warned.html (accessed December 2013).

35 Soltani-Arabshahi, R., Wong, J.W., Duffy, K.L., *et al.* (2013). Facial allergic granulomatous reaction and systemic hypersensitivity associated with microneedle therapy for skin rejuvenation. *JAMA Dermatol.* 150: 68–72.

36 Roby, K.D. and Nardo, A.D. (2013). Innate immunity and the role of the antimicrobial peptide cathelicidin in inflammatory skin disease. *Drug Discovery Today: Dis. Mech.* 10: 3–4.

37 Donnelly, R.F., Thakur, R.R.S., Tunney, M.M., *et al.* (2009). Microneedle arrays allow lower microbial penetration than hypodermic needles *in vitro. Pharm. Res.* 26: 2513–2522.

38 Wei-Ze, L., Mei-Ronga, H. and Jian-Pinga, Z. (2010). Super-short solid silicon microneedles for transdermal drug delivery applications *Int. J. Pharm.* 389: 122–129.

39 Vicente-Perez, E.M., Larrañeta, E. and McCrudden, M.T.C. (2017). Repeat application of microneedles does not alter skin appearance or barrier function and causes no measurable disturbance of serum biomarkers of infection, inflammation or immunity in mice *in vivo. Eur. J. Pharm. Biopharms.*117, 400–407.

40 Rouphael, N.G., Paine, M., Mosley, R., *et al.* (2017). The safety, immunogenicity, and acceptability of inactivated influenza vaccine delivered by microneedle patch (TIV-MNP 2015): A randomised, partly blinded, placebo-controlled, phase 1 trial. *Lancet* 390: 649–658.

41 Fujifilm (2012). Delivery of solids/liquids. http://www.fujifilm.com/innovation/technologies/delivery-of-solids-or-liquids/ (accessed October 2017).

42 Inside Philanthropy (2017). Leapfrogging the cold chain. Why Gates is big into microneedles. http://www.insidephilanthropy.com/home/2017/3/31/leapfrogging-the-cold-chain-why-gates-is-big-into-microneedles (accessed October 2017).

43 United States Food and Drug Administration (2017). Regulatory Considerations for Microneedling Devices: Draft Guidance for Industry and Food and Drug Administration Staff. U.S. Food & Drug Administration, Center for Devices & Radiological Health.

Index

Microneedles for Drug and Vaccine Delivery and Patient Monitoring, First Edition.
Edited by Ryan F. Donnelly, Thakur Raghu Raj Singh, Eneko Larrañeta, and Maelíosa T.C. McCrudden.

n

Printed and bound by CPI Group (UK) Ltd, Croydon, CR0 4YY
17/09/2026
14964302-0003